# Groups and Symmetry

Bijan Davvaz

# Groups and Symmetry

## Theory and Applications

 Springer

Bijan Davvaz
Department of Mathematics
Yazd University
Yazd, Iran

ISBN 978-981-16-6110-5      ISBN 978-981-16-6108-2   (eBook)
https://doi.org/10.1007/978-981-16-6108-2

This Springer imprint is published by the registered company Springer Nature Singapore Pte Ltd.
The registered company address is: 152 Beach Road, #21-01/04 Gateway East, Singapore 189721,
Singapore

# Preface

Group theory is an extensive topic, and in this book, I have tried to select the important and representative items and to organize them in a coherent way. Symmetry occurs not only in geometry, but also in group theory. Group theory is the study of symmetry. Indeed, it appears in any situation where symmetry plays a part. When we are dealing with an object that appears symmetric, group theory can help with the analysis. We apply the label symmetric to anything which stays invariant under some transformations. As numbers measure size, we can say groups measure symmetry. Symmetry is a type of invariance: the property that something does not change under a set of transformations. Given a structured object of any sort, a symmetry is a mapping of the object onto itself which preserves the structure. In group theory, an automorphism is an isomorphism from a group to itself. It is, in some sense, a symmetry of the object and a way of mapping the object to itself while preserving all of its structure. The set of all automorphisms of a group forms a group, called the automorphism group. It is, loosely speaking, the symmetry group of the object. Although there are many excellent texts about group theory, this is the new one that covers all the topics of groups and symmetry.

The reader is supposed to be familiar with the elementary basic facts about groups; we refer the readers to Davvaz "A First Course in Group Theory."

The book is organized into ten chapters. Chapter 1 discussed $G$-sets; these are sets which admit an action by elements of a group $G$. The basic results here are orbit-stabilizer theorem, Burnside's lemma, and Pólya enumeration. Sylow's theorems and several examples of their applications are discussed in Chap. 2. Simple groups are presented in Chap. 3. In this chapter, we investigate the simplicity of alternating groups and projective special linear groups. In Chap. 4, we study the notion of products of groups. The next objective in Chap. 5 is the Jordan–Hölder theorem. This result shows the importance of the composition factors in determining the structure of finite groups. In continue, we study the notions of solvable and nilpotent groups. The main objective in Chap. 6 is free groups. We show that free groups exist, and we use these to develop a method of describing groups in terms of generators and relations. In Chap. 7, we investigate isometry between Euclidean spaces. In Chap. 8, we review Platonic solids, and then, we give a classification of finite groups of rotations in space.

A wallpaper group is a mathematical device used to describe any classify repetitive designs on two-dimensional spaces. The designs are classified into groups by looking at the type of symmetries presents. The main subject of Chap. 9 is wallpaper groups, and we shall see there are precisely seventeen wallpaper groups. The aim of the last chapter (i.e., Chap. 10) is to introduce the reader to the theory of representations of groups by linear transformations over a vector space or, equivalently, by matrices over a field. The study focuses on character theory and how character theory can be used to extract information from a group.

Each section ends in a collection of exercises. The purpose of these exercises is to allow students to test their assimilation of the material and to challenge their knowledge and ability. Moreover, at the end of each chapter, we have two special sections: 1. Worked-out problems; 2. Supplementary exercises.

In each worked-out problems section, I decided to try teach by examples, by writing out to solutions to problems. In choosing problems, three major criteria have been considered to be challenging, interesting, and educational. Moreover, there are many exercises at the end of each chapter as supplementary exercises. They are harder than before and serve to present the interesting concepts and theorems which are not discussed in the text.

The list of references at the end of the book is confined to works actually used in the text.

Yazd, Iran                                                                      Bijan Davvaz

# Contents

# About the Author

**Bijan Davvaz** is Professor at the Department of Mathematics, Yazd University, Iran. He earned his Ph.D. in Mathematics with a thesis on "Topics in Algebraic Hyperstructures" from Tarbiat Modarres University, Iran, and completed his M.Sc. in Mathematics from the University of Tehran, Iran. He also has served as Head of the Department of Mathematics (1998–2002), Chairman of the Faculty of Science (2004–2006), and Vice-President for Research (2006–2008) at Yazd University, Iran. His areas of interest include algebra, algebraic hyperstructures, rough sets, and fuzzy logic. On the editorial boards for 25 mathematical journals, he has authored six books and over 600 research papers, especially on algebra, fuzzy logic, algebraic hyperstructures, and their applications.

# Notations

| | |
|---|---|
| $G, H, \ldots$ | Sets, groups, … |
| $\in, a \in A$ | Membership |
| $\varnothing$ | Empty set |
| $\{x \mid p(x)\}$ | Set of all $x$ that $p(x)$ |
| $A \subseteq B$ | $A$ is a subset of $B$ |
| $A \subset B$ | $A$ is a proper subset of $B$ |
| $A_1 \cap A_2 \cap \ldots \cap A_n$ and $\bigcap_{i \in I} A_i$ | Intersection of sets |
| $A_1 \cup A_2 \cup \ldots \cup A_n$ and $\bigcup_{i \in I} A_i$ | Union of sets |
| $A - B$ | Difference set of $A$ and $B$ |
| $\mathbb{N}$ | Natural numbers |
| $\mathbb{Z}$ | Integers, additive group, and ring of integers |
| $\mathbb{Q}$ | Field of rational numbers |
| $\mathbb{R}$ | Field of real numbers |
| $\mathbb{C}$ | Field of complex numbers |
| $\mathbb{F}$ | Field |
| $\mathbb{F}^*$ | If $\mathbb{F}$ is a field, $\mathbb{F}^* = \mathbb{F} - \{0\}$ |
| $R[x]$ | Ring of polynomials over a ring $R$ |
| $\mathbb{Z}_n$ | Ring and group of integers modulo $n$ |
| $U_n$ | Group of units modulo $n$ |
| $a \mid b$ | $a$ divides $b$ |
| $a \nmid b$ | $a$ does not divide $b$ |
| $a \equiv b \pmod{n}$ | Congruence modulo $n$ |
| $(a, b)$ | Greatest common divisor of integers $a$ and $b$ |
| $f : A \rightarrow B$ | A function of $A$ to $B$ |
| $f(a)$ | Image of $a$ under $f$ |
| $f(A)$ | Image of $A$ under $f$ |
| $Im f$ | Image of $f$ |
| $f^{-1}(B)$ | Inverse image of $B$ under $f$ |
| $id_X$ | The identity function $X \rightarrow X$ |

| | |
|---|---|
| $\varphi$ | Euler function |
| $\mu$ | Möbius function |
| $f^{-1}$ | The inverse of the function $f$ |
| $f \circ g$ | Composite function |
| $\begin{pmatrix} n \\ k \end{pmatrix}$ | Binomial coefficient $n!/k!(n-k)!$ |
| $\lvert A \rvert$ | Number of elements in a finite set $A$ |
| $\lvert G \rvert$ | Order of the group $G$ |
| $\circ(a)$ | Order of element $a$ |
| $H \leqslant G$ | $H$ is a subgroup of $G$ |
| $H \trianglelefteq G$ | $H$ is a normal subgroup of $G$ |
| $a^{-1}$ | The inverse of $a$ |
| $Z(G)$ | Center of $G$ |
| $C_G(a)$ | Centralizer of $a$ in $G$ |
| $C_G(X)$ | Centralizer of the set $X$ in $G$ |
| $N_G(X)$ | Normalizer of the $X$ in $G$ |
| $\langle a \rangle$ | Cyclic group generated by $a$ |
| $\langle X \rangle$ | Subgroup generated by a set $X$ |
| $AB$ | $\{ab \ \ a \in A$ and $b \in B\}$ |
| $A = (a_{ij})$ | Matrix whose entry in row $i$ and column $j$ is $a_{ij}$ |
| $\det(A)$ | Determinant of matrix $A$ |
| $I_n$ | The identity $n \times n$ matrix |
| $A^t$ | Transpose of the matrix $A$ |
| $tr(A)$ | Trace of the matrix $A$ |
| $Mat_{m \times n}(\mathbb{F})$ | The set of all $m \times n$ matrices over $\mathbb{F}$ |
| $aH$ | Left coset |
| $Ha$ | Right coset |
| $G/H$ | Factor group |
| $[G : H]$ | Index of the subgroup $H$ in the group $G$ |
| $G \times H$ | Direct product of groups |
| $\prod_{i \in I} G_i$ | Direct product of groups |
| $K \ltimes_{\varphi} H$ | Semidirect product |
| $RG$ | Group ring of a group $G$ over a ring $R$ |
| $G \cong H$ | $G$ is isomorphic to $H$ |
| $C_n$ | Cyclic group of order $n$ |
| $S_X$ | Symmetric group on $X$ |
| $S_n$ | Symmetric group of degree $n$ |
| $A_n$ | Alternating group |
| $D_n$ | Dihedral group of order $2n$ |
| $Q_8$ | Quaternion group |
| $GL_n(\mathbb{F})$ | General linear group |
| $SL_n(\mathbb{F})$ | Special linear group |
| $O_n(\mathbb{F})$ | Orthogonal group |
| $SO_n(\mathbb{F})$ | Special orthogonal group |

| $UT_n(\mathbb{F})$ | The set of upper triangular matrices such that all entries on the diagonal are nonzero |
| $LT_n(\mathbb{F})$ | The set of lower triangular matrices such that all entries on the diagonal are nonzero |
| $PGL_n(\mathbb{F})$ | Projective general linear group |
| $PSL_n(\mathbb{F})$ | Projective special linear group |
| $\Gamma L(V, \mathbb{F})$ | Semi-linear group |
| $P\Gamma L(V, \mathbb{F})$ | Projective semi-linear group |
| $\langle X, Y \rangle$ | Inner product of two vectors |
| $[x, y]$ | $x^{-1}y^{-1}xy$ |
| $G'$ | Derived subgroup of a group $G$ |
| $Kerf$ | Kernel of the homomorphism $f$ |
| $End(G)$ | The set of all endomorphism from $G$ to itself |
| $Aut(G)$ | Automorphism group |
| $Inn(G)$ | Inner automorphism group |
| $\mathrm{Stab}_G(X)$ | Stabilizer of $X$ in $G$ |
| $\mathrm{Orb}_G(x)$ | Orbit of $x$ |
| $\mathrm{Fix}_G(Y)$ | Fixed point of $Y$ |
| $F_X(A)$ | Fixed point subset under $A$ |
| $\mathrm{Syl}_p(G)$ | The set of all distinct Sylow $p$-subgroups of $G$ |
| $\langle X \vert Y \rangle$ | Group presented by generators $X$ and relations $Y$ |

# Chapter 1
# Group Actions on Sets

Groups often arise as the set of permutations of some mathematical object. Some examples of this would be the symmetric group $S_n$, the alternating group $A_n$, or the dihedral group $D_n$. We can formalize this notion with the concept of a group action. To do this, we begin with developing the necessary tools we need to investigate group actions. We begin by discussing the definition of group actions and simple examples. A group action is not the same thing as a binary operation. In a binary operation, we combine two elements of $G$ to get a third element of $G$. In a group action, we combine an element of a group $G$ with an element of a set $X$ to get an element of $X$.

## 1.1  Group Actions and $G$-Sets

We will give two equivalent definitions of a group $G$ acting on a set $X$.

**Definition 1.1** Let $G$ be a group and $X$ be a non-empty set. A (*right*) *action* of $G$ on $X$ is a function $\theta : X \times G \to X$ such that

(1) $\theta(x, ab) = \theta\,(\theta(x, a), b)$, for all $a, b \in G$ and $x \in X$;
(2) $\theta(x, e) = x$, for all $x \in X$.

Often the function $\theta$ is suppressed in the notation, and one may write $x^a$ or $x \cdot a$ for $\theta(x, a)$. Similarly, a left action of $G$ on $X$ is a function $\theta : G \times X \to X$ satisfying (in the evident juxtaposition notation) $\theta(ab, x) = \theta\,(a, \theta(b, x))$ and $\theta(e, x) = x$. Of course the distinction between left and right actions does not depend on whether we write the domain of $\theta$ as $X \times G$ or $G \times X$. The distinction is that in a right action $ab$ acts by $a$ first, then $b$, whereas in a left action $b$ acts first, then $a$. From now on, unless specified otherwise, "action" means "right action". We think of an action as a way of making the group $G$ move the points in $X$ around. Thus the first condition states that applying two elements $a$ and $b$ in sequence has the same effect as applying the

© The Author(s), under exclusive license to Springer Nature Singapore Pte Ltd. 2021
B. Davvaz, *Groups and Symmetry*, https://doi.org/10.1007/978-981-16-6108-2_1

product $ab$, while the second condition is that the identity element fixes every point, and so acts in the same way as the identity permutation of $X$.

A set $X$ with an action of a group $G$ is often known as a *G-set*. If $|X| = n$, then $n$ is called the *degree* of the $G$-set $X$. We often regard geometrically $X$ as a set of points, and the elements of $X$ are called *point*.

**Definition 1.2** We say that $G$ acts on a non-empty set $X$ if there is a homomorphism $f : G \to S_X$.

Before giving examples, we need to show that Definitions 1.1 and 1.2 actually define the same notion.

**Theorem 1.3** *Definitions 1.1 and 1.2 are equivalent.*

***Proof*** Suppose that $G$ and $X$ satisfy Definition 1.2, so that we have a homomorphism $f : G \to S_X$. We show that $G$ and $X$ satisfy Definition 1.1 too. We define a function $\theta :$ $X \times G \to X$ by $\theta(x, a) = x^{f(a)}$, for all $a \in G$ and $x \in X$. Since $f$ is a homomorphism, it follows that

$$\theta(x, ab) = x^{f(ab)} = x^{f(a)f(b)}$$
$$= \left(x^{f(a)}\right)^{f(b)} = \theta\left(\theta(x, a), b\right),$$

for all $a, b \in G$ and $x \in X$. So, $\theta$ satisfies the first condition of Definition 1.1. Also, since $f$ is a homomorphism, it follows that $f(e) = \mathrm{id}_X$, the identity permutation. This yields that $\theta(x, e) = x^{f(e)} = x$, which is the second condition of Definition 1.1. Consequently, $G$ and $X$ satisfy Definition 1.1.

Now, suppose that $G$ and $X$ satisfy Definition 1.1. Hence, we have an action $\theta : X \times G \to X$. We define a map $f : G \to S_X$ such that $x^{f(a)} = \theta(x, a)$, for all $a \in G$ and $x \in X$. First, we prove that $f$ is well defined, i.e., $f(a)$ is a bijective function from $X$ to itself. To show that $f(a)$ is onto, let $x \in X$, and consider $\theta(x, a^{-1}) \in X$. Then, we have

$$\theta(x, a^{-1})^{f(a)} = \theta\left(\theta(x, a^{-1}), a\right) = \theta(x, a^{-1}a) = \theta(x, e) = x.$$

This yields that $f(a)$ is onto. Now, we show that $f(a)$ is one to one. If $x^{f(a)} = y^{f(a)}$, for some $x, y \in X$, then $\theta(x, a) = \theta(y, a)$. So, we have $\theta\left(\theta(x, a), a^{-1}\right) = \theta\left(\theta(y, a), a^{-1}\right)$. By the first condition of Definition 1.1, we obtain $\theta(x, aa^{-1}) = \theta(y, aa^{-1})$ or $\theta(x, e) = \theta(y, e)$. Next, by the second condition of Definition 1.1, we conclude that $x = y$. Finally, we show that $f$ is a homomorphism. For every $x, y \in G$ and $x \in X$, we have

$$x^{f(ab)} = \theta(x, ab) = \theta(\theta(x, a), b) = \theta(x, a)^{f(b)} = x^{f(a)f(b)}.$$

This shows that $G$ and $X$ satisfy Definition 1.2.                                        ∎

Theorem 1.3 shows that the notion of a $G$-set is equivalent to the notion of representation of $G$ by permutations on the set $X$.

**Definition 1.4** Let $G$ be a group. A *permutation representation* of $G$ is a homomorphism from $G$ to $S_X$, for some set $X$.

Now that we have a few ways of thinking about group actions, let's see some examples.

**Example 1.5** The symmetric group $S_n$ acts on the set $X = \{1, 2, \ldots, n\}$ by $(x, \sigma) \mapsto x\sigma$, the effect of applying the permutation $\sigma$ to the number $x \in X$.

We call this action the *natural action* of $S_n$.

**Example 1.6** If $\mathbb{F}$ is a field, then $S_n$ acts on $\mathbb{F}[x_1, \ldots, x_n]$ by $f^\sigma(x_1, \ldots, x_n) = f(x_{1\sigma}, \ldots, x_{n\sigma})$, for all $\sigma \in S_n$ and $f \in \mathbb{F}[x_1, \ldots, x_n]$.

**Example 1.7** Let $V$ be a vector space of dimension $n$ over a field. Let $T$ and $T'$ be two linear transformations in $GL_n(\mathbb{F})$. Similar to the notation for symmetric groups, $TT'$ means that first apply $T$ and then $T'$. Now, if we define

$$x^T := xT,$$

for all $T \in GL_n(\mathbb{F})$ and $x \in V$, then it is easy to verify that this is a group action. Hence, $GL_n(\mathbb{F})$ acts on $V$ as linear transformations.

**Example 1.8**

(1) Let $G$ be a group and $H$ a subgroup of $G$. An action of the group $H$ on the set $G$ is $(g, h) \mapsto hg$ (product in $G$), where $h \in H$ and $g \in G$. This action is called a *left translation*. In particular, if $G = H$, then $G$ acts on itself;

(2) Let $K$ be another subgroup of $G$ and $\mathcal{L}$ be the set of all left cosets of $K$ in $G$. Then, $H$ acts on $\mathcal{L}$ by translation $(gK, h) \mapsto hgK$, where $gK \in \mathcal{L}$ and $h \in H$.

**Example 1.9** Every group $G$ acts on itself by conjugation $x^a := a^{-1}xa$, for $x, a \in G$. To see that this is an action, note that for every $a, b \in G$, we have

$$x^{ab} = (ab)^{-1}x(ab) = b^{-1}a^{-1}xab = b^{-1}(x^a)b = (x^a)^b.$$

Since clearly $x^e = x$, conjugation does give an action of $G$ on itself. This action is the trivial if and only if $G$ is abelian.

**Example 1.10** Every group $G$ acts on the family of all subgroups by conjugation.

**Example 1.11** Let $G = \{1, -1\}$ and $X = \mathbb{R}$.

(1) Define $\theta : X \times G \to X$ by

$$\theta(x, a) = \begin{cases} x & \text{if } a = 1 \\ -x & \text{if } a = -1. \end{cases}$$

This defines an action of $G$ on $X$.

(2) Define $\theta : X \times G \to X$ by

$$\theta(x, a) = \begin{cases} x & \text{if } a = 1 \text{ and } x \neq 0 \\ x^{-1} & \text{if } a = -1 \text{ and } x \neq 0 \\ 0 & \text{if } x = 0. \end{cases}$$

Also, this defines an action of $G$ on $X$.

Therefore, it is possible for a group to act on a set via different actions.

**Example 1.12** The dihedral group $D_n$ acts on the vertices of a regular $n$-gon. Note that $D_n$ can be considered as a subgroup of $S_n$ generated by the following two permutations:

$$\alpha = (1\ 2\ 3\ \dots\ n) \text{ and } \beta = \begin{pmatrix} 1 & 2 & 3 & \dots & n \\ 1 & n & n-1 & \dots & 2 \end{pmatrix}.$$

Recall that this has the following properties: $o(\alpha) = n$, $o(\beta) = 2$ and $\beta\alpha = \alpha^{-1}\beta$. Now, we consider a regular $n$-gon with vertices labeled from 1 to $n$ as in Fig. 1.1. Applying $\alpha$ to the vertices induces a rotation of the regular $n$-gon. Applying $\beta$ produces a reflection in the axis through vertex 1. Hence both $\alpha$ and $\beta$ induce transformations of the regular $n$-gon and consequently any product of them does so also.

**Definition 1.13** Let $G$ be a group and $\theta$ be an action of $G$ on a non-empty set $X$. The *kernel* of the action $\theta$ is the set of all group elements that fix every element in $X$, i.e.,

$$\mathrm{Ker}\theta = \{a \in G \mid x^a = x, \text{ for all } x \in X\}.$$

**Theorem 1.14** $\mathrm{Ker}\theta$ *is a normal subgroup of* $G$.

**Proof** Since $x^e = x$, for all $x \in X$, it follows that $e \in \mathrm{Ker}\theta$, and so it is non-empty. Further, if $a, b \in \mathrm{Ker}\theta$, then $x^{ab} = (x^a)^b = x^b = x$, for all $x \in X$, and so $ab \in \mathrm{Ker}\theta$. Also, if $a \in \mathrm{Ker}\theta$, then $x = x^e = x^{aa^{-1}} = (x^a)^{a^{-1}} = x^{a^{-1}}$, for all $x \in X$, and hence $a^{-1} \in \mathrm{Ker}\theta$. This shows that $\mathrm{Ker}\theta$ is a subgroup of $G$. For normality, let $a \in \mathrm{Ker}\theta$ and $g \in G$. Then, we can write

**Fig. 1.1** Dihedral group $D_n$ acts on the vertices of a regular $n$-gon

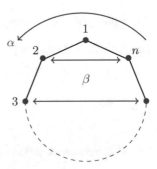

$$x^{g^{-1}ag} = \left(\left(x^{g^{-1}}\right)^a\right)^g = \left(x^{g^{-1}}\right)^g = x^{g^{-1}g} = x^e = x,$$

for all $x \in X$ and so $g^{-1}ag \in \text{Ker}\theta$. Thus, $\text{Ker}\theta \trianglelefteq G$.                                     ∎

**Definition 1.15** In Definition 1.13, if $\text{Ker}\theta = \{e\}$, then we say that the action is *faithful*.

Thus an action is faithful if $x^a = x$ for all $x \in X$ implies that $a = e$. In other words, the only element of $G$ that fixes everything in $X$ is the identity.

The fundamental question on group actions is to classify, for a given group $G$, all possible sets with a $G$-action. The first step, as always with such classification problems, is to decide when we wish two consider to $G$-sets equivalent.

**Definition 1.16** Let $G$ be a group and $X_1$, $X_2$ be $G$-sets. A function $\psi : X_1 \to X_2$ is said to be *$G$-equivalent* if $\psi(x^a) = \psi(x)^a$, for all $x \in X_1$ and $a \in G$. In terms of the associated homomorphisms $\alpha : G \to S_{X_1}$ and $\beta : G \to S_{X_2}$ we can write the condition as follows: For each $a \in G$, the diagram (Fig. 1.2) commutes.

We say that two $G$-sets $X_1$ and $X_2$ are *isomorphic* if there is a $G$-equivalent bijection $\psi : X_1 \to X_2$.

**Exercises**

1. Let $G$ be a group and $H$ be a subgroup of $G$. Suppose that $\mathcal{R}$ is the set of distinct right cosets of $H$ in $G$ and $\mathcal{L}$ is the set of distinct left cosets of $H$ in $G$.

   (a) Show that how each of these sets is a $G$-set in a natural way;
   (b) Show that $G$-sets $\mathcal{R}$ and $\mathcal{L}$ are isomorphic.

2. Let $X = \{1, 2, 3\}$ and $G = \mathbb{Z}_2$.

   (a) How many different actions are there of $G$ on $X$?
   (b) Up to isomorphism how many different $G$-sets do these give?

3. The *extended complex numbers* consist of the complex numbers $\mathbb{C}$ together with $\infty$. Geometrically, the set of extended complex numbers is referred to as the *Riemann sphere*. Show that the group $SL_2(\mathbb{C})$ of $2 \times 2$ complex matrices with determinant 1 acts (from left) on the Riemann sphere by

$$\left(\begin{bmatrix} a & b \\ c & d \end{bmatrix}, z\right) \mapsto \frac{az + b}{cz + d}.$$

**Fig. 1.2** $G$-equivalent function

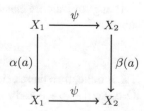

4. Let $X = \{1, 2, 3\}$ and $G = \mathbb{Z}_2$.

    (a) How many different actions are there of $G$ on $X$?
    (b) Up to isomorphism how many different $G$-sets do these give?

5. Let $X$ be a $G$-set. Show that $G$ acts faithfully on $X$ if and only if no two distinct elements of $G$ have the same action on each element of $X$.

## 1.2   Orbits and Stabilizers

We now define a few important terms relevant to group actions.

Let $G$ be a group and $X$ be a $G$-set. For every $x, y \in X$, we define

$$x \sim y \iff \text{ there is } a \in G \text{ such that } x^a = y.$$

**Lemma 1.17** $\sim$ *is an equivalence relation on* $X$.

*Proof* We investigate the three conditions of an equivalence relation.

(1) Suppose that $x \in X$. Since $x^e = x$, it follows that $x \sim x$.
(2) If $x, y \in X$ such that $x \sim y$, then by the definition, there is $a \in G$ such that $x^a = y$. Hence, we obtain $y^{a^{-1}} = (x^a)^{a^{-1}} = x^{aa^{-1}} = x^e = x$. This shows that $y \sim x$.
(3) If $x, y, z \in X$ such that $x \sim y$ and $\sim z$, then by the definition, there are $a, b \in G$ such that $x^a = y$ and $y^b = z$. Hence, we can write $z = y^b = (x^a)^b = x^{ab}$. Since $ab \in G$, it follows that $x \sim z$.     ∎

**Definition 1.18** The corresponding equivalence classes are called the *orbits* of $G$ on $X$. For $x \in X$,

$$\{x^a \mid a \in G\}$$

is the *orbit* that contains $x$; we write it as $\mathrm{Orb}_G(x)$ or $x^G$. The number of elements in $\mathrm{Orb}_G(x)$ is called the *length* of the $\mathrm{Orb}_G(x)$. A subset $Y$ of $X$ is said to be $G$-invariant if for all $a \in G$,

$$y \in Y \implies y^a \in Y.$$

Being equivalence classes, the orbits partition $X$ into a union of pairwise disjoint subsets:

$$X = \bigcup_{x \text{ in disjoint orbits}} \mathrm{Orb}_G(x).$$

If $X$ is finite, then there exist disjoint orbits $\mathrm{Orb}_G(x_1), \ldots, \mathrm{Orb}_G(x_k)$ such that $|X| = |\mathrm{Orb}_G(x_1)| + \cdots + |\mathrm{Orb}_G(x_k)|$.

Orbits are $G$-invariant subsets of $X$ (in the sense defined above of $G$-invariance), and are therefore themselves $G$-sets: in fact, a subset of $X$ is $G$-invariant precisely when it is a union of orbits.

**Definition 1.19** Let $G$ be a group which acts on a set $X$. For $x \in X$, the *stabilizer* of $x$ in $G$, written

$$\mathrm{Stab}_G(x) = \{a \in G \mid x^a = x\}.$$

Thus, the stabilizer of $x$ is the set of all group elements which fix $x$. More generally, for every subset $Y \subseteq X$, we can consider elements of $G$ which fix $Y$:

$$\mathrm{Fix}_G(Y) = \{a \in G \mid x^a = x, \text{ for all } x \in Y\}.$$

***Example 1.20*** Consider $S_n$ acts on $\{1, \ldots, n\}$ by $k^\sigma := k\sigma$, for $\sigma \in S_n$ and $1 \le k \le n$. Then there is just one orbit. Moreover, $\mathrm{Stab}_{S_n}(n) = \{\sigma \in S_n \mid n^\sigma = n\} \cong S_{n-1}$.

***Example 1.21*** Consider a group $G$ acting on itself by conjugation as in Example 1.9; in other words, $x^a := a^{-1}xa$, for all $x, a \in G$. The orbit of an element can be either just that element, or it can be many elements. It is easy to check that the following are equivalent:

(1) $a \in Z(G)$;
(2) $\mathrm{Stab}_G(x) = G$;
(3) $\mathrm{Orb}_G(x) = \{x\}$;
(4) $|\mathrm{Orb}_G(x)| = 1$.

***Example 1.22*** Let $S_3$ act on itself by conjugation. Its orbits are the conjugacy classes, which are

$$\mathrm{id}, \{(1\ 2), (1\ 3), (2\ 3)\}, \{(1\ 2\ 3), (1\ 3\ 2)\}.$$

The conjugacy class of $(1\ 2)$ has size 3, and the stabilizer of $(1\ 2)$ is its centralizer $\{\mathrm{id}, (1\ 2)\}$, which has index 3 in $S_3$.

**Lemma 1.23** *Let $G$ be a group acting on a set $X$ and suppose that $x \in X$ and $Y \subseteq X$. Then, both $\mathrm{Stab}_G(x)$ and $\mathrm{Fix}_G(Y)$ are subgroups of $G$.*

**Proof** First, we prove that $\mathrm{Stab}_G(x)$ is a subgroup of $G$. Since $x^e = x$, it follows that $e \in \mathrm{Stab}_G(x)$ and hence $\mathrm{Stab}_G(x)$ is non-empty. Now, assume that $a$ and $b$ are two arbitrary elements of $\mathrm{Stab}_G(x)$. Then, we have $x^{ab} = (x^a)^b = x^b = x$, and so $ab \in \mathrm{Stab}_G(x)$, while $x^{a^{-1}} = (x^a)^{a^{-1}} = x^{aa^{-1}} = x^e = x$, and hence $a^{-1} \in \mathrm{Stab}_G(x)$. This yield that $\mathrm{Stab}_G(x)$ is a subgroup of $G$.

We leave it to reader to show that $\mathrm{Fix}_G(Y)$ is a subgroup, but the argument is almost exactly the same. ∎

**Lemma 1.24** *With notation as above, we have the following equality of sets:*

$$\mathrm{Fix}_G(Y) = \bigcap_{x \in Y} \mathrm{Stab}_G(x).$$

***Proof*** It is straightforward.                                                            ∎

The next result is the most important basic result in the theory of group actions.

**Theorem 1.25  (Orbit-stabilizer theorem).** *If $G$ is a group which acts on a set $X$, then for any $x \in X$, we have*

$$|\text{Orb}_G(x)| = [G : \text{Stab}_G(x)].$$

***Proof*** We demonstrate the existence of a bijection from the set of right cosets of the $\text{Stab}_G(x)$ in $G$ to the orbit of $x$. We define $\varphi : \text{Stab}_G(x)a \mapsto x^a$, for all $a \in G$. First, we show that $\varphi$ is well defined. Suppose that $\text{Stab}_G(x)a = \text{Stab}_G(x)b$, for some $a, b \in G$. Then, we have $ab^{-1} \in \text{Stab}_G(x)$, which implies that $x^{ab^{-1}} = x$. Apply $b$, we obtain $x^a = x^b$. Consequently, $\varphi$ is well defined. Now, we show that $\varphi$ is one to one. Let $\varphi\big(\text{Stab}_G(x)a\big) = \varphi(\text{Stab}_G(x)b)$, for some $a, b \in G$. Then, we have $x^a = x^b$, which implies that $x^{ab^{-1}} = x$, and so $ab^{-1} \in \text{Stab}_G(x)$. This yields that $\text{Stab}_G(x)a = \text{Stab}_G(x)b$, and hence $\varphi$ is one to one. Finally, if $y \in \text{Orb}_G(x)$, then there is $a \in G$ such that $y = x^a$. Thus, we have $\varphi\big(\text{Stab}_G(x)a\big) = y$, and so $\varphi$ is onto. Therefore, $\varphi$ is a one-to-one correspondence between the set of all right cosets of $\text{Stab}_G(x)$ to the orbit of $x$, and this completes the proof.                                        ∎

**Corollary 1.26** *Let a group $G$ act on a set $X$, where $X$ is finite. Let the distinct orbits of $X$ be represented by $x_1, \ldots, x_k$. Then,*

$$|X| = \sum_{i=1}^{k} |\text{Orb}_G(x_i)| = \sum_{i=1}^{k} [G : \text{Stab}_G(x_i)].$$

***Proof*** The set $X$ can be written as the union of its orbits, which are mutually disjoint. The Orbit-Stabilizer theorem tells us how large each orbit is.                        ∎

One thing to consider is the following observation: Let $G$ acts on a set $X$ and suppose that $x, y \in X$ are two points that lie in the same orbit. We know that orbits are either disjoint or equal, so we have $\text{Orb}_G(x) = \text{Orb}_G(y)$. Hence, by the Orbit-Stabilizer theorem, we conclude that $[G : \text{Stab}_G(x)] = [G : \text{Stab}_G(y)]$. In particular, if $G$ is a finite group, we can deduce immediately that $|\text{Stab}_G(x)| = |\text{Stab}_G(y)|$.

**Theorem 1.27** *Let $G$ be a group which acts on a set $X$ and suppose that $x, y \in X$. If $x$ and $y$ lie in the same orbit of $G$ on $X$, then the stabilizers $\text{Stab}_G(x)$ and $\text{Stab}_G(y)$ are conjugate subgroups of $G$. In particular, we have*

$$\text{Stab}_G(x^a) = \big(\text{Stab}_G(x)\big)^a,$$

*for some $a \in G$.*

***Proof*** Since $y \in \text{Stab}_G(x)$, it follows that there is $a \in G$ such that $y = x^a$. We show that

$$\text{Stab}_G(y) = \big(\text{Stab}_G(x)\big)^a = a^{-1}\text{Stab}_G(x)a.$$

Suppose that $b$ is an arbitrary element of $\text{Stab}_G(x)$. Then, we have $a^{-1}ba \in \big(\text{Stab}_G(x)\big)^a$, and so $y^{a^{-1}ba} = (x^a)^{a^{-1}ba} = x^{ba} = x^a = y$. Hence, we conclude that $a^{-1}ba \in \text{Stab}_G(y)$, and this shows that $\big(\text{Stab}_G(x)\big)^a \subseteq \text{Stab}_G(y)$.

For the converse of inclusion, since $y = x^a$, it follows that $y^{a^{-1}} = x$. Hence, reasoning in the same way as we did from the equation $x^a = y$, we obtain $\big(\text{Stab}_G(y)\big)^{a^{-1}} \subseteq \text{Stab}_G(x)$, or equivalently, $\text{Stab}_G(y) \subseteq a^{-1}\text{Stab}_G(x)a = \big(\text{Stab}_G(x)\big)^a$.

Therefore, we proved that $\text{Stab}_G(y) = \big(\text{Stab}_G(x)\big)^a$, for some $a \in G$, as required. ■

**Definition 1.28** Let $G$ be a group acting on a set $X$. The collection of all fixed points of $a \in G$ is denoted by $F_X(a)$ and is called the *fixed point subset* of $a$, i.e.,

$$F_X(a) = \{x \in X \mid x^a = x\}.$$

More generally, for a subset $A$ of $G$, let $F_X(A)$ represent the set of elements of $X$ that are fixed by all elements of $A$, i.e.,

$$F_X(A) = \{x \in X \mid x^a = x, \text{ for all } x \in A\} = \bigcap_{a \in A} F_X(a).$$

$F_X(A)$ is called the *fixed point subset* under $A$.

**Theorem 1.29  (Burnside's lemma).** *If $G$ is a finite group acting on a set $X$, then the number of orbits of $G$ on $X$ is*

$$k = \frac{1}{|G|} \sum_{a \in G} |F_X(a)|.$$

**Proof** Let $n$ denote the number of pairs $(x, a) \in X \times G$ with $x^a = x$. We begin by counting these pairs in two different ways. The first way would be to look at $F_X(a)$. For every $a \in G$, the number of pairs $(x, a)$, where $x^a = x$, would be $|F_X(a)|$. Then, in order to obtain the total number of pairs, we need to sum $|F_X(a)|$, for all $a \in G$. Consequently, we obtain

$$n = \sum_{a \in G} |F_X(a)|.$$

The second way to count the number of pairs would be to look at $\text{Stab}_G(x)$. For each $x \in X$, the number of pairs $(x, a)$, where $x^a = x$, would be $|\text{Stab}_G(x)|$. Then, in order to get the total number of pairs, we need to sum $|\text{Stab}_G(x)|$, for all $x \in X$. Hence, we have

$$n = \sum_{x \in X} |\text{Stab}_G(x)|.$$

Therefore, we have two different expressions that allow us to count the number of pairs, $n$, of the form $(x, a)$, where $x^a = x$. Consequently, we are able to set them equal to each other:

$$n = \sum_{a \in G} |F_X(a)| = \sum_{x \in X} |\text{Stab}_G(x)|. \tag{1.1}$$

By the Orbit-Stabilizer theorem, we have

$$\frac{|G|}{|\text{Orb}_G(x)|} = |\text{Stab}_G(x)|.$$

Now, we focus on a single orbit of $G$. If $y$, $z$ are in the same orbit of $G$, then $\text{Orb}_G(y) = \text{Orb}_G(z)$. Further, we conclude that $|\text{Stab}_G(y)| = |\text{Stab}_G(z)|$. For each orbit, we sum the order of the stabilizer over the elements in the orbit:

$$\sum_{t \in \text{Orb}_G(y)} |\text{Stab}_G(t)| = |\text{Orb}_G(y)| \, |\text{Stab}_G(y)| = |G|.$$

Suppose that $\text{Orb}_G(x_1), \ldots, \text{Orb}_G(x_k)$ are distinct orbits. So, by (1.1), we have

$$n = \sum_{a \in G} |F_X(a)| = \sum_{x \in X} |\text{Stab}_G(x)|$$

$$= \sum_{x \in \text{Orb}_G(x_1)} |\text{Stab}_G(x)| + \cdots + \sum_{x \in \text{Orb}_G(x_k)} |\text{Stab}_G(x)|$$

$$= k|G|.$$

From the second and the last part of this equation, we deduce that

$$k = \frac{1}{|G|} \sum_{a \in G} |F_X(a)|,$$

as desired.                                                                                        ∎

Burnside's Lemma can be described as finding the number of distinct orbits by taking the average size of the fixed sets.

**Exercises**

1. Let $G = \{\text{id}, (1\ 2)(3\ 4), (1\ 2\ 3\ 4)(5\ 6), (1\ 3)(2\ 4), (1\ 4\ 3\ 2)(5\ 6), (5\ 6)(1\ 3), (1\ 4)(2\ 3), (2\ 4)(5\ 6)\}$.

   (a) Find the stabilizer of 1 and the orbit of 1;
   (b) Find the stabilizer of 3 and the orbit of 3;
   (c) Find the stabilizer of 5 and the orbit of 5.

2. Find the number of orbits in $\{1, 2, 3, 4, 5, 6, 7, 8\}$ under the subgroup of $S_8$ generated by $(1\ 3)$ and $(2\ 4\ 7)$.

**Fig. 1.3** A torus

3. Let $H, K \leq S_8, H = \langle (1\,2\,3)(4\,5\,6)(7\,8) \rangle$ and $K = \langle (3\,8) \rangle$. Determine the orbits of $H, K$ and $\langle H, K \rangle$ on $\{1, \ldots, 8\}$.
4. Let $G$ be the additive group of real numbers. Let the action of $\theta \in G$ on the real plane $\mathbb{R}^2$ be given by rotating the plane counterclockwise about the origin through $\theta$ radians. Let $P$ be a point other than the origin in the plane.

   (a) Show $\mathbb{R}^2$ is a $G$-set;
   (b) Describe geometrically the orbit containing $P$;
   (c) Find $\mathrm{Stab}_G(P)$.

5. Let $\psi : X_1 \to X_2$ be a $G$-equivalent. Show that if $O_i$ is an orbit in $X_1$, then $\psi(O_i)$ is an orbit in $X_2$.
6. Let $X$ be a countably infinite set. Show that $S_X$ has countably many orbits on the power set $\mathcal{P}(X)$, and describe these orbits.
7. In geometry, a *torus* is a surface of revolution generated by revolving a circle in three-dimensional space about an axis that is coplanar with the circle, see Fig. 1.3. We can consider a torus as a product of two circles. Describe the orbits of the following actions of $\mathbb{R}$ on the torus.

   (a) The real number $t$ sends $(e^{ix}, e^{iy})$ to $(e^{i(x+t)}, e^{iy})$;
   (b) The real number $t$ sends $(e^{ix}, e^{iy})$ to $(e^{i(x+t)}, e^{i(y+t)})$;
   (c) The real number $t$ sends $(e^{ix}, e^{iy})$ to $(e^{i(x+t)}, e^{i(y+t\sqrt{2})})$.

## 1.3 Transitive $G$-Sets

We begin this section with the definition of a transitive action.

**Definition 1.30** Suppose that a group $G$ acts on a set $X$. We say that the action is *transitive*, or that $G$ *acts transitively* on $X$ if there is only one orbit, namely, for every $x, y \in X$, there exists $a \in G$ such that $x^a = y$. The set $X$ is then called a *homogeneous G-set*.

   A permutation representation $f$ on $X$ of a group $G$ is said to be *transitive* if the $G$-set $X$ via the action $f$ is transitive.

**Example 1.31**  The symmetric group $S_n$ acts transitively on the set $\{1, \ldots, n\}$, i.e., there is only one orbit.

**Example 1.32**  For any subgroup $H$ of a group $G$, $G$ acts transitively on $G/H$, but the action of $G$ on itself is never transitive if $G \neq \{e\}$ because $\{e\}$ is always a conjugacy class.

**Example 1.33**  The dihedral group $D_n$ acts transitively on the set of vertices $V$ of the $n$-gon. Let $v_1$ and $v_2$ be vertices, then certainly there exists some rotation $\rho$ for which $v_1 = v_2^\rho$. Also, the action is faithful since if an element leaves all the vertices fixed, it must be the identity.

**Example 1.34**  Orthogonal transformations preserve length, so if $X$ is the unit sphere in $\mathbb{R}^3$, we have a natural action of $SO_3(\mathbb{R})$ on $X$. The orthogonal transformation $T$ sends the unit vector $v_1$ to the unit vector $v_1 T$. The orbit of any vector is the whole sphere. Two unit vectors $v_1$ and $v_2$ determine a plane and a suitable rotation of $\mathbb{R}^3$ about the axis through the origin perpendicular to this plane will bring $v_1$ to the position of $v_2$. The matrix of this rotation is an element of $SO_3(\mathbb{R})$ which sends $v_1$ to $v_2$. If $T$ fixes the first element of the standard basis for $\mathbb{R}^3$, then the matrix of $T$ has the form

$$\begin{bmatrix} 1 & 0 & 0 \\ 0 & & \\ 0 & A & \end{bmatrix}$$

where $A \in SO_2(\mathbb{R})$. Points in the same orbit always have conjugate stabilizers. Therefore, the stabilizer of every unit vector is isomorphic to $SO_2(\mathbb{R})$. This is an example of a transitive action.

**Definition 1.35**  We say that a group $G$ acts $k$-*transitively* on a set $X$ if for any two ordered tuples $(x_1, \ldots, x_k)$ and $(y_1, \ldots, y_k)$ of elements of $X$, where $x_i \neq x_j$ and $y_i \neq y_j$ for $i \neq j$, then there exists an element $a$ in $G$ such that $x_i^a = y_i$, for all $1 \leq i \leq k$.

Of course, 1-transitivity is ordinary transitivity. If $k > 1$, then every $k$-transitive $G$ set is $(k-1)$-transitive. A $k$-transitive $G$-set $X$ is called *doubly transitive* if $k \geq 2$, *triply transitive* if $k \leq 3$, and so forth. It is quite common to say that a group $G$ is $k$-transitive if there exists a $k$-transitive $G$-set.

**Example 1.36**  The symmetric group $S_n$ acts on $\{1, \ldots, n\}$, $n$-transitively, and the subgroup $A_n$ consisting of all even permutations acts on $\{1, \ldots, n\}$, $(n-2)$-transitively, for $n > 3$. The first statement is evident. The second follows from the fact that if a permutation $\sigma$ carries the symbols $i_1, \ldots, i_{n-2}$ to the symbols $j_1, \ldots, j_{n-2}$, then the permutation $\sigma(i_{n-1} \, i_n)$ does too. Note that one of these permutations is even.

**Theorem 1.37**  *If a finite group $G$ acts transitively on $X$, then $|X| \, | \, |G|$.*

**Proof** For each $x \in X$, we have $\text{Orb}_G(x) = X$. Now, by Orbit-Stabilizer theorem we conclude that

$$|X| = [G : \text{Stab}_G(x)] = \frac{|G|}{|\text{Stab}_G(x)|}.$$

This completes the proof. ∎

**Theorem 1.38** *Let $G$ be a group. The following three conditions are equivalent:*

*(1) There is a transitive permutation representation of $G$ on a set of $n$ elements;*
*(2) There is a homomorphism from $G$ into $S_n$ such that the image of $G$ is transitive;*
*(3) The group $G$ has a subgroup of index $n$.*

*If a subgroup $H$ of index $n$ is given, then an action on the set of right cosets of $H$ is determined. The kernel of this action is the intersection of all conjugate subgroups of $H$.*

**Proof** If there is a transitive permutation representation $f$ of $G$ on $X = \{1, \ldots, n\}$, then the $G$-set $X$ via the action $f$ is transitive. Hence, the stabilizer $H = \text{Stab}_G(e)$ is a subgroup of index $n$. Conversely, if $H$ is a subgroup of index $n$, then the set $X$ of right cosets of $H$ in $G$ is a $G$-set by the action $Hx \mapsto Hxa$. So, there is a homomorphism $f : G \to S_X$, and this representation is transitive. Since $S_X \cong S_n$, it follows that there is a transitive homomorphism from $G$ to $S_n$.

Now, suppose that $K$ is the kernel of $f$. Then, $K$ fixes all the right cosets $Hx$. Hence, we have

$$K = \bigcap_{x \in G} \text{Stab}_G(Hx) = \bigcap_{x \in G} x^{-1}Hx.$$

This completes the proof. ∎

**Definition 1.39** Let $H$ be a subgroup of a group $G$. Define

$$\text{Core}_G(H) = \bigcap_{x \in G} H^x,$$

where $H^x = x^{-1}Hx$. This subgroup is called the *core* of $H$.

**Theorem 1.40** *The core of $H$ is the largest normal subgroup of $G$ which is contained in $H$.*

**Proof** Clearly, we have $H \trianglelefteq G$. Assume that $K$ is a normal subgroup of $G$ which contained in $H$. We have $K = K^x \subseteq H^x$, for all $x \in G$. Hence, we get

$$K \subseteq \bigcap_{x \in G} H^x = \text{Core}_G(H) \subseteq H.$$

Consequently, $\text{Core}_G(H)$ is the largest subgroup of $G$ contained in $H$. ∎

**Definition 1.41** The action of a group $G$ on a set $X$ is *regular* if for every pair $(x, y) \in X \times X$, there exists exactly one $a \in G$ such that $x^a = y$. If $N$ is a normal subgroup of $G$ that acts regularity on $X$, then $N$ is called a *regular normal subgroup* of $G$.

**Example 1.42**  The action of a group $G$ on itself by multiplication (on the left or on the right) is a regular action, called the (*left* or *right*) *regular representation* of $G$.

**Theorem 1.43**  *The following statements are equivalent:*

*(1)  $G$ acts regularity on $X$;*
*(2)  $G$ acts transitively on $X$ and $\mathrm{Stab}_G(x) = \{e\}$, for some $x \in X$.*

**Proof**  $(1 \Rightarrow 2)$: It holds by definition.

$(2 \Rightarrow 1)$: Suppose that $x, y \in X$ and $a, b \in G$ such that $x^a = x^b = y$. Then, we have $ab^{-1} \in \mathrm{Stab}_G(x)$. By Theorem 1.27, $\mathrm{Stab}_G(x)$ is conjugate to $\mathrm{Stab}_G(y)$. This shows that $x = y$.  ∎

**Corollary 1.44**  *Let $G$ be a transitive abelian permutation group on $X$. Then, $G$ acts regularity on $X$.*

**Proof**  Since $G$ is abelian, it follows that $\mathrm{Stab}_G(x)^a = \mathrm{Stab}_G(x)$, for all $a \in G$ and $x \in X$. Hence, $\mathrm{Stab}_G(x)$ fixes every element in $x^G = X$. This gives $\mathrm{Stab}_G(x) = \{e\}$. Now, by Theorem 1.43, we conclude that $G$ acts regularity on $X$.  ∎

**Corollary 1.45**  *Let $x \in X$ and $N$ be a regular normal subgroup of $G$. For $y \in X$, let $n_y \in N$ be the unique element of $N$ such that $x^{n_y} = y$. Then, for all $y \in X$ and $a \in \mathrm{Stab}_G(x)$, we have*

$$(n_y)^a = n_{y^a}.$$

*In particular, the action of $\mathrm{Stab}_G(x)$ on $X - \{e\}$ is equivalent to the action of $\mathrm{Stab}_G(x)$ on $N - \{e\}$ by conjugation.*

**Proof**  We have

$$y^a = \left(x^{n_y}\right)^a = x^{aa^{-1}n_y a} = \left(x^a\right)^{a^{-1}n_y a} = x^{(n_y)^a},$$

and we are done.  ∎

**Exercises**

1. Let $G$ act transitively on a set $X$. Let $N$ be a normal subgroup of $G$, and suppose that $Y$ is the set of orbits of $N$ in $X$. Prove that

   (a) There is a natural action of $G$ on $Y$ which is transitive and shows that every orbit of $N$ on $X$ has the same length;
   (b) Show by example that if $N$ is not normal, then its orbits need not have the same length.

2. Let $G$ be a finite group acting transitively on the set $X$. Prove that

$$\frac{1}{|G|} \sum_{a \in G} |F_X(a)| = 1.$$

   This says that in a transitive action, the average number of fixed points of an element is 1.

3. Let $X$ be a transitive $G$-set, and let $x_0 \in X$. Show that $X$ is isomorphic to the $G$-set $\mathcal{L}$ of all left cosets of $\operatorname{Stab}_G(x_0)$.
   *Hint:* For $x \in X$, suppose that $x = x_0^a$, for some $a \in G$ and define $\psi : X \to \mathcal{L}$ by $\psi(x) = a\operatorname{Stab}_G(x_0)$.

4. Let $G$ be a finite group acting doubly transitively on the set $X$. Prove that

$$\frac{1}{|G|} \sum_{a \in G} |F_X(a)| = 2.$$

5. Let $G$ be a non-trivial finite group acting transitively on the set $X$. Show that $G$ must contain at least one element that fixes no element of $X$.

6. Let $G$ be a non-trivial finite group and let $H$ be a proper subgroup of $G$. Use Exercise 1.3 to prove that $G \neq \bigcup_{a \in G} a^{-1}Ha$. Give a counter example to this result if the finiteness assumption is dropped.

7. Let $\psi : X_1 \to X_2$ be a $G$-equivalent. If the action is transitive on $X_2$, prove that $|X_2| \,|\, |X_1|$.

8. For which $n$ is the special linear group $SL_n(\mathbb{F})$ transitive on $\mathbb{F}^n - \{0\}$?

9. Let $n \geq 5$ and $2 < t < n$.

   (a) Show that $S_n$ cannot act transitively on a set with $t$ elements. Conclude that every orbit of an $S_n$-set with more than two elements has at least $n$ elements;

   (b) Show that $S_n$ has no subgroups of index $t$.

10. Let $X$ be a transitive $G$-set, and suppose that $H$ is a subgroup of $G$ with $[G : H] = n$. Show that if $H$ has $k$ orbits in $X$, then $k \leq n$.

11. Let $N$ be a normal subgroup of $G$ and $Y$ be the set of orbits of $N$ on $X$. Show that $G$ acts transitively on $Y$.

12. Let $X$ be a $G$-set and $k \geq 2$. Prove that $X$ is $k$-transitive if and only if for each $x \in X$, the $\operatorname{Stab}_G(x)$-set $X - \{x\}$ is $(k-1)$-transitive.

13. A permutation $\sigma \in S_n$ is *regular* if either $\sigma$ has no fixed points and it is the product of disjoint cycles of the same length or $\sigma = \operatorname{id}$. Prove that $\sigma$ is regular if and only if $\sigma$ is a power of a cycle $\tau$ of length $n$, i.e., $\sigma = \tau^m$, for some $m$.

14. Prove that the image of the regular representation of a group $G$ is a regular permutation group.

15. If $G$ is a transitive group of degree $n$, prove that each non-identity element $a$ in the center $Z(G)$ moves every letter. In particular, $|Z(G)| \leq n$.

## 1.4  Primitive $G$-Sets

If $G$ is a group of order $n$, then Cayley's theorem gives a faithful permutation representation of $G$ of order $n$, i.e., $G$ may be regarded as a subgroup of $S_n$. The coming discussion can often give representations of smaller degree.

**Fig. 1.4** Block example

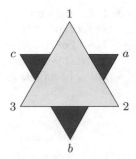

**Definition 1.46** Let $X$ is a $G$-set. A *block* is a subset $B$ of $X$ such that for each $a \in G$, either $B^a = B$ or $B^a \cap B = \emptyset$, where $B^a = \{x^a \mid x \in B\}$.

Trivial examples of blocks are $\emptyset$, $X$ and one point subsets.

**Example 1.47** Look at Fig. 1.4. Suppose that

$$G = \{\text{id}, \ (1\ 2\ 3)(a\ b\ c), \ (1\ 3\ 2)(a\ c\ b), \ (1\ b)(2\ a)(3\ c),$$
$$(1\ a)(2\ c)(3\ b), \ (1\ c)(2\ b)(3\ a)\}.$$

It is easy to check that $G \cong S_3$. We observe that either triangle $\{1, 2, 3\}$ or $\{a, b, c\}$ is a block. In addition, $\{1, a\}$, $\{2, b\}$ and $\{3, c\}$ are also blocks.

**Definition 1.48** A transitive $G$-set $X$ is *primitive* if it contains non-trivial block; otherwise, it is *imprimitive*.

Every partition $\{B_1, \ldots, B_m\}$ of a set $X$ determines an equivalence relation on $X$ whose equivalence classes are the $B_i$. In particular, if $X$ is a $G$ set and $B$ is a block, then there is an equivalence relation on $X$ given by $x \sim y$ if there is some $B^{a_i}$ containing both $x$ and $y$. This equivalence relation is $G$-*invariant*, i.e., if $x \sim y$, then $x^a \sim y^a$, for all $a \in G$. Conversely, if $x \sim y$ is a $G$-invariant equivalence relation on a $G$-set $X$, then any one of its equivalence classes is a block. Thus, a $G$-set is primitive if and only if it admits no non-trivial $G$-invariant equivalence relations.

**Theorem 1.49** *Every doubly transitive $G$-set $X$ is primitive.*

**Proof** If $X$ has a non-trivial block $B$, then there are elements $x, y, z \in X$ with $x, y \in B$ and $z \notin B$. Since $X$ is doubly transitive, it follows that there exists $a \in G$ such that $x^a = x$ and $y^a = z$. So, we obtain $x \in B \cap B^a$ and $B \neq B^a$, a contradiction. ∎

**Theorem 1.50** *Let $X$ be a transitive $G$-set of degree $n$ and let $B$ be a non-trivial block of $X$.*

*(1) If $a \in G$, then $B^a$ is a block;*

(2) *There are elements $a_1, \ldots, a_m$ of $G$ such that*

$$Y = \{B^{a_1}, \ldots, B^{a_m}\}$$

*is a partition of $X$;*

(3) *$|B|$ divides $n$, and $Y$ is a transitive $G$-set of degree $m = n/|B|$.*

**Proof** (1) If $B^a \cap B^{ab} \neq \emptyset$, for some $b \in G$, then $B \cap B^{aba^{-1}} \neq \emptyset$. Since $B$ is a block, it follows that $B = B^{aba^{-1}}$ or $B^a = B^{ab}$.

(2) Choose $y \in B$ and $x_1 \notin B$. Since $G$ acts transitively, it follows that there is $a_1 \in G$ such that $y^{a_1} = x_1$. Now, since $B$ is a block and $B \neq B^{a_1}$, it follows that $B \cap B^{a_1} = \emptyset$. If $X = B \cup B^{a_1}$, we are done. Otherwise, choose $x_2 \notin B \cup B^{a_1}$ and choose $a_2 \in G$ such that $y^{a_2} = x_2$. It is easy to see that $B^{a_2}$ is distinct from $B$ and $B^{a_1}$. Hence, since $B$ and $B^{a_1}$ are blocks, it follows that $B^{a_2} \cap (B \cup B^{a_1}) = \emptyset$. The proof is completed by iterating this procedure.

(3) Since $|B| = |B^{a_i}|$, for all $i$, it follows that $n = |X| = m|B|$ and $|Y| = m = n/|B|$. In order to check that $Y$ is a transitive $G$-set, let $B^{a_i}, B^{a_j} \in Y$ and choose $x \in B$. Since $X$ is a transitive $G$-set, it follows that there exists $a \in G$ such that $x^{a_i a} = x^{a_j}$, i.e., $B^{a_i a} \cap B^{a_j} = B^{a_j(a_j^{-1}a_i a)} \cap B^{a_j} \neq \emptyset$. Since $B^{a_j}$ is a block, it follows that $B^{a_i a} = B^{a_j}$, as desired.

∎

If $B$ is a block of a transitive $G$-set and if $Y = \{B^{a_1}, \ldots, B^{a_m}\}$ is a partition of $X$ (where $a_1, \ldots, a_m \in G$), then $Y$ is called the *imprimitive system* generated by $B$.

**Corollary 1.51** *A transitive $G$-set of prime degree is primitive.*

**Proof** This follows from Theorem 1.50 (3). ∎

**Theorem 1.52** *Let $X$ be a transitive $G$-set. Then, $X$ is primitive if and only if for each $x \in X$, the stabilizer $\mathrm{Stab}_G(x)$ is a maximal subgroup of $G$.*

**Proof** If $\mathrm{Stab}_G(x)$ is not maximal, then there is a subgroup $H$ such that $\mathrm{Stab}_G(x) \subset H \subset G$. We show that $x^H = \{x^h \mid h \in H\}$ is a non-trivial block, i.e., $X$ is imprimitive. If $a \in G$ and $x^H \cap x^{Ha} \neq \emptyset$, then $x^h = x^{h'a}$, for some $h, h' \in H$. This implies that $x = x^{h'ah^{-1}}$. Since $h'ah^{-1}$ fixes $x$, it follows that $h'ah^{-1} \in \mathrm{Stab}_G(x) \subset H$. Hence, we conclude that $a \in H$. Therefore, $x^{Ha} = x^H$, and $x^H$ is a block. It remains to show that $x^H$ is non-trivial. Obviously, $x^H$ is non-empty. Suppose that $a \in G$ with $a \notin H$. If $x^H = X$, then for every $y \in X$, there is $h \in H$ such that $y = x^h$. In particular, $x^a = x^h$, for some $h \in H$. This yields that $ha^{-1} \in \mathrm{Stab}_G(x) \subset H$. So, we obtain $a \in H$, which is a contradiction. Finally, if $x^H$ is a singleton, then $H \leq \mathrm{Stab}_G(x)$, contracting $\mathrm{Stab}_G(x) \subset H$. Consequently, $X$ is imprimitive.

Conversely, suppose that each $\mathrm{Stab}_G(x)$ is a maximal subgroup of $G$, yet there exists a non-trivial block $B$ in $X$. Define a subgroup $H$ of $G$ as follows:

$$H = \{a \in G \mid B^a = B\}.$$

Let $x \in B$ and suppose that $a \in \text{Stab}_G(x)$ is arbitrary. Then, we have $x^a = x$, and hence $x \in B \cap B^a$. Since $B$ is a block, it follows that $B^a = B$, which implies that $a \in H$. So, we conclude that $\text{Stab}_G(x) \leq H$. Since $B$ is non-trivial, it follows that there exists $y \in B$ with $y \neq x$. Since $X$ is a transitive $G$-set, it follows that $x^a = y$, for some $a \in G$. Hence, we obtain $y \in B \cap B^a$, and so $B = B^a$. This shows that $a \in H$, while $a \notin \text{Stab}_G(x)$, i.e., $\text{Stab}_G(x)$ is a proper subgroup of $H$. If $H = G$, then $B^a = B$, for all $a \in G$, and this contradicts $X \neq B$ being a transitive $G$-set. Therefore, $\text{Stab}_G(x) \subset H \subset G$, contradicting the maximality of $\text{Stab}_G(x)$.                      ∎

**Theorem 1.53** *Let $G$ be a primitive permutation group on $X$ and $\{e\} \neq N \trianglelefteq G$. Then, $N$ acts transitively on $G$. In addition, if $N$ is regular on $X$, then $N$ is a minimal normal subgroup of $G$.*

***Proof*** Suppose that $x \in X$. By Theorem 1.52, $\text{Stab}_G(x)$ is a maximal subgroup of $G$. If $N$ is a subgroup of $\text{Stab}_G(x)$, then $N$ acts trivially on $X = x^G$, which contradicts $N \neq \{e\}$. Thus, we have $\text{Stab}_G(x) \subset \text{Stab}_G(x)N = G$ and $X = x^G = x^N$ follows. Now, assume that $N$ acts regularity on $X$. Then, every normal subgroup $\{e\} \neq K \trianglelefteq G$ with $K \leq N$ is also regular on $X$. This implies that $|N| = |x^N| = |X| = |x^K| = |K|$. Therefore, we conclude that $N = K$.                      ∎

***Example 1.54*** The action of $A_4$ on $\{1, 2, 3, 4\}$ is doubly transitive and the normal subgroup $\{\text{id}, (1\ 2)(3\ 4), (1\ 3)(2\ 4), (1\ 4)(2\ 3)\} \trianglelefteq A_4$ acts transitively on $\{1, 2, 3, 4\}$.

**Theorem 1.55** *Let $X$ be a $G$-set and suppose that $x, y \in X$.*

*(1) If $H$ is a subgroup of $G$, then*

$$x^H \cap y^H \neq \emptyset \Rightarrow x^H = y^H;$$

*(2) If $H$ is a normal subgroup of $G$, then the subsets $x^H$ are blocks of $X$.*

***Proof*** (1) It is easy to check that $x^H = y^H$ if and only if $y \in x^H$. If $x^H \cap y^H \neq \emptyset$, then there exist $h, h' \in H$ such that $y^h = x^{h'}$. This implies that $y = x^{h'h^{-1}} \in x^H$, and so $x^H = y^H$.

(2) Suppose that $x^{Ha} \cap x^H \neq \emptyset$. Since $H \trianglelefteq G$, it follows that $x^{Ha} \cap x^H = x^{aH} \cap x^H \neq \emptyset$. So, there exist $h, h' \in H$ such that $x^{ah} = x^{h'}$. This yields that $x^a = x^{h'h^{-1}} \in x^H$. Consequently, we obtain $x^{Ha} = x^H$.

                      ∎

### Exercises

1. Show that if the action of $G$ on $X$ is primitive and faithful, then the action of any normal subgroup $N \neq \{e\}$ of $G$ is transitive.
2. Prove that the symmetric group of any degree and the alternating groups of degree greater than 2 are primitive.
3. Find all primitive and imprimitive subgroups of the symmetric group $S_4$.
4. Prove that if a proper transitive subgroup $H$ of the symmetric group $S_n$ contains a transposition, then it is imprimitive.
5. Suppose that $G$ is primitive on $X$ and contains a transposition. Prove that $G \cong S_X$.

## 1.5 Applications of Burnside's Lemma

To demonstrate Burnside's Lemma for counting colorings of an object with symmetries, we take a look at the following examples.

Two squares will be considered distinct if they cannot be made identical by rotating and/or flipping one of the squares.

***Example 1.56*** How many distinct squares are there with red and black beads at the corners? Let $G$ be the dihedral group,

$$D_4 = \{R_0, \ R_1, \ R_2, \ R_3, \ S_0, \ S_1, \ S_2, \ S_3\},$$

where

$R_0 =$    Rotation of $0°$,
$R_1 =$    Rotation of $90°$,
$R_2 =$    Rotation of $180°$,
$R_3 =$    Rotation of $270°$,
$S_0 =$    Reflection about a horizontal axis,
$S_1 =$    Reflection about a vertical axis,
$S_2 =$    Reflection about the main diagonal,
$S_3 =$    Reflection about the other diagonal.

and let $X$ be the set of not necessarily distinct squares that can be created with red and black beads. Since there are 4 corners and 2 colors, it follows that there are 16 elements in $X$. Now we must find how many square configurations each element of $D_4$ fixes.

The identity, $R_0$, fixes all 16 configurations. Both $R_1$ and $R_3$ only fix two elements, namely black with all black dots and the square with all red dots. Note that every element in $D_4$ fixes these two monochrome squares. In addition to these two, $R_2$ also fixes the squares shown in Fig. 1.5. $S_0$ also fixes the squares in Fig. 1.6 and $S_1$ fixes those in Fig. 1.7. The six additional squares fixed by $S_2$ and $S_3$ are shown in Figs. 1.8 and 1.9, respectively.

**Fig. 1.5** Fixed by $R_2$

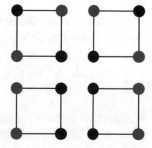

**Fig. 1.6** Fixed by $S_0$

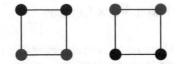

**Fig. 1.7** Fixed by $S_1$

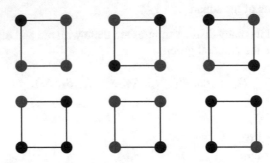

**Fig. 1.8** Fixed by $S_2$

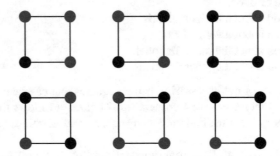

**Fig. 1.9** Fixed by $S_3$

**Table 1.1** Number of elements of $F_X(a)$, for all elements of $D_4$

| $a \in D_4$ | $R_0$ | $R_1$ | $R_2$ | $R_3$ | $S_0$ | $S_1$ | $S_2$ | $S_3$ |
|---|---|---|---|---|---|---|---|---|
| $|F_X(a)|$ | 16 | 2 | 4 | 2 | 4 | 4 | 8 | 8 |

In Table 1.1, we indicate the number of elements of $F_X(a)$, for all elements of $D_4$: Now we use Burnside's Lemma to count the number of orbits of $D_4$ on $X$. We have

$$\frac{1}{|G|} \sum_{a \in G} |F_X(a)| = \frac{1}{8}(16 + 2 + 4 + 2 + 4 + 4 + 8 + 8) = 6.$$

We conclude that it is possible to create 6 distinct squares with red and black beads at the corners. These 6 orbits can be seen in Fig. 1.10, where each row shows a different orbit (Fig. 1.11).

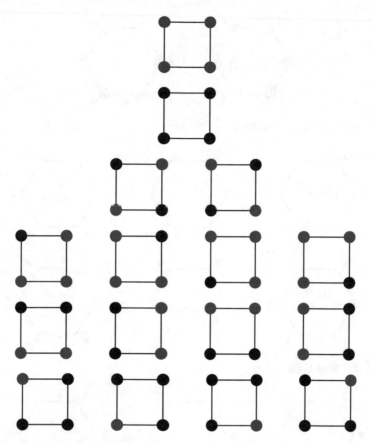

**Fig. 1.10**  All possible square configurations arranged into equivalent rows

*Example 1.57*  A jeweler would like to make necklaces containing six beads, each of which could be black or brown. How many unique necklaces can they make? To solve the problem using Burnside's Lemma, we consider each of the beads on the necklace as a vertex on a hexagon, and so our group can be the dihedral group $D_6$, which consists 6 rotations (Fig. 1.12) and 6 reflections (Fig. 1.13) such that each reflection is across the red line.

If the vertices are labeled from 1 to 6 counterclockwise, then the elements of $D_6$ can be represented by permutations in $D_6$. The set $X$ on which $D_6$ acts is the set of all functions from the set of vertices of the hexagon to the set {black, red}. We observe that $|X| = 2^6$. Numbering the beads 1 till 6 allow us to keep track of which placement of a bead we are looking at on the necklace since there is no clasp or start point on the type of necklace that we are discussing. Therefore the number of unique necklaces and the number of distinct colorings is equal to the number of orbits of

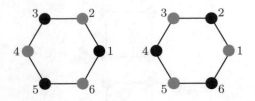

**Fig. 1.11**  Equivalent necklace colorings

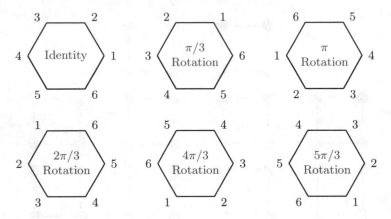

**Fig. 1.12**  Rotations in $D_6$

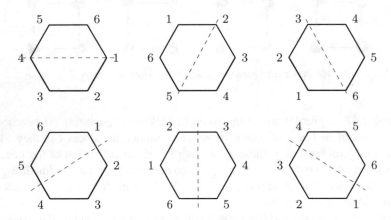

**Fig. 1.13**  Reflections in $D_6$

$D_6$ on $X$. This is because for example the necklace with beads in the odd positions colored brown and beads in the even positions colored black is the same as the one with beads in the odd positions colored black and beads in the even positions colored brown. This can be seen in Fig. 1.11.

Now, for each $\rho \in D_6$ we determine the number of elements of $F_X(\rho)$.

- If $\rho$ is the identity element, then each one of the vertices is fixed under this action. Thus, no matter what each vertex is colored, the entire coloring will be fixed. Therefore, $F_X(\rho) = 2^6$.
- If $\rho$ is the $\pi/3$ rotation, then none of the vertices are fixed and there are no subsets of vertices that get sent to one another. Thus for an entire coloring to be fixed, all of the vertices must be the same color and since there are 2 colors, $F_X(\rho) = 2^1$. The same argument can be given if $\rho$ is the $5\pi/3$.
- If $\rho$ is the $2\pi/3$ rotation, then the vertices 1, 3, and 5 rotate to one another and the vertices 2, 4, and 6 rotate to one another. Thus as long as the vertices in the first set are colored the same and the vertices in the second set are colored the same, the coloring will be fixed. Therefore having two color choices for each set, $F_X(\rho) = 2^2$. The same argument can be given if $\rho$ is the $4\pi/3$ rotation.
- If $\rho$ is the $\pi$ rotation, then the vertices 1 and 4 will be switched with one another, as will 2 and 5 with one another and 3 and 6 with one another. Thus, as long as a vertex is the same color as the one that it is being switched with, the coloring will be fixed. Therefore having two color choices for each pairing, $F_X(\rho) = 2^3$.
- If $\rho$ is any one of the three reflections on the bottom of Fig. 1.13, then it is a reflection across a line that is not through vertices. Thus, no vertices get fixed and each vertex is switched with one other. Thus as long as a vertex is the same color as the one that it is being switched with, the coloring will be fixed. Therefore having two colors choices for each pairing will give each of these reflections, $F_X(\rho) = 2^3$.
- If $\rho$ is any one of the three reflections on the top of Fig. 1.13, then it is a reflection across a line that is through two vertices. Thus, two vertices get fixed, while the other four have a partner that they are switched with. Thus, the color of the fixed vertices do not matter, but the other vertices must be the same color as the one it is being switched with. Therefore, having two color choices, for each of these reflections, $F_X(\rho) = 2^4$.

The previous calculations can be seen in Table 1.2. Therefore, using Burnside's Lemma, the number of unique colorings of the vertices and hence the number of unique necklaces that can be made is

$$\frac{1}{|D_6|} \sum_{\rho \in D_6} |F_X(\rho)| = \frac{1}{12} \left( 2^6 + (2^1 \cdot 2) + (2^2 \cdot 2) + 2^3 + (2^3 \cdot 3) + (2^4 \cdot 3) \right) = 13.$$

These 13 options for unique necklaces can be seen in Fig. 1.14.

***Example 1.58*** The *Möbius strip* is a surface with only one side and only one boundary component (see Fig. 1.15). The Möbius strip is the simplest non-orientable surface. It was codiscovered independently by the German mathematicians August Ferdinand Möbius and Johann Benedict Listing in 1858. A model can easily be created by taking a paper strip and giving it a half-twist, and then merging the ends of the strip together to form a single strip.

**Table 1.2**  Result of calculations of fixed points

| Description of $\rho$ | Permutation | $|F_X(\rho)|$ |
|---|---|---|
| Identity | id | $2^6$ |
| $\pi/3$ Rotation | (1 2 3 4 5 6) | $2^1$ |
| $2\pi/3$ Rotation | (1 3 5)(2 4 6) | $2^2$ |
| $\pi$ Rotation | (1 4)(2 5)(3 6) | $2^3$ |
| $4\pi/3$ Rotation | (1 5 3)(2 4 6) | $2^2$ |
| $5\pi/3$ Rotation | (1 6 5 4 3 2) | $2^1$ |
| Reflection through 1 and 4 | (2 6)(3 5) | $2^4$ |
| Reflection through 2 and 5 | (1 3)(4 6) | $2^4$ |
| Reflection through 3 and 6 | (1 5)(2 4) | $2^4$ |
| Reflection through 1 and 2 | (1 2)(3 6)(4 5) | $2^3$ |
| Reflection through 2 and 3 | (1 4)(2 3)(5 6) | $2^3$ |
| Reflection through 3 and 4 | (1 6)(2 5)(3 4) | $2^3$ |

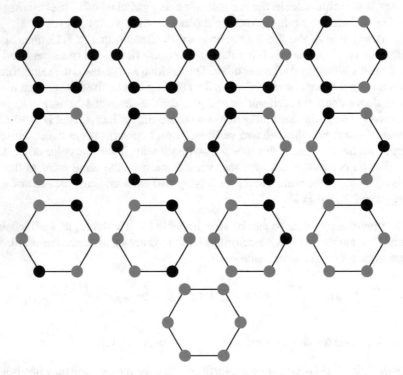

**Fig. 1.14**  Unique necklaces with six beads

**Fig. 1.15**  Möbius strip

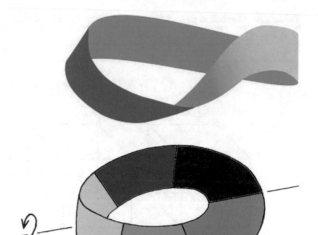

**Fig. 1.16**  Two types of
symmetry in Möbius strip

A $5 \times 1$ rectangular strip of paper is marked off on both sides into five unit squares.
The ends of the strip are then joined with a half-twist to produce a Möbius strip $M$.
How many different strips can result if we have three colors with which to paint the
squares? There are ten squares and three colors, giving a total of $3^{10}$ painted strips.
"Different" now means different up to the "natural" symmetry of the Möbius strip.
Make a model of $M$ and run it through our fingers so that the squares move along
one position, call this symmetry $R$. After ten moves we get back to where we started,
so $R^{10}$ is the identity. Note that $R$ is not induced by a rotation of $\mathbb{R}^3$; it is a movement
of the Möbius strip in itself. This type of symmetry shows up well if the strip is used
as a belt drive connecting two pulleys. There is another natural symmetry $S$; just turn
$M$ over as in Fig. 1.16. Both $R$ and $S$ together generate a group which is isomorphic
to the dihedral group $D^{10}$ and which acts on the set of painted strips in the obvious
way. The conjugacy classes of the dihedral group $D_{10}$ are as follows:

(1)  Two conjugacy classes of size 1: $\{I\}$ and $\{R^5\}$;
(2)  Four conjugacy classes of size 2: $\{R, \ R^9\}$, $\{R^2, \ R^8\}$, $\{R^3, \ R^7\}$ and $R^4, \ R^6\}$;
(3)  Two conjugacy classes of size 5 related to reflections:

$$\{S, \ R^2S, \ R^4S, \ R^6S, \ R^8S\} \text{ and } \{RS, \ R^3S, \ R^5S, \ R^7S, \ R^9S\}.$$

Taking the first element of each of these conjugacy classes and working out how
many painted strips it leaves fixed gives (Table 1.3).

For example, $S$ leaves two squares invariant, those which are pierced at their center
by its axis, and interchanges the other eight in pairs, so it fixes $3^6$ painted strip. By

**Table 1.3** Number of fixed points on conjugacy classes of $D_{10}$

| $a \in D_{10}$ | I | R | $R_2$ | $R^3$ | $R^4$ | $R^5$ | S | RS |
|---|---|---|---|---|---|---|---|---|
| $|F_X(a)|$ | $3^{10}$ | 3 | $3^2$ | 3 | $3^2$ | $3^5$ | $3^6$ | $3^5$ |

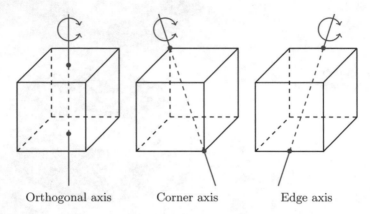

Orthogonal axis          Corner axis          Edge axis

**Fig. 1.17** Axes of symmetry for a cube

Burnside's Lemma, the number of distinct painted Möbius strips is

$$\frac{1}{20}\left(3^{10} + (3 \cdot 2) + (3^2 \cdot 2) + (3 \cdot 2) + (3^2 \cdot 2) + 3^5 + (3^6 \cdot 5) + (3^5 \cdot 5)\right) = 3210.$$

***Example 1.59*** Let $G$ be the group of rotational symmetries of the cube in three dimensions. Using Burnside's Lemma, we want to determine a formula for the number of different ways to color the faces of the cube with $n$ given colors. Two colorings are equivalent if they differ only by a rotation of the cube. The axes of rotational symmetries of a cube are given in Fig. 1.17.

We label a number from 1 to 6 to each face of the cube as in Fig. 1.18.

Since any rotation of the cube must carry each face of the cube to exactly one other face of the cube and different rotations induce different permutations of the faces, $G$ can be viewed as a group of permutations on the set $\{1, 2, 3, 4, 5, 6\}$. The group $G$ acts on $X$, the set of faces of the cube. In order to determine distinct coloring, we compute the number of orbits. First, for each $\rho \in G$ we determine the number of elements of $F_X(\rho)$.

- If $\rho$ is the identity rotation, then the number of different coloring of the cube is $n^6$.
- If $\rho = (2\ 3\ 4\ 5)$, then it is a $\pi/2$ rotation around the orthogonal axis. Since $\rho$ makes faces 1 and 6 fixed, and each can be colored with $n$ colors, it follows that there are $n^2$ distinct colorings for them. The faces 2, 3, 4, and 5 are not fixed, but they are in the same cycle, so they must all be the same color to be invariant under $\rho$. Hence, they can be colored in $n$ possible ways. Consequently, the total number of colorings under $\rho$ is $n^2 n = n^3$.

**Fig. 1.18** Labeling of the cube's faces

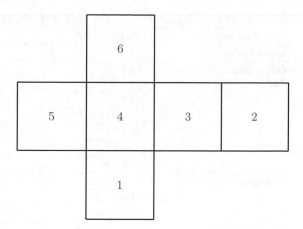

Similarly, we can obtain the number of fixed colorings of all 24 permutations in $G$. In Table 1.4, we present all 24 permutations in $G$ and the number of colorings which they keep fixed.

Finally, by Burndide's Lemma, the total number of distinct colorings of faces with $n$ colors of the cube under rotation is

$$\frac{1}{|G|} \sum_{\rho \in G} |F_X(\rho)| = \frac{1}{24} \left( n^6 + 6n^3 + 3n^4 + 6n^3 + 8n^2 \right). \tag{1.2}$$

**Exercises**

1. Find the number of different squares with vertices colored red, white, or blue.
   *Solution*: 21.
2. A wheel is divided evenly into six different compartments. Each compartment can be painted red or white. The back of the wheel is black. How many different color wheels are there?
   *Solution*: 14.
3. A rectangular design consists of 11 parallel strips of equal width. If each stripe can be painted red, blue, or green, find the number of possible patterns.
   *Solution*: 88938.
4. Each side of an equilateral triangle is divided into two equal parts, and each part is colored red or green. Find the number of patterns.
   *Solution*: 16.
5. The interior of an equilateral triangle is divided into six parts by the medians. Each part is painted with one of $m$ colors. Find the number of patterns.

**Table 1.4** All 24 permutations in $G$ and the number of colorings which they keep fixed

| Axis of rotations | Permutation | $|F_X(\rho)|$ |
|---|---|---|
| Identity rotation | id | $n^6$ |
| Orthogonal axis ($\pi/2$ and $3\pi/2$) | (2 3 4 5) | $n^3$ |
| | (1 5 6 3) | |
| | (1 2 6 4) | |
| | (5 4 3 2) | |
| | (3 6 5 1) | |
| | (4 6 2 1) | |
| Orthogonal axis ($\pi$) | (2 4)(3 5) | $n^4$ |
| | (1 6)(3 5) | |
| | (2 4)(1 6) | |
| Edge axis ($\pi$) | (1 4)(2 6)(3 5) | $n^3$ |
| | (1 2)(4 6)(3 5) | |
| | (1 6)(4 5)(2 3) | |
| | (1 6)(3 4)(2 5) | |
| | (2 4)(1 3)(5 6) | |
| | (2 4)(1 5)(3 6) | |
| Corner axis ($2\pi/3$ and $4\pi/3$) | (1 5 4)(2 6 3) | $n^2$ |
| | (1 2 3)(4 5 6) | |
| | (1 5 2)(3 4 6) | |
| | (1 4 3)(2 5 6) | |
| | (4 5 1)(3 6 2) | |
| | (3 2 1)(6 5 4) | |
| | (2 5 1)(6 4 3) | |
| | (3 4 1)(6 5 2) | |

## 1.6 Pólya's Enumeration

In this section, we introduce a powerful enumeration technique generally referred to as Pólya's Enumeration Theorem. Pólya's approach to counting allows us to use symmetries (such as those of geometric objects like polygons) to form generating functions. These generating functions can then be used to answer combinatorial questions.

We know that a permutation $\sigma \in S_n$ can be represented uniquely as a product of disjoint cycles. We say that $\sigma$ is of type $(k_1(\sigma), k_2(\sigma), \ldots, k_n(\sigma))$, where $k_i(\sigma)$ $(1 \leq i \leq n)$ is the number of cycles of length $i$ in the factorization of $\sigma$.

***Example 1.60*** For instance, the following permutation

$$\begin{pmatrix} 1 & 2 & 3 & 4 & 5 & 6 & 7 & 8 & 9 \\ 2 & 1 & 3 & 9 & 6 & 8 & 7 & 5 & 4 \end{pmatrix} = (3)(7)(1\ 2)(4\ 9)(5\ 6\ 8)$$

is of type $(2, 2, 1, 0, 0, 0, 0, 0, 0)$.

**Definition 1.61** Let $G$ be a group of permutations. The *cycle index* is the polynomial

$$P_G(x_1, x_2, \ldots, x_n) = \frac{1}{|G|} \sum_{\sigma \in G} x_1^{k_1(\sigma)}, x_2^{k_2(\sigma)}, \ldots, x_n^{k_n(\sigma)},$$

with $n$ variables $x_1, x_2, \ldots, x_n$, where $x_1^{k_1(\sigma)}, x_2^{k_2(\sigma)}, \ldots, x_n^{k_n(\sigma)}$ is formed for each $\sigma \in G$ from its type $(k_1(\sigma), k_2(\sigma), \ldots, k_n(\sigma))$.

The cycle index depends on the structure of the group action $G$ on the set $X$. Thus, for each structure of $G$ and different sizes of $X$ we must find a new cycle index.

***Example 1.62*** We consider the group $G$ of rotational symmetries of the cube, and assume that it acts on the faces of the cube. We summarize all computations in Table 1.5. Therefore, the cycle index equals to

$$P_G(x_1, x_2, x_3, x_4) = \frac{1}{24} \left( x_1^6 + 6x_1^2 x_4 + 3x_1^2 x_2^2 + 6x_2^3 + 8x_3^2 \right). \tag{1.3}$$

Substituting $x_i = n$ (for all $i$) in (1.3) gives (1.2).

Let $n$ be the number of color choices and let $k(\sigma)$ be the number of cycles within the permutation $\sigma \in G$ when it is represented in disjoint cycles form.

**Theorem 1.63** *If $G$ is a finite group of permutations acting on a set $X$ of coloring of $n$ elements in which the element can be any of $n$ color choices and $\sigma \in G$, then*

$$|F_G(\sigma)| = n^{k(\sigma)}.$$

***Proof*** By the definition, $|F_G(\sigma)|$ is the number of coloring which are unchanged when acted on by $\sigma$. A coloring remains the same when $\sigma$ acts on it, if all vertices within a vertex cycle have same color. Hence, there is $n$ color choices for each of the vertex cycles. This completes the proof. ∎

**Theorem 1.64** *The cycle index of the symmetric group acting on a set of $n$ elements is given by*

$$P_{S_n}(x_1, x_2, \ldots, x_n) = \sum \frac{1}{(1^{k_1} k_1!)(2^{k_2} k_2!) \ldots (n^{k_n} k_n!)} (x_1^{k_1}, x_2^{k_2}, \ldots, x_n^{k_n}),$$

*where the summation is over all $n$-tuple $(k_1, k_2, \ldots, k_n)$ of non-negative integers $k_i$ satisfying $k_1 + 2k_2 + 3k_3 + \cdots + nk_n = n$.*

**Table 1.5** Group of rotational symmetries of the cube, and assume that it acts on the faces of the cube

| Permutation | Type | Monomial |
|---|---|---|
| id | $(6, 0, \ldots, 0)$ | $x_1^6$ |
| $(2\ 3\ 4\ 5)$ | $(2, 0, 0, 1, 0, \ldots, 0)$ | $x_1^2 x_4$ |
| $(1\ 5\ 6\ 3)$ | | |
| $(1\ 2\ 6\ 4)$ | | |
| $(5\ 4\ 3\ 2)$ | | |
| $(3\ 6\ 5\ 1)$ | | |
| $(4\ 6\ 2\ 1)$ | | |
| $(2\ 4)(3\ 5)$ | $(2, 2, 0, \ldots, 0)$ | $x_1^2 x_2^2$ |
| $(1\ 6)(3\ 5)$ | | |
| $(2\ 4)(1\ 6)$ | | |
| $(1\ 4)(2\ 6)(3\ 5)$ | $(0, 3, 0, \ldots, 0)$ | $x_2^3$ |
| $(1\ 2)(4\ 6)(3\ 5)$ | | |
| $(1\ 6)(4\ 5)(2\ 3)$ | | |
| $(1\ 6)(3\ 4)(2\ 5)$ | | |
| $(2\ 4)(1\ 3)(5\ 6)$ | | |
| $(2\ 4)(1\ 5)(3\ 6)$ | | |
| $(1\ 5\ 4)(2\ 6\ 3)$ | $(0, 0, 2, 0, \ldots, 0)$ | $x_3^2$ |
| $(1\ 2\ 3)(4\ 5\ 6)$ | | |
| $(1\ 5\ 2)(3\ 4\ 6)$ | | |
| $(1\ 4\ 3)(2\ 5\ 6)$ | | |
| $(4\ 5\ 1)(3\ 6\ 2)$ | | |
| $(3\ 2\ 1)(6\ 5\ 4)$ | | |
| $(2\ 5\ 1)(6\ 4\ 3)$ | | |
| $(3\ 4\ 1)(6\ 5\ 2)$ | | |

**Proof** Suppose that $\sigma \in S_n$ is of type $(k_1, k_2, \ldots, k_n)$. First we determine the number of elements of $S_n$, $m$, having the above type. We know that each permutation $\sigma \in S_n$ is associated with a partition $k_1 + 2k_2 + 3k_3 + \cdots + nk_n = n$. Among the permutations in $S_n$, the $k_i$ cycles of length $i$ contribute a rearrangement which can be done by interchanging cycles of the same length. They make a contribution of $k_1! k_2! \ldots k_n!$ arrangement by the product rule. On the other hand, a cycle of length $i$ can be represented by $i$ different ways. Hence, we have to consider the contribution where they make to the $n!$ rearrangements. But there are $1^{k_1} 2^{k_2} \ldots n^{k_n}$ of them because there are $k_i$ cycles of length $i$, for each $i$. Consequently, we obtain

$$m = \frac{n!}{(1^{k_1} k_1!)(2^{k_2} k_2!) \ldots (n^{k_n} k_n!)}.$$

This completes the proof.                                                                                  ∎

**Theorem 1.65** *The cycle index of the alternating group acting on a set of $n$ elements is given by*

$$P_{A_n}(x_1, x_2, \ldots, x_n) = P_{S_n}(x_1, x_2, \ldots, x_n) + P_{S_n}(x_1, -x_2, \ldots, (-1)^{n+1}x_n)$$

**Proof** Assume that $\sigma \in S_n$ and $\sigma$ is of type $(k_1, k_2, \ldots, k_n)$. If the sum $k_2 + k_4 + \cdots$ is even, then $\sigma \in A_n$, because this sum totals all disjoint cycles having lengths $2, 4, 6, \ldots$. ∎

A cyclic group, $C_n$ can be considered as the group of rotations of a regular $n$-gon, i.e., $n$ elements equally spaced around a circle.

**Theorem 1.66** *The cycle index of $C_n$, the cyclic group of permutations of order $n$ is given by*

$$P_{C_n}(x_1, x_2, \ldots, x_n) = \frac{1}{n} \sum_{d \mid n} \varphi(d) x_d^{n/d}.$$

*where $\varphi$ the Euler function.*

**Proof** Suppose that $C_n$ is generated by $\sigma = (1, 2, \ldots, n)$ acting on a set $X$ of $n$ elements, and let $\tau_k = \sigma^k$ ($k = 0, 1, \ldots, n - 1$). We know that $\tau_k$ and $\tau_{(k,n)}$ generate the same cyclic subgroup. Moreover, they are of the same type, because if $x \in X$, then $x$ belongs to the cycle obtained from

$$x \longrightarrow x\tau_k \longrightarrow x\tau_k^2 \longrightarrow \cdots \longrightarrow x\tau_k^{n/d} \longrightarrow x.$$

This shows that $\tau_k$ is the product of $d$ cycles of length $n/d$. Next, in order to determine the cycle index of $C_n$, we find the number of solutions of $d = (k, n)$ or $(k/d, n/d) = 1$. Note that for all such $k$, $\tau_k$'s have the same type. Finally, we know that the number of solutions of $(k/d, n/d) = 1$ equals to $\varphi(n/d)$. This completes the proof. ∎

**Theorem 1.67** *The cycle index of the dihedral group $D_n$ acting on a set of $n$ elements is given by*

$$P_{D_n}(x_1, x_2, \ldots, x_n) = \frac{1}{2} P_{C_n}(x_1, x_2, \ldots, x_n)$$
$$+ \begin{cases} \dfrac{1}{2} x_1 x_2^{(n-1)/2} & \text{if } n \text{ is odd} \\[2mm] \dfrac{1}{4}\left(x_1^2 x_2^{(n-2)/2} + x_2^{n/2}\right) & \text{if } n \text{ is even.} \end{cases}$$

**Proof** The cycle index of the dihedral group is like the cyclic group but also includes reflections. ∎

Now, we consider the functions between two finite sets $X$ and $Y$. The set of all function from $X$ to $Y$ is denoted by $Y^X$.

**Definition 1.68** Let $X$ and $Y$ be finite sets, with the group $G$ acting on the set $X$. Let $\sim$ be the equivalence relation on $Y^X$ given by

$$f_1 \sim f_2 \Leftrightarrow \text{ there exists } a \in G \text{ such that } f_1(x^a) = f_2(x), \text{ for all } x \in X.$$

The equivalence class of an element $f$ in $Y^X$ is called *configuration*. The next theorem shows how the number of configurations is related to the group $G$.

**Theorem 1.69** *If $C$ denotes the set of configurations of $Y^X$, and $G$ is a permutation group acting on $X$, then the number of configurations of $Y^X$ is given by*

$$|C| = \frac{1}{|G|} \sum_{\alpha \in G} |\{f : X \to Y \mid f \circ \alpha = f\}|.$$

***Proof*** We begin by using the group $G$ acting on $X$ to define a group $\mathcal{G}$ acting on $Y^X$. Consider the set $\{\psi_\alpha : Y^X \to Y^X \mid \psi_\alpha(f) = f \circ \alpha^{-1}, \ \alpha \in G\}$ with the composition of functions operation. We show that this is a group acting on $Y^X$, denoting it by $\mathcal{G}$. Clearly, for every $\alpha, \beta \in G$, we have $\psi_\alpha \circ \psi_\beta = \psi_{\alpha \circ \beta} \in \mathcal{G}$. Also, associativity holds. The identity element of $\mathcal{G}$ is $\psi_{\mathrm{id}}$, where id is the identity element of $G$. Also, obviously, associativity holds. The identity element of $\mathcal{G}$ is $\psi_{\mathrm{id}}$, where id is the identity element of $G$. In addition, the inverse of $\psi_\alpha$ in $\mathcal{G}$ is $\psi_{\alpha^{-1}}$ because $\psi_\alpha \circ \psi_{\alpha^{-1}} = \psi_{\alpha \circ \alpha^{-1}} = \psi_{\mathrm{id}} = \psi_{\alpha^{-1} \circ \alpha} = \psi_{\alpha^{-1}} \circ \psi_\alpha$. This yields that $\mathcal{G}$ is a group of permutations on $Y^X$. Finally, $\mathcal{G}$ acts on $Y^X$ because for every $\psi_\alpha, \ \psi_\beta \in \mathcal{G}$ and $f \in Y^X$ we have

$$(\psi_\alpha \circ \psi_\beta)(f) = \psi_{\alpha \circ \beta}(f) = f \circ (\alpha \circ \beta)^{-1} = f \circ \beta^{-1} \circ \alpha^{-1} = \psi_\alpha\big(\psi_\beta(f)\big),$$

and $\psi_{\mathrm{id}}(f) = f$. Now, we can write

$$\begin{aligned}
f_1, f_2 \in \mathrm{Orb}_{\mathcal{G}}(Y^X) &\Leftrightarrow f_1 = \psi_\alpha(f_2) = f_2 \circ \alpha^{-1}, \text{ for some } \alpha \in G \\
&\Leftrightarrow f_1 \circ \alpha = f_2, \text{ for some } \alpha \in G \\
&\Leftrightarrow f_1, f_2 \text{ belong to a configuration in } C.
\end{aligned}$$

Therefore, we conclude that $|C| = |\mathrm{Orb}_{\mathcal{G}}(Y^X)|$. Next, by using Burnside's Lemma we get

$$|C| = \frac{1}{|\mathcal{G}|} \sum_{\psi_\alpha \in \mathcal{G}} |F_{\mathcal{G}}(\psi_\alpha)|. \tag{1.4}$$

Note that $\psi_\alpha \in \mathcal{G}$ if and only if $\alpha \in G$. On the other hand, we have

$$F_{\mathcal{G}}(\psi_\alpha) = \{f : X \to Y \mid \psi_\alpha(f) = f\} = \{f : X \to Y \mid f = f \circ \alpha\}. \tag{1.5}$$

Now, if we substitute (1.5) in (1.4), then we obtain

$$|C| = \frac{1}{|G|} \sum_{\alpha \in G} \left| \{f : X \to Y \mid f \circ \alpha = f\} \right|,$$

as desired.    ∎

**Definition 1.70** Let $Y$ be a finite set, and suppose that $R = \mathbb{Q}[x_1, \ldots, x_n]$ is the polynomial ring. If $w : Y \to R$ is a function, then the *weight* of each $y \in Y$ is the element $w(y)$. The *weight of a function* $f : X \to Y$, denoted by $W(f)$, is the product

$$W(f) = \prod_{x \in X} w\big(f(x)\big).$$

**Theorem 1.71** *Two functions in the same configuration have the same weight.*

**Proof** Suppose that $f_1$ and $f_2$ are in the same configuration. By the definition, there is $\sigma \in G$ such that $f_1(x^\sigma) = f_2(x)$, for all $x \in X$. Then, we can write

$$W(f_1) = \prod_{x \in X} w\big(f_1(x)\big) = \prod_{x \in X} w\big(f_1(x^\sigma)\big) = \prod_{x \in X} w\big(f_2(x)\big) = W(f_2),$$

and we are done.    ∎

**Definition 1.72** The *weight of a configuration* is defined to be the common weight of all functions in the configuration. If $c$ is a configuration, the weight of $c$ is denoted by $W(c)$.

***Example 1.73*** Suppose that $\{1,\ 2\}$, $Y = \{a,\ b\}$ and $G = \{\text{id},\ (1\ 2)\}$. There are four functions $\{f_1,\ f_2,\ f_3,\ f_4\}$ from $X$ to $Y$ as follows:

$$f_1(1) = f_1(2) = a, \qquad f_1(1) = a \text{ and } f_2(2) = b,$$
$$f_3(1) = b \text{ and } f_3(2) = a, \quad f_4(1) = f_4(2) = b.$$

It is easy to see that $\{f_1\}$, $\{f_2,\ f_3\}$, and $\{f_4\}$ are distinct configurations. Now, assume that $w(a) = x_1$ and $w(b) = x_2$. Then, we have

$$W(f_1) = w\big(f_1(1)\big)w\big(f_1(2)\big) = w(a)w(a) = x_1^2,$$
$$W(f_2) = w\big(f_2(1)\big)w\big(f_2(2)\big) = w(a)w(b) = x_1 x_2,$$
$$W(f_3) = w\big(f_3(1)\big)w\big(f_3(2)\big) = w(b)w(a) = x_1 x_2,$$
$$W(f_4) = w\big(f_4(1)\big)w\big(f_4(2)\big) = w(b)w(b) = x_2^2.$$

***Example 1.74*** Let $X = \{1,\ 2,\ 3,\ 4,\ 5\}$ be the set of labeled corners of a pentagon (see Fig. 1.19), $Y$ be the set containing the colors "green" and "blue", and the group be the cyclic group $C_5$. There are eight distinct configurations. The eight configurations can be represented by the symbols $c_1, \ldots, c_8$ as in Fig. 1.20.

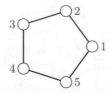

**Fig. 1.19** Pentagon with labeled corners

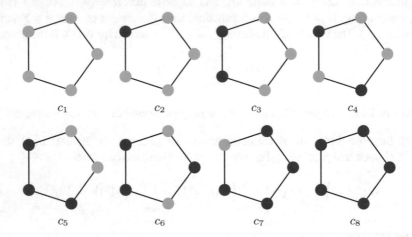

**Fig. 1.20** Eight distinct configurations

We define $f_1, f_2 : X \to Y$ by

$$f_1(x) = \begin{cases} \text{blue} & \text{if } x \in \{3,\ 5\} \\ \text{green} & \text{if } x \in \{1,\ 2,\ 4\} \end{cases} \text{ and } f_2(x) = \begin{cases} \text{blue} & \text{if } x \in \{2,\ 4\} \\ \text{green} & \text{if } x \in \{1,\ 3,\ 5\}. \end{cases}$$

We observe that $f_1$ and $f_2$ belong to the configuration $c_4$, because if we take $\sigma = (1\ 2\ 3\ 4\ 5)$, then we have $f_1(x^\sigma) = f_2(x)$, for all $1 \leq x \leq 5$. Now, if we define $w : Y \to \mathbb{Q}[B, G]$ by $w(\text{green}) = G$ and $w(\text{blue}) = B$, then

$$W(f_1) = w\big(f_1(1)\big) \ldots w\big(f_1(5)\big) = GGBGB = B^2 G^3.$$

Hence, the weight of $c_3$ is $B^2 G^3$.

**Theorem 1.75** *Let $X_1, \ldots, X_n$ be a partition of a set $X$ and $Y = \{y_1, \ldots, y_m\}$. If $S \subseteq Y^X$ is the set of functions which are constant on each $X_i$, then*

$$\sum_{f \in S} W(f) = \prod_{i=1}^{n} \left( w(y_1)^{|X_i|} + \cdots + w(y_m)^{|X_i|} \right). \tag{1.6}$$

**Proof** Multiplying out the right side of (1.6) gives us a sum. Since $f \in S$ is constant on each $X_i$, it follows that each term in this sum corresponds to a single function $f \in S$. Consequently, the sum can be compacted as follows:

$$\sum_{f \in S} w(f(x_1))^{|X_1|} \ldots w(f(x_n))^{|X_n|}, \tag{1.7}$$

where $x_i \in X_i$ (for $1 \le i \le n$). Finally, the sum (1.7) can be written as follows:

$$\sum_{f \in S} \left( \prod_{x \in X_1} w(f(x)) \right) \ldots \left( \prod_{x \in X_n} w(f(x)) \right) = \sum_{f \in S} \prod_{x \in X} w(f(x)) = \sum_{f \in S} W(f).$$

This completes the proof. ∎

**Definition 1.76** For a set $C$ of configurations on $Y^X$, the *configuration generating function* is defined by

$$\mathcal{F}(C) = \sum_{c \in C} W(c).$$

**Example 1.77** Consider Example 1.74. Then, the set of configurations is $\{c_1, \ldots, c_8\}$ and the configuration generating function is

$$\begin{aligned}
\sum_{c \in C} W(c) &= G^5 + G^4 B + G^3 B^2 + G^3 B^2 + G^2 B^3 + G^2 B^3 + G B^4 + B^5 \\
&= G^5 + G^4 B + 2 G^3 B^2 + 2 G^2 B^3 + G B^4 + B^5.
\end{aligned}$$

Hence, the configuration generating function lists all possible weights of configurations with the coefficient representing the number configurations with the given weight. For example, the weight $G^2 B^3$ has coefficient 2, so we know that there are two distinct configurations with weight $G^2 B^3$. In other words there are two distinct colorings of the corners of the pentagon such that two corners are green and three are blue. Thus, knowing the configuration generating function, we know how many distinct colorings there are for each set of weights. To find the total number of colorings, we can simply assign weights $w(\text{blue}) = w(\text{blue}) = 1$ which gives $W(c_i) = 1$, for all $i$. Now, the configuration generating function gives $\mathcal{F}(C) = 8$, the number of distinct colorings.

**Theorem 1.78 (Pólya's enumeration theorem).** *Let $X$ and $Y$ be finite sets and $|X| = n$. Suppose that $G$ is a finite group acting on $X$ and $P_G$ is the cycle index polynomial. If $w$ is the weight function on $Y$, then the configuration generating function is given by*

$$P_G \left( \sum_{y \in Y} w(y), \sum_{y \in Y} w(y)^2, \ldots, \sum_{y \in Y} w(y)^n \right).$$

*In particular, if $w(y) = 1$, for all $y \in Y$, then the total number of configurations equals to $P_G(|Y|, |Y|, \ldots, |Y|)$.*

**Proof**  Assume that $\mathcal{S}_w \subseteq Y^X$ is the set of functions with weight $w$. Take $\psi_w(\alpha) = \{f : X \to Y \mid f = f \circ \alpha$ and $W(f) = w\}$. By Theorem 1.69, the number of configurations in $\mathcal{S}_w$ is given by

$$|\mathcal{S}_w| = \frac{1}{|G|} \sum_{\alpha \in G} |\psi_w(\alpha)|.$$

This implies that

$$\sum_{c \in C} W(c) = \sum_w w|\mathcal{S}_w| = \frac{1}{|G|} \sum_w \sum_{\alpha \in G} |\psi_w(\alpha)|$$

$$= \frac{1}{|G|} \sum_{\alpha \in G} \sum_w w|\psi_w(\alpha)|. \tag{1.8}$$

If $\psi(\alpha)$ denotes the set $\{f : X \to Y \mid f = f \circ \alpha\}$, then we have

$$\sum_w w|\psi_w(\alpha)| = \sum_{f \in \psi(\alpha)} W(f). \tag{1.9}$$

Therefore, from (1.8) and (1.9), we conclude that

$$\sum_{c \in C} W(c) = \frac{1}{|G|} \sum_{\alpha \in G} \sum_{f \in \psi(\alpha)} W(f). \tag{1.10}$$

On the other hand, we know that each $\alpha \in G$ partitions $X$ into a disjoint union of cycles, say $X_1, \ldots, X_m$, where $m \leq n$. If $f \in \mathcal{S}_w$, then $f = f \circ \alpha = f \circ \alpha^2 = \cdots$, which implies that $f$ is constant on each $X_i$. Conversely, if $f$ is constant on each $X_i$, then $f = f \circ \alpha$, and so $f \in \mathcal{S}_w$. Now, by Theorem 1.75 we obtain

$$\sum_{f \in \psi(\alpha)} W(f) = \prod_{i=1}^n \sum_{y \in Y} w(y)^{|X_i|} = \left( \sum_{y \in Y} w(y)^{|X_1|} \right) \ldots \left( \sum_{y \in Y} w(y)^{|X_m|} \right).$$

If the cycle type of $\alpha$ is $(k_1, \ldots, k_n)$, then in $|X_1|, \ldots, |X_m|$, the number $i$ occurs $k_i$ times. Consequently, we have

$$\sum_{f \in \psi(\alpha)} W(f) = \left( \sum_{y \in Y} w(y) \right)^{k_1} \ldots \left( \sum_{y \in Y} w(y) \right)^{k_n}. \tag{1.11}$$

Substituting (1.11) in (1.10) gives the configuration generating function

$$\frac{1}{|G|} \sum_{\alpha \in G} \left( \left( \sum_{y \in Y} w(y) \right)^{k_1} \cdots \left( \sum_{y \in Y} w(y) \right)^{k_n} \right). \tag{1.12}$$

Finally, we observe that (1.11) is precisely the cycle index

$$P_G\left( \sum_{y \in Y} w(y), \sum_{y \in Y} w(y)^2, \ldots, \sum_{y \in Y} w(y)^n \right).$$

This completes the proof of Pólya's Enumeration Theorem.  ∎

***Example 1.79*** In how many ways can the six faces of a cube be colored such that three faces are red, two are blue, and one is purple?

Assume that $X$ is the set of six faces of the cube, and $Y$ the set of colors with weights $w(\text{red}) = R$, $w(\text{blue}) = B$ and $w(\text{purple}) = P$. The appropriate permutation group is the group of rotational symmetries of the cube. According to Example 1.62 is

$$P_G(x_1, x_2, x_3, x_4) = \frac{1}{24}\left(x_1^6 + 6x_1^2 x_4 + 3x_1^2 x_2^2 + 6x_2^3 + 8x_3^2\right).$$

Now, Pólya's Enumeration Theorem gives the configuration generating function:

$$\begin{aligned}
\mathcal{F}(R, B, P) &= P_G\big((R + B + P), (R^2 + B^2 + P^2), \ldots, (R^4 + B^4 + P^4)\big) \\
&= \frac{1}{24}\big((R + B + P)^6 + 6(R + B + P)^2(R^4 + B^4 + P^4) \\
&\quad + 3(R + B + P)^2(R^2 + B^2 + P^2)^2 + 6(R^2 + B^2 + P^2)^3 \\
&\quad + 8(R^3 + B^3 + P^3)^2\big) \\
&= R^6 + R^5 B + R^5 P + 2R^4 B^2 + 2R^4 BP + 2R^4 P^2 + 2R^3 B^3 \\
&\quad + 3R^3 B^2 P + 3R^3 BP^2 + 2R^3 P^3 + 2R^2 B^4 + 3R^2 B^3 P \\
&\quad + 6R^2 B^2 P + 3R^2 BP^3 + 2R^2 P^4 + RB^5 + 2RB^4 P \\
&\quad + 3RB^3 P^2 + 3RB^2 P^3 + 2RBP^4 + RP^5 + B^6 + B^5 P \\
&\quad + 2B^4 P^2 + 2B^3 P^3 + 2B^2 P^4 + BP^5 + P^6.
\end{aligned}$$

The coefficient of $R^3 B^2 P$ is 3, thus there are 3 distinct colorings of the cube such that three faces are red, two are blue and one is purple. The total number of 3-colorings of the cube is found by

$$\begin{aligned}
P_G\big(|Y|, |Y|, \ldots, |Y|\big) &= P_G(3, 3, \ldots, 3) \\
&= \frac{1}{24}\left(3^6 + 6 \cdot 3^2 \cdot 3 + 3 \cdot 3^2 \cdot 3^2 + 6 \cdot 3^3 + 8 \cdot 3^2\right) = 57.
\end{aligned}$$

**Exercises**

1. If $a$ and $n$ are positive integers, prove that

$$\sum_{d|n} \varphi(d) a^{n/d} \equiv 0 (\text{mod } n).$$

2. Show that

   (a) Fermat's Little Theorem: For any prime number $p$ and any integer $a$, $a^p \equiv a(\text{mod } p)$, and

   (b) Gauss' Theorem: If $n$ is a positive integer then $\sum_{d|n} \varphi(d) = n$,

   are simple consequences of Exercise 1.6.

3. Suppose that we are painting the faces of a cube and we have white, gold, and blue paints available. Two painted cubes are equivalent if you can rotate one of them so that all corresponding faces are painted the same color. Determine the number of non-equivalent ways that we can paint the faces of the cube as well as the number having two faces of each color.

4. Suppose that we are given 6 similar spheres in three different colors, say, three red, two blue and one yellow (spheres of the same color being indistinguishable). In how many ways can we distribute the six spheres on the 6 vertices of an octahedron freely movable in space?

## 1.7   Worked-Out Problems

**Problem 1.80** Let $G$ be a group and $X_1, X_2$ be $G$-sets (with actions $\theta_1, \theta_2$).

(1) We define
$$X = X_1 + X_2 = (X_1 \times \{1\}) \cup (X_2 \times \{2\})$$

   by $\theta\big((x, i), a)\big) \mapsto (\theta_i(x, a), i)$, for $i = 1, 2$, $x \in X_i$ and $a \in G$. Show that this makes $X$ into a $G$-set (the sum of $X_1$ and $X_2$);

(2) Similarly, we define $\theta : (X_1 \times X_2) \times G \to X_1 \times X_2$ by $(x_1, x_2)^a := (x_1^a, x_2^a)$, for all $(x_1, x_2) \in X_1 \times X_2$ and $a \in G$. Show that this is an action.

*Solution* (1) Clearly, $\theta$ is a function from $(X_1 + X_2) \times G$ into $X_1 + X_2$. We need to check the conditions specifying an action. Since $\theta_i$ is an action, it follows that $\theta_i(x, e) = x$ for all $x \in X_i$ and that $\theta_i\big(\theta_i(x, a), b\big) = \theta_i(x, ab)$, for all $x \in X_i$, $a, b \in G$. Elements of $X_1 + X_2$ are of the forms $(x, 1)$ with $x \in X_1$ and $(x, 2)$ with $x \in X_2$.
Suppose that $a, b \in G$, $i = 1, 2$ and $x \in X_i$. Then, we have

$$\theta\big((x, i), ab\big) = \big(\theta_i(x, ab), i\big) = \big(\theta_i(\theta_i(x, a), b), i\big)$$
$$= \theta\big((\theta_i(x, a), i), b\big) = \theta\big((\theta((x, i), a), b\big).$$

Moreover, we have $\theta\big((x,i),e\big) = \big(\theta_i(x,e),i\big) = (x,i)$, where $i = 1,2$ and $x \in X_i$. Therefore, $\theta$ as given does satisfy the conditions for an action, as required.

(2) For the product set we have

$$(x_1,x_2)^{ab} = (x_1^{ab}, x_2^{ab}) = \big((x_1^a)^b, (x_2^a)^b\big)$$
$$= (x_1^a, x_2^a)^b = \big((x_1,x_2)^a\big)^b,$$

for all $(x_1,x_2) \in X_1 \times X_2$ and $a,b \in G$. Also, we have $(x_1,x_2)^e = (x_1^e, x_2^e) = (x_1,x_2)$, for all $(x_1,x_2) \in X_1 \times X_2$. Consequently, $\theta$ is indeed an action. ∎

**Problem 1.81** Let $G$ be a finite group and suppose that $H$ and $K$ are subgroups of $G$ of orders $m$ and $n$, respectively. If none of the elements of $K - \{e\}$ are conjugate to elements of $H$, prove that $mn \big| |G|$.

*Solution* Suppose that $X$ is the set of right cosets of $K$ in $G$, i.e., $X = \{Ka \mid a \in G\}$. Obviously, $H$ acts on $X$. If $h \in H$ leaves $Ka$ fixed, then $(Ka)^h = Ka$ or $Kah = Ka$. This shows that $aha^{-1} \in K$, and so there exists $b \in K$ such that $aha^{-1} = b$. Since the elements of $K - \{e\}$ cannot be conjugate to the elements of $H$, it follows that $b = e$, and consequently $h = e$. Thus, we conclude that the stabilizer of each elements of $X$ under $H$ is trivial. So, the length of each orbit is equal to $|H|$. If $k$ is the number of orbits of $H$ on $X$, then

$$\frac{|G|}{n} = \frac{|G|}{|K|} = [G:K] = |X| = k|H| = km.$$

Therefore, we obtain $|G| = kmn$, and this completes the proof. ∎

**Problem 1.82** Let a finite group $G$ acts transitively on a set $X$ contains $n$ elements. If $H$ is a subgroup of $G$ and $[G : N_G(H)] = m$ such that $(m,n) = 1$, prove that

(1) $N_G(H)$ acts transitively on $X$;
(2) $|\mathrm{Orb}_G(x)| = |\mathrm{Orb}_G(y)|$, for all $x, y \in X$.

*Solution* (1) Since $G$ acts transitively on $X$, it follows that $[G : \mathrm{Stab}_G(x)] = |X| = n$, for all $x \in X$. As $(m,n) = 1$, we deduce that $G = \mathrm{Stab}_G(x)N_G(H)$. Now, assume that $y$ is an arbitrary element of $X$. Since $G$ acts transitively on $X$, it follows that there is $a \in G$ such that $x^a = y$. On the other hand, $a \in G = \mathrm{Stab}_G(x)N_G(H)$ implies that $a = bc$, for some $b \in \mathrm{Stab}_G(x)$ and $c \in N_G(H)$. So, we have $y = x^a = x^{bc} = (x^b)^c = x^c$. This shows that $N_G(H)$ acts transitively on $X$.

(2) First, we prove that

$$a^{-1}\mathrm{Stab}_H(x)a = \mathrm{Stab}_H(x^a), \tag{1.13}$$

for all $x \in X$ and $a \in N_G(H)$. Clearly, we have $a^{-1}\mathrm{Stab}_H(x)a \subseteq \mathrm{Stab}_H(x^a)$. For the converse inclusion, let $h \in \mathrm{Stab}_H(x^a)$ be arbitrary. Then, we get $(x^a)^h = x^a$ or $x^{ah} = x^a$. This yields that $x^{aha^{-1}} = x$, and so $aha^{-1} \in \mathrm{Stab}_G(x)$. Since $H \trianglelefteq$

**Table 1.6**  Results of using the Orbit-Stabilizer theorem

| $a$ | Conjugate | Element fixed by $a$ | $F_X(a)$ | Contribution |
|---|---|---|---|---|
| id | 1 | $(x_1, x_2, x_3)$ | 4851 | 4851 |
| (1 2) | 3 | $(x_1, x_1, x_3)$ | 49 | 147 |
| (1 2 3) | 2 | $(x_1, x_1, x_1)$ | 0 | 0 |

$N_G(H)$, it follows that $aha^{-1} \in H$. Hence, we have $aha^{-1} \in \text{Stab}_G(x) \cap H = \text{Stab}_H(x)$, which implies that $h \in a^{-1}\text{Stab}_H(x)a$. Therefore, we conclude that $a^{-1}\text{Stab}_H(x)a \subseteq \text{Stab}_H(x^a)$.

Now, suppose that $y \in X$ is arbitrary. By (1), there is $a \in N_G(H)$ such that $x^a = y$. By (1.13), we conclude that $|\text{Stab}_H(x)| = |\text{Stab}_H(y)|$. This yields that $|\text{Orb}_H(x)| = |\text{Orb}_H(y)|$.

∎

**Problem 1.83** How many different triples $(x_1, x_2, x_3)$ of positive integers are there such that $x_1 + x_2 + x_3 = 100$ and $x_1 \leq x_2 \leq x_3$.

*Solution*  Suppose that

$$X = \{(x_1, x_2, x_3) \mid x_i \geq 1 \text{ and } x_1 + x_2 + x_3 = 100\}.$$

Then, $S_3$ acts naturally on $X$ and in each orbit of this action there is a unique $(x_1, x_2, x_3)$ such that $x_1 \leq x_2 \leq x_3$. Hence, the question is equivalent to finding the number of orbits of this action. Note that $x_1$ can be any number from 1 to 98, and $x_2$ any number from 1 to $99 - x_1$, with $x_3$ then determined by the choices of $x_1$ and $x_2$. Consequently, we obtain

$$|X| = \sum_{x_1=1}^{98}(99 - x_1) = 99 \times 98 - \frac{1}{2} \times 98 \times 99 = 49 \times 99 = 4851.$$

Now, we apply the Orbit-Stabilizer theorem as Table 1.6. Therefore, the number of orbits equals $(4851 + 147)/6 = 833$.                                                          ∎

**Problem 1.84** Suppose that $G$ acting transitively on the finite set $X$. If $|X| > 1$, show that there exist elements of $G$ that have no fixed point in $X$.

*Solution*  There is nothing lose if we assume that $G$ acts faithfully on $X$. Then, $G$ must be finite and we can apply Burnside's Lemma. Clearly, we have $|F_X(e)| = |X| > 1$. If we had $|F_X(a)| > 1$, for all $a \in G$, then $\sum_{a \in G} |F_X(a)|$ would exceed $|G|$. Therefore, there must exist $a \in G$ with $F_X(a) = \emptyset$.

∎

**Problem 1.85** How many essentially different ways are there to make a bracelet which has three red beads, two blue beads, and two white beads?

*Solution* These seven beads can be considered to occupy the vertices of a regular heptagon and then two different colorings would be considered indistinguishable if some element of $D_7$ connects the two. So, we are again being asked to determine the number of orbits of this action of $D_7$ on the set of (labeled) colorings. The total number of different colorings (while the positions are labeled) is

$$\frac{7!}{3!2!2!} = 210.$$

Let $R$ denote a rotation by $2\pi/7$ and $S$ denote a (fixed) reflection. Note that a rotation (in this case) would only fix those colorings that are monochromatic (not possible here). All reflections are in an axis that goes through a vertex and the opposite edges midpoint. A coloring would be fixed if vertices and their mirror images were of the same color; with the given beads this is only possible if we color the vertex on the axis red and the other six in pairs opposite one another (one pair red, one blue, one white). There are $6 = 3!$ ways of doing this. Hence our answer is

$$\frac{\overbrace{210}^{\text{id}} + \overbrace{6 \times 10}^{\text{rotations}} + \overbrace{7 \times 3!}^{\text{reflections}}}{14} = 18,$$

and we are done. ∎

**Problem 1.86** A manufactures turns out necklaces consisting of 18 colored glass beads on a circular $n$-gon string whose ends are joined with an invisible weld. There are 6 white beads, 6 pink beads and 6 coral beads in each necklace. How many essentially different necklaces can she produce?

*Solution* Two necklaces can be compared by arranging them on a flat surface with the string arranged as a circle and the beads equally spaced round it. They are then deemed to be essentially the same if the one can be rotated through some angle $k \cdot 2\pi/18$ or can be lifted up, turned over and replaced, so that it matches the other. The group $G$ of possible operations here is the dihedral group of order 36, and the number of essentially different necklaces that can be produced is the number of its orbits in the set of all conceivable necklaces. We calculate this number using Burnside's Lemma.

Rotations of a necklace form a cyclic group of order 18 containing elements of orders 1, 2, 3, 6, 9, 18. An operation of turning a necklace over, which we call a reflection, is one of two kinds: Either two diametrically opposite beads are fixed or the line of reflection meets the string midway between two neighboring beads. We denote a rotation of order $k$ by $R_k$, reflections of the first kind by $S_2$ and reflections of the second kind by $S_2'$. Thinking of these symmetry operations as permutations of the beads, a necklace is fixed by the operation if and only if each cycle of the permutation consists of beads of the same color. Thus, rotations $R_9$ and $R_{18}$ cannot fix any necklaces. The identity rotation $R_1$ fixes all the necklaces, of which there are $18!/(6!)^3$. A necklace is fixed by the rotation $R_2$ if and only if three of its 2-cycles

**Table 1.7** Essentially different necklaces

| $a \in G$ | Number of $a \in G$ | $|F_X(a)|$ | Contribution |
|---|---|---|---|
| $R_1$ | 1 | $18!/(6!)^3$ | 17153136 |
| $R_2$ | 1 | $9!/(3!)^3$ | 1680 |
| $R_3$ | 2 | $6!/(2!)^3$ | 180 |
| $R_6$ | 2 | $3!$ | 12 |
| $R_9$ | 6 | 0 | 0 |
| $R_{18}$ | 6 | 0 | 0 |
| $S_2$ | 9 | $9!/(3!)^3$ | 15120 |
| $S_2'$ | 9 | $9!/(3!)^3$ | 15120 |
| Totals | 36 | | 17185248 |

consist of white beads, three consist of pink beads, and three consist of coral beads. Thus, the number of necklaces fixed by $R_2$ is $9!/(3!)^3$. Very similar calculations give the remaining entries in Table 1.7. Consequently, by Burnside's Lemma, the number of essentially distinct necklaces is $17185248/36 = 477368$.

∎

**Problem 1.87** Give a general formula for the number of necklaces that can be made with $k_1$ blue beads and $k_2$ red beads, where $k_1 + k_2$ is an odd prime and $k_1, k_2 \neq 0$.

*Solution* First, assume that $k_1$ and $k_2$ be positive integers such that $k_1 + k_2 = n$, where $n$ is an odd prime. Since $n$ is odd, we can assume without loss of generality that $k_1$ is even. We assign weights $B$ and $R$ to the blue and red beads, respectively. Now, the relevant group acting on the necklace is the dihedral group and since $n$ is odd, by Theorem 1.67, we find the cycle index as follows:

$$P_{D_n}(x_1, x_2, \ldots, x_n) = \frac{1}{2}P_{C_n}(x_1, x_2, \ldots, x_n) + \frac{1}{2}x_1 x_2^{(n-1)/2}. \qquad (1.14)$$

Substituting the cycle index $P_{C_n}$ from Theorem 1.66 in (1.14) gives

$$P_{D_n}(x_1, x_2, \ldots, x_n) = \frac{1}{2n}\sum_{d|n}\varphi(d)x_d^{n/d} + \frac{1}{2}x_1 x_2^{(n-1)/2}$$
$$= \frac{1}{2n}\left(x_1^n - (n-1)x_n\right) + \frac{1}{2}x_1 x_2^{(n-1)/2}. \qquad (1.15)$$

Using the cycle index in (1.15), we apply Pólya's Enumeration Theorem to obtain the configuration generating function:

$$\mathcal{F}(B, R) = \frac{1}{2n}(B + R)^n - \frac{1}{2n}(n - 1)(B^n + R^n)$$
$$+ \frac{1}{2}(B + R)(B^2 + R^2)^{(n-1)/2}. \qquad (1.16)$$

We need not expand the polynomials since we are only interested in the coefficient of the term $B^{k_1} R^{k_2}$. We look separately at the three summands of (1.16). First we notice that the coefficient of $B^{k_1} R^{k_2}$ in $1/(2n)(B + R)^n$, by the binomial theorem, is given as $\frac{1}{2n}\binom{n}{k_1}$. The second summand $\frac{1}{2n}(n - 1)(B^n + R^n)$ does not affect the coefficient of $B^{k_1} R^{k_2}$, since it is only adding to the coefficients of $B^n$ and $R^n$. In the last summand, using the binomial theorem we observe that $(B^2 + R^2)^{(n-1)/2}$ contributes $\binom{(n - 1)/2}{k_1/2}$ to the coefficient of $B^{k_1} R^{k_2}$. Then multiplying through by $(1/2)(B + W)$, the coefficient of $B^{k_1} R^{k_2}$ becomes $\frac{1}{2}\binom{(n - 1)/2}{k_1/2}$. Having now accounted for each of the summands, the coefficient of $B^{k_1} R^{k_2}$ term is given by the sum

$$\frac{1}{2n}\binom{n}{k_1} + \frac{1}{2}\binom{(n - 1)/2}{k_1/2}.$$

Therefore, the general formula for the number of necklaces that can be formed using $k_1$ blue beads and $k_2$ red beads, where $k_1 + k_2$ is an odd prime, is given by

$$\frac{1}{2(k_1 + k_2)}\binom{k_1 + k_2}{k_1} + \frac{1}{2}\binom{(k_1 + k_2 - 1)/2}{k_1/2},$$

and we are finished.                                                      ∎

## 1.8  Supplementary Exercises

1. If $G$ acts on $X_1$ and on $X_2$, we know that $G$ acts on $X_1 \times X_2$ (see the second part of Problem 1.80). Show that the stabilizer of $(x_1, x_2)$ is the intersection of $\mathrm{Stab}_G(x_1)$ and $\mathrm{Stab}_G(x_2)$. Give an example which shows this action need not be transitive even if $G$ acts transitively on both $X_1$ and $X_2$.
2. Let $X_i$ for $i \in I$ be $G$-sets for the same group $G$, and suppose the sets $X_i$ are not necessarily disjoint. Let $X_i' = \{(x, i) \mid x \in X_i\}$ for all $i \in I$. Then the sets $X_i'$ are disjoint, and each can still be regarded as a $G$-set in an obvious way (see the first part of Problem 1.80). The elements of $X_i$ have simply been tagged by $i$ to distinguish them from the elements of $X_j$ for $i \neq j$. The $G$-set $\bigcup_{i \in I} X_i'$ is the disjoint union of the $G$-sets $X_i$. Show that every $G$-set is isomorphic to a disjoint union of left coset $G$-sets.
3. Exercise 1.8 shows that every $G$-set $X$ is isomorphic to a disjoint union of left coset $G$-sets. The question then arises whether left coset $G$-sets of distinct subgroups $H$ and $K$ of $G$ can themselves be isomorphic. If $\mathrm{Stab}_G(x_0) \neq \mathrm{Stab}_G(x_0^a)$, for some $x_0 \in X$ and $a \in G$, then the collections $\mathcal{L}_H$ of left cosets

of $H = \text{Stab}_G(x_0)$ and $\mathcal{L}_K$ of left cosets of $K = \text{Stab}_G(x_0^a)$ form distinct $G$-sets that must be isomorphic, since both $\mathcal{L}_H$ and $\mathcal{L}_H$ are isomorphic to $X$.

(a) Let $X$ be a transitive $G$-set and let $x_0 \in X$ and $a \in G$. If $H = \text{Stab}_G(x_0)$, describe $K = \text{Stab}_G(x_0^a)$ in terms of $H$ and $a$;

(b) Based on part (a), conjecture conditions on subgroups $H$ and $K$ of $G$ such that the left coset $G$-sets of $H$ and $K$ are isomorphic;

(c) Prove your conjecture in part (b).

4. Let $n \leq 4$. Show that the action of $D_n$ on the vertices of a regular $n$-gon is transitive but not 2-transitive.

5. Let $X = \{1, \ldots, n\}$ and $G = S_n$. Consider the induced action of $G$ in turn on each of the following. Determine in each case the orbits, stabilizers, and their cardinalities.

(a) The power set of $X$;

(b) Product $X \times X$;

(c) Complex valued functions on $X$;

(d) Functions from $X$ to $X$.

6. Up to isomorphism, how many transitive $\mathbb{Z}_6$-sets $X$ are there?

7. Up to isomorphism, how many transitive $S_3$-sets $X$ are there?

8. Let $X$ be a $G$-set. We define

$$X^k = \{(x_1, \ldots, x_k) \mid x_i \in X\},$$
$$X^{(k)} = \{(x_1, \ldots, x_k) \mid x_i \in X, \text{ all different}\},$$
$$X^{\{k\}} = \{A \mid A \subseteq X \text{ and } |A| = k\},$$
$$\mathcal{P}(X) = \{A \mid A \subseteq X\}.$$

Show that how each of these is a $G$-set in a natural way. Show also $X^{(k)}$ is a $G$-subset of $X^k$ and that $X^{\{k\}}$ is a $G$-subset of $\mathcal{P}(X)$.

9. Show that the following two conditions on a group $G$ are equivalent:

(a) There is a homomorphism $\varphi : G \to S_n$ such that $\varphi(a) \neq \text{id}$, for some $a \in G$;

(b) The group $G$ contains a proper subgroup of index at most $n$.

10. Suppose that a group $G$ acts on a set $X$. Show that

(a) $G$ is 2-transitive on $X$ if and only if $G$ has two orbits in its action on $X^2$;

(b) $G$ is 3-transitive on $X$ if and only if $G$ has five orbits in its action on $X^3$;

(c) In general, $G$ is $k$-transitive on $X$ if and only if $G$ has $r_k$ orbits in its action on $X^k$, where $r_k$ is the number of equivalence relations on the set $\{1, \ldots, k\}$ (so $r_1 = 1, r_2 = 2, r_3 = 5, r_4 = 15, \ldots$).

11. Let $n$ be a positive integer, and suppose that $\tau(n)$ is the number of divisors of $n$. Show that

$$\sum_{\substack{m=0 \\ (m,n)=1}}^{n-1} (m - 1, n) = \varphi(n)\tau(n),$$

where $\varphi$ is the Euler function. Hint: Let $\mathbb{Z}_n \cong \langle x \rangle$ be the cyclic group of order $n$ and let $G = \text{Aut}(\mathbb{Z}_n)$. What is the order of $G$? How many orbits does $G$ produce in $\mathbb{Z}_n$? If $a \in G$ has the effect $x \mapsto x^a$, what is $|F_{\mathbb{Z}_n}(a)|$?

12. Let $s$ and $n$ be two integers such that $s \geq 0$ and $n > 0$. Prove that

$$\sum_{m=0}^{n-1} s^{(m,n)} \equiv 0 (\text{mod } n).$$

*Hint:* Suppose that $X$ is a set with $|X| = s$, and let $G$ be a group acting on the set $Y$. Let $X^Y = \{f \mid f : X \to Y \text{ is a function}\}$, and suppose that $G$ acts on the set $X^Y$ by

$$X^Y \times G \to X^Y$$
$$(f, a) \mapsto f^a,$$

where $f^a : X \to Y$ is defined by $f^a(x) := \left(f(x)\right)^{a^{-1}}$, for all $x \in X$. If $\Omega = X^Y$ and $k(a)$ is the number of orbits of $\langle a \rangle$ on $Y$, show that $|F_\Omega(a)| = s^{k(a)}$. Next, assume that $G = \langle a \rangle$ is a cyclic group of order $n$ and take $Y = G$, where the action is multiplication. Note that if $0 \leq m \leq n - 1$, then $o(a^m) = n/(m, n)$. Hence, the number of orbits of $\langle a \rangle$ on $G$ is $n/o(a^m) = (m, n)$.

13. A traditional soccer ball has 20 faces that are regular hexagons and 12 faces that are regular pentagons (see Fig. 1.21). The technical term for this solid is *truncated icosahedron*. How many rotation symmetries does a soccer ball have? Use Orbit-Stabilizer theorem to explain why a soccer ball cannot have a $\pi/3$ rotational symmetry about a line through the center of a hexagonal face.

14. Let $X$ be an imprimitive $G$-set and let $B$ be a maximal non-trivial block of $X$; that is, $B$ is not a proper subset of a non-trivial block. Show that the imprimitive system $Y$ generated by $B$ is a primitive $G$-set. Give an example with $X$ faithful and $Y$ not faithful.

15. If $n \geq 2$, prove that $n - 2$ transpositions cannot generate a transitive group of degree $n$.

16. Let $G$ be a group acting on a set $X$ and $H$ be a subgroup of $G$. Suppose the action of $H$ on $X$ is transitive. Pick $x \in X$ and show that $G = H\text{Stab}_G(x)$.

17. Let $X$ be the set of all functions $f : \mathbb{R}^3 \to \mathbb{R}$ all of whose partial derivatives of any order exist everywhere. Let $G$ be the group of all transformations

$$T_{U,b} : a \mapsto aU + b \ (a \in \mathbb{R}^3),$$

where $U \in O_3(\mathbb{R})$ and $b \in \mathbb{R}^3$. For $a \in G$ and $f \in X$, we define

$$f^a(x, y, z) := f\left((x, y, z)a^{-1}\right).$$

**Fig. 1.21** A soccer ball

(a) Show that the prescription $(f, a) \mapsto f^a$ gives an action of $G$ on $X$;

(b) Let $\nabla$ be the Laplace operator on $X$,

$$\nabla f := \frac{\partial^2 f}{\partial x^2} + \frac{\partial^2 f}{\partial y^2} + \frac{\partial^2 f}{\partial z^2}.$$

Show that $\nabla(f^a) = (\nabla f)^a$, for all $f \in X$ and all $a \in G$.

# Chapter 2
# Sylow's Theorems and Applications

Sylow's theorems do provide us with a sort of partial converse to Lagrange's theorem, by asserting the existence of certain subgroups (called Sylow $p$-subgroups) of any finite group with a given order. In this chapter, we wish to build up to the proofs of Sylow's theorems by using of group actions and the Orbit-Stabilizer theorem. The study of finite non-abelian groups is much more complicated. Sylow's theorems give us some important and useful information about them.

## 2.1  $p$-Groups

The order of a group $G$ has consequences for its structure. Our aim is fundamental for understanding the structure of a finite group. A rough rule of thumb is that the more complicated the prime factorization of $|G|$, the more complicated the group. Now, we begin with the "local" case when only one prime divides $|G|$.

**Definition 2.1**  If $p$ is a prime, then a *$p$-group* is a group in which every element has order a power of $p$.

A subgroup of a group $G$ is called a *$p$-subgroup* if its order is a power of $p$.

***Example 2.2***  Both the quaternion group and the dihedral group of order 8 are 2-groups.

The following theorem gives a simple characterization of finite $p$-groups.

**Theorem 2.3**  *A finite group $G$ is a p-group if and only if $|G|$ is a power of p.*

***Proof***  If $|G| = p^n$, then Lagrange's theorem shows that $G$ is a $p$-group. Conversely, suppose that there is a prime $q$ ($\neq p$) which divides $|G|$. By Cauchy's theorem, $G$ contains an element of order $q$, and this contradicts $G$ being a $p$-group. ∎

© The Author(s), under exclusive license to Springer Nature Singapore Pte Ltd. 2021
B. Davvaz, *Groups and Symmetry*, https://doi.org/10.1007/978-981-16-6108-2_2

**Lemma 2.4** *Let G be a p-group. Any subgroup of G is also a p-group; its index is a power of p. If N is a normal subgroup of G, then the factor group $G/N$ is a p-group.*

**Proof** It is straightforward.                                                                                            ∎

**Theorem 2.5** *Let G be a p-group, and let X be a non-empty finite G-set. If $|X| \not\equiv 0(\mathrm{mod}\ p)$, then there is a fixed point of the action of G on X.*

**Proof** Suppose that $X = \mathrm{Orb}_G(x_1) \cup \cdots \cup \mathrm{Orb}_G(x_m)$, the union of disjoint orbits. For each $i$, we have $|\mathrm{Orb}_G(x_i)| = [G : \mathrm{Stab}_G(x_i)]$. By assumption, $G$ is a $p$-group, so by Lemma 2.4, $[G : \mathrm{Stab}_G(x_i)]$ is a power of $p$. Since $|X| \not\equiv 0(\mathrm{mod}\ p)$, it follows that

$$\sum_{i=1}^{m}[G : \mathrm{Stab}_G(x_i)] \not\equiv 0(\mathrm{mod}\ p).$$

Hence, for at least one $i$, $[G : \mathrm{Stab}_G(x_i)]$ is not divisible by $p$. A power $p^n$ is not divisible by $p$ only when $n = 0$. This yields that $|\mathrm{Orb}_G(x_i)| = 1$. Consequently, $\mathrm{Orb}_G(x_i)$ consists a fixed point of the action of $G$ on $X$.                                                                    ∎

**Corollary 2.6** *Suppose that a p-group G acts on another p-group H. If $H \neq \{e\}$, then there is a fixed point of H other than the identity.*

**Proof** Let $X = H - \{e\}$. By the assumption, $X$ is a non-empty $G$-set such that $|X| \not\equiv 0(\mathrm{mod}\ p)$. Hence, by Theorem 2.5, there is a fixed point of $X$.                     ∎

**Theorem 2.7** *Let G be a p-group and let N be a normal subgroup of G. If $N \neq \{e\}$, then $N \cap Z(G) \neq \{e\}$.*

**Proof** By Lemma 2.4, $N$ is a $p$-subgroup. Since $N \trianglelefteq G$, it follows that the $p$-group $G$ acts on the $p$-group $N \neq \{e\}$ by conjugation. Theorem 2.5 says that $N$ contains a fixed point $x$ other than $e$. By the definition of action, we have $x = x^a = a^{-1}xa$, for all $a \in G$. This shows that $x$ is contained in the center of $G$. Hence, we have $x \in N \cap Z(G) \neq \{e\}$. This completes the proof.                                                       ∎

**Theorem 2.8** *Let H be a subgroup of a p-group G. Then, either $H \trianglelefteq G$, or a conjugate subgroup $H^a$ different from H is contained in $N_G(H)$.*

**Proof** Suppose that $H$ is not normal, and let $X$ be the all conjugates $H^a$ of $H$ different from $H$. Then, $X$ is a non-empty set on which $H$ acts by conjugation. We know that the number of conjugates of $H$ equals to $[G : N_G(H)]$. Hence by Lemma 2.4 and assumption, we conclude that

$$|X| = [G : N_G(H)] - 1 \not\equiv 0(\mathrm{mod}\ p).$$

Thus, by Theorem 2.5, $X$ contains a fixed point, say $H^b$. Since $H^b$ is a fixed point, it follows that $H \subseteq N_G(H^b)$. Take $a = b^{-1}$, then we obtain $H \neq H^a \subseteq N_G(H)$.    ∎

**Theorem 2.9** *Let G be a finite p-group.*

*(1)* If $H$ is a proper subgroup of $G$, then $H$ is a proper subgroup of $N_G(H)$ too;
*(2)* Each maximal subgroup of $G$ is normal.

**Proof**

(1) If $H \trianglelefteq G$, then $N_G(H) = G$ and the result follows. If $X$ is the set of all conjugate of $H$, then we may assume that $|X| = [G : N_G(H)] > 1$. Now, $G$ acts on $X$ by conjugation. Since $G$ is a $p$-group, it follows that every orbit of $X$ has length a power of $p$. Since $\{H\}$ is an orbit of length 1, it follows that there exist at least $p - 1$ other orbits of length 1. Hence, there is at least one conjugate $aHa^{-1} \neq H$ with $\{aHa^{-1}\}$ also an orbit of length 1. Now, we have $baHa^{-1}b^{-1} = aHa^{-1}$, for all $b \in H$. This implies that $a^{-1}baH(a^{-1}ba)^{-1} = H$, and hence $a^{-1}ba \in N_G(H)$, for all $a \in H$. But $aHa^{-1} \neq H$ gives at least one $a \in H$ such that $a^{-1}ba \notin H$, and consequently $H \subset N_G(H)$.
(2) If $H$ is a maximal subgroup of $G$, then $H \subset N_G(H)$ implies that $N_G(H) = G$. This shows that $H$ is a normal subgroup of $G$. ∎

**Lemma 2.10** *In each finite p-group, distinct subgroups of order p have trivial intersection.*

**Proof** Suppose that $H$ and $K$ are two distinct subgroups of a $p$-group $G$. By Lagrange's theorem we have $|H \cap K| \,|\, |G|$. This forces that $H \cap K| = 1$ or $|H \cap K| = p$. If $|H \cap K| = p$, it follows that $H = K$, a contradiction. Hence, $|H \cap K| = 1$ or equivalently $H \cap K = \{e\}$. ∎

**Theorem 2.11** *Let $G$ be a finite p-group. If $k$ is the number of subgroups of $G$ having order $p$, then $k \equiv 1 \,(\mathrm{mod}\ p)$.*

**Proof** First, we count the number of elements of order $p$. Since $Z(G)$ is abelian, it follows that all its elements of order $p$ together with $e$ form a subgroup $H$ whose order is power of $p$. So, the number of central elements of order $p$ is $|H| - 1 \equiv -1 \,(\mathrm{mod}\ p)$. If $a$ is an element of $G$ such that $o(a) = p$ and it is not central, then its conjugacy class $a^G$ consists of several elements of order $p$. Indeed, we have $|a^G| = [G : C_G(a)] > 1$, and so $|a^G|$ is a power of $p$. This yields that the number of elements in $G$ of order $p$ is congruent to $-1$ modulo $p$. In other words, there are $mp - 1$ elements of order $p$ in $G$. Now, since the intersection of any distinct pairs of subgroups of order $p$ is trivial, it follows that the number of elements of order $p$ is $k(p - 1)$. Hence, we have $k(p - 1) = mp - 1$, which implies that $k \equiv 1 \,(\mathrm{mod}\ p)$. ∎

**Theorem 2.12** *If $G$ is a p-group acting on a finite set $X$, then*

$$|X| \equiv |F_X(G)| \,(\mathrm{mod}\ p),$$

*where $F_X(G) = \{x \in X \mid x^a = x, \text{ for all } a \in G\}$.*

***Proof*** Suppose that distinct orbits of $X$ is represented by $x_1, \ldots, x_k$. Then, we have $|X| = |\mathrm{Orb}_G(x_1)| + \cdots + |\mathrm{Orb}_G(x_k)|$. Since $|\mathrm{Orb}_G(x_i)| \, \big| \, |G|$, it follows that $|\mathrm{Orb}_G(x_i)|$ is a power of $p$. Assume that $r$ orbits are singleton, where $0 \le r \le k$. So, we obtain $F_X(G)| = r$, and so

$$|X| = r + \text{a sum of positive powers of } p.$$

Hence, we conclude that $|X| \equiv F_X(G)(\mathrm{mod}\, p)$.                          ∎

**Exercises**

1. Let $G$ be a group. If $N \trianglelefteq G$ and $N$, $G/N$ are both $p$-groups, prove that $G$ is a $p$-group.
2. Let $G$ be a finite $p$-group. If $M$ is a maximal subgroup of $G$, prove $[G : M] = p$.
3. Give an example of a finite group $G$ generated by two elements of order 2 but where $G$ is not a 2-group.
4. If $G$ is an infinite group, prove that either $G$ has a subgroup of order $p^n$ for each $n \ge 1$ or there exists a positive integer $m$ such that for every finite subgroup $H \le G$ we have $|H| \le p^m$.

## 2.2  Sylow's Theorems

As we have seen, the converse to Lagrange's theorem is false in general: If $G$ is a finite group of order $n$ and $d$ divides $n$, then there need not exist a subgroup of $G$ whose order is $d$. (For example, $|A_4| = 12$ and $A_4$ has no subgroup of order 6). Already we observed that he converse is true for primed, it was Cauchy's theorem. Sylow's theorems say that such a subgroup exists in one special but very important case: when $d$ is the largest power of a prime which divides $n$. (It then turns out that $G$ has a subgroup of every order which is a prime power dividing $n$, not necessarily the largest such). In fact, Sylow's theorems tell us much more about such subgroups, by giving information on how many such subgroups can exist. As we shall see, this will sometimes enable us to show that $G$ has a non-trivial proper normal subgroup.

**Definition 2.13** Let $G$ be a finite group. We write

$$|G| = p^n m \ (p, m) = 1.$$

A subgroup of $G$ is said to be a *Sylow p-subgroup* if its order is exactly $p^n$. In other words, a Sylow $p$-subgroup is a maximal $p$-subgroup.

We may denote the set of all Sylow $p$-subgroups of $G$ by $\mathrm{Syl}_p(G)$.

We begin with a combinatorial lemma.

**Lemma 2.14** *If* $p \nmid m$, *then* $p \nmid \displaystyle\binom{p^n m}{p^n}$.

**Proof** If we expand the binomial coefficient, we obtain

$$\binom{p^n m}{p^n} = \frac{(p^n m)!}{(p^n)!(p^n m - p^n)!}$$
$$= \frac{p^n m(p^n m - 1)\ldots(p^n m - k)\ldots(p^n m - p^n + 1)}{p^n(p^n - 1)\ldots(p^n - k)\ldots(p^n - p^n + 1)}.$$

Now, we show that if $p$ divides a term $p^n m - k$ in the numerator, then it divides the corresponding term $p^n - k$ in the denominator. If we take $k = p^l s$, where either $l < n$ and $p \nmid s$ or $l = n$, then

$$\frac{p^n m - k}{p^n - k} = \frac{p^{n-l} m - s}{p^{n-l} - s}.$$

which is not divisible by $p$. Therefore, we conclude that the whole product is not divisible by $p$. ∎

**Theorem 2.15** (**Sylow's first theorem**). *If $p$ is a prime number and $p \,\|\, |G|$, then there exists a Sylow $p$-subgroup of $G$.*

**Proof** Let $|G| = p^n m$ with $(p, m) = 1$, and suppose that $X$ is the set of all subsets of order $p^n$. Then, we have $|X| = \binom{p^n m}{p^n}$. Assume that $G$ acts on $X$ by right multiplication. Let the distinct orbits of $X$ be represented by $A_1, \ldots, A_k$. Then, we have $|X| = |\mathrm{Orb}_G(A_i)| + \cdots |\mathrm{Orb}_G(A_i)|$. So, by Lemma 2.14, we deduce that $p \nmid |X|$, and so there exist at least one $1 \leq j \leq k$ such that $p \nmid |\mathrm{Orb}_G(A_j)|$. Now, we set $H := \mathrm{Stab}_G(A_j)$. Then, by Orbit-Stabilizer theorem, we have $|\mathrm{Orb}_G(A_j)| = [G : H]$. Since $|G| = p^n m = [G : H]|H|$ and $p \nmid [G : H]$, it follows that $p^n \,\|\, |H|$. If $A_j = \{a_1, \ldots, a_{p^n}\}$, then we have $A_j^h = A_j$, for all $h \in H$. This implies that $a_1^h = a_i$, for some $1 \leq i \leq p^n$. Hence, $a_1 h = a_i$ or $h = a_i a_1^{-1}$. Consequently, we obtain $|H| \leq p^n$. Therefore, we conclude that $|H| = p^n$, and so $H$ is a Sylow $p$-subgroup of $G$. ∎

**Theorem 2.16** (**Sylow's second theorem**). *For a given prime $p$, all Sylow $p$-subgroups of $G$ are conjugate to each other.*

**Proof** Consider two Sylow $p$-subgroups of $G$, call $H$ and $K$. We show that they are conjugate. Let $X$ be the set of right cosets of $H$ in $G$, and consider the action of $K$ on $X$ by right multiplication. Since $K$ is a finite $p$-group, by Theorem 2.12, it follows that

$$|X| \equiv |F_X(K)| \pmod{p}, \tag{2.1}$$

The left side of (2.1) is $[G : H] = m$, which is nonzero modulo $p$. Hence, $|F_X(K)|$ cannot be zero. Consequently, there exist a fixed point in $X$, say $Ha$. So, we get $Hab = Ha$, for all $b \in K$, which implies that $b \in a^{-1}Ha$. Therefore, we conclude that $K \subseteq a^{-1}Ha$. Since $|K| = |a^{-1}Ha|$, it follows that $K = a^{-1}Ha$, thereby giving the desired result. ∎

**Lemma 2.17** *If $P \in \mathrm{Syl}_p(G)$ and $H$ is any $p$-group of $G$, then $H \cap P = H \cap N_G(P)$.*

**Proof** Since $P \leq N_G(P)$, it follows that $H \cap P \subseteq H \cap N_G(P)$. Now, we prove the converse inclusion. Since $H \cap N_G(P) \leq H$ and $H$ is a $p$-subgroup, it follows that either $H \cap N_G(P) = \{e\}$ or $H \cap N_G(P)$ is a $p$-subgroup. Note that $P \trianglelefteq N_G(P)$ and $[N_G(P) : P]$ has no factor of $p$. If we consider the restrict canonical homomorphism $\varphi : N_G(P) \to N_G(P)/P$ to $H \cap N_G(P)$, then we have $\varphi(a) = aP$, for all $a \in H \cap N_G(P)$. Since $N_G(P)/P$ has no elements whose orders are power of $p$, it follows that each element of $H \cap N_G(P)$ whose order is a power of $p$ must map to the identity element of $N_G(P)/P$. Therefore, we obtain $aP = P$, for all $a \in H \cap N_G(P)$, which implies that $a \in P$. This proves that $H \cap N_G(P) \leq P$. Therefore, we obtain $H \cap P = H \cap N_G(P)$.                                                                ∎

**Lemma 2.18** *If $P \in \mathrm{Syl}_p(G)$ and $X$ is the set of all distinct conjugations of $P$, then $p \nmid |X|$.*

**Proof** Suppose that $|G| = p^n m$ with $(p, m) = 1$. We know the action of $G$ on $X$ is transitive. Hence, we have

$$|X| = |\mathrm{Orb}_G(P)| = [G : N_G(P)] = \frac{|G|}{|N_G(P)|}.$$

Since $P \leq N_G(P) \leq G$, it follows that $p^n \big| |N_G(P)|$. So, $|N_G(P)|$ contains $p^n$ as a factor, which is the maximum power of $p$ that it can have. Consequently, $[G : N_G(P)]$ contains no factor of $p$, and so $p \nmid |X|$.                                                                ∎

**Theorem 2.19** *Any $p$-subgroup of a group $G$ is contained in a Sylow $p$-subgroup.*

**Proof** Let $H$ be any $p$-subgroup of $G$. We want to show that $H \leq P$, for some $P \in \mathrm{Syl}_p(G)$. Assume that $H$ acts on $\mathrm{Syl}_p(G)$ by conjugation, and distinct orbits is represented by $P_1, \ldots, P_k$. Then, we have $|\mathrm{Syl}_p(G)| = |\mathrm{Orb}_H(P_1)| + \cdots + |\mathrm{Orb}_H(P_k)|$. By Lemma 2.18, $p \nmid |\mathrm{Syl}_p(G)|$, so $p$ cannot dividing all of the terms $|\mathrm{Orb}_H(P_1)|, \ldots, |\mathrm{Orb}_H(P_k)|$. Hence, we can assume that $|\mathrm{Orb}_H(P_i)| = 1$, for some $1 \leq i \leq k$. On the other hand, by Orbit-Stabilizer theorem, we have

$$|\mathrm{Orb}_H(P_i)| = [H : \mathrm{Stab}_H(P_i)] = [H : H \cap N_G(P_i)].$$

Thus, we must have $[H : H \cap N_G(P_i)] = 1$, which implies that $H = H \cap N_G(P_i)$. Since $H$ is a $p$-subgroup, by Lemma 2.17, we conclude that $H \cap N_G(P_i) = H \cap P_i$. Consequently, we have $H = H \cap P_i$. This shows that $H \leq P_i$.                                                                ∎

**Theorem 2.20** (**Sylow's third theorem**). *We denote the number of Sylow $p$-subgroups of $G$ by $n_p$. Then, the following statements hold:*

*(1)  $n_p \equiv 1 \pmod{p}$;*
*(2)  If $|G| = p^n m$ such that $p \nmid m$, then $n_p | m$;*

**Proof**

(1) Suppose that $\mathrm{Syl}_p(G) = \{P_1, \ldots, P_{n_p}\}$ is the set of all distinct Sylow $p$-subgroups of $G$. Let $P_1$ acts on $\mathrm{Syl}_p(G)$ by conjugation. Since $aP_1a^{-1} = P_1$, for all $a \in P_1$, it follows that $|\mathrm{Orb}_{P_1}(P_1)| = 1$. If $k \neq 1$, then by Orbit-Stabilizer theorem we have

$$|\mathrm{Orb}_{P_1}(P_k)| = [P_1 : P_1 \cap N_G(P_k)] = [P_1 : P_1 \cap P_k]. \tag{2.2}$$

On the other hand, since $P_1$ and $P_k$ are distinct subgroups of $G$, it follows that

$$[P_1 : P_1 \cap P_k] = \frac{|P_1|}{|P_1 \cap P_k|} > 1. \tag{2.3}$$

By (2.2) and (2.3), we conclude that $|\mathrm{Orb}_{P_1}(P_k)| > 1$, for all $k \neq 1$. Moreover, since $|P_1| = p^n$, it follows that $|\mathrm{Orb}_{P_1}(P_k)|$ (for $k \neq 1$) is a power of $p$. Now, we have

$$n_p = |\mathrm{Syl}_p(G)| = |\mathrm{Orb}_{P_1}(P_1)| + \sum_{k=2}^{n_k} |\mathrm{Orb}_{P_1}(P_k)| = 1 + rp,$$

for some positive integer $r$. This implies that $n_p \equiv 1 (\mathrm{mod}\ p)$.

(2) To show that $n_p | m$, we consider the action of $G$ on $\mathrm{Syl}_p(G)$ by conjugation. Since the Sylow $p$-subgroups are conjugate to each other, it follows that the action is transitive. So, by Theorem 1.37, we conclude that $|\mathrm{Syl}_p(G)| \big| |G|$, or equivalently $n_p | p^n m$. From $n_p \equiv 1 (\mathrm{mod}\ p)$, it follows that $n_p$ and $p$ are relatively prime. Since $n_p | p^n m$ and $(n_p, p) = 1$, it follows that $n_p | m$. ∎

**Corollary 2.21** *A finite group $G$ has a unique Sylow $p$-subgroup $P$, for some prime $p$, if and only if $P \trianglelefteq G$.*

**Proof** If $G$ has only one Sylow $p$-subgroup $P$, then $1 = [G : N_G(P)]$. This implies that $G = N_G(P)$, and so $P \trianglelefteq G$. Conversely, if $P$ is a normal Sylow $p$-subgroup of $G$, then it is unique, because all Sylow $p$-subgroups of $G$ are conjugate. ∎

**Corollary 2.22** *Let $p$ and $q$ are different prime factors of $|G|$. If $n_p = 1$ and $n_q = 1$, then the elements of the Sylow $p$-subgroup commute with the elements of the Sylow $q$-subgroup.*

**Proof** Suppose that $P$ is the Sylow $p$-subgroup and $Q$ is the Sylow $q$-subgroup. Since $(|P|, |Q|) = 1$, by Lagrange's theorem, we conclude that $P \cap Q = \{e\}$. Moreover, by Corollary 2.21, the subgroups $P$ and $Q$ are normal in $G$. Hence, for $a \in P$ and $b \in Q$, we can write

$$aba^{-1}b^{-1} = (aba^{-1})b^{-1} = a(ba^{-1}b^{-1}) \in P \cap Q = \{e\}.$$

This shows that $ab = ba$. ∎

**Theorem 2.23** *If $P \in \text{Syl}_p(G)$ and $H$ is a subgroup of $G$ that contains $N_G(P)$, then $N_G(H) = H$.*

**Proof** Clearly, we have $H \subseteq N_G(H)$. For the converse inclusion, suppose that $a \in N_G(H)$ is an arbitrary element. Then, we have $aHa^{-1} = H$. Since $P \subseteq N_G(P) \subseteq H$, it follows that $aPa^{-1} \subseteq aN_G(P)a^{-1} \subseteq aHa^{-1} = H$. Hence, $P$ and $aPa^{-1}$ are Sylow $p$-subgroups of $H$. So, we conclude that $aPa^{-1} = hPh^{-1}$, for some $h \in H$, which implies that $h^{-1}aP(h^{-1}a)^{-1} = P$. This yields that $h^{-1}a \in N_G(P) \subseteq H$, and so $a \in hH = H$. This completes the proof. ∎

**Corollary 2.24** *If $P \in \text{Syl}_p(G)$, $N_G(N_G(P)) = N_G(P)$.*

**Proof** Take $H = N_G(P)$ in Theorem 2.23. ∎

**Theorem 2.25** *If $P \in \text{Syl}_p(G)$ and $N$ is a normal subgroup of $G$, then $P \cap N$ is a Sylow p-subgroup of $N$ and $PN/N$ is a Sylow p-subgroup of $G/N$.*

**Proof** By the second isomorphism theorem, we find out $[N : P \cap N] = [PN : P]$, and it is relatively prime to $p$. Hence, since $P \cap N$ is a $p$-subgroup, it must be a Sylow $p$-subgroup of $N$.

Also, as $PN/N \leq G/N$, by the second isomorphism theorem, we have $|PN/N| = |P/(P \cap N)|$. This implies that $|PN/N|$ is a power of $p$. On the other hand, we have $[G/N : PN/N] = [G : PN]$ is relatively prime to $p$. Therefore, we deduce that $PN/N$ is a Sylow $p$-subgroup of $G/N$. ∎

**Theorem 2.26** *(**Frattini argument***). Let $H$ be a normal subgroup of a finite group $G$. If $P \in \text{Syl}_p(H)$, then $G = HN_G(P)$.*

**Proof** Suppose that $a \in G$ is an arbitrary element. Since $H \trianglelefteq G$, it follows that $aPa^{-1} \leq aHa^{-1} = H$. This yields that $aPa^{-1}$ is a Sylow $p$-subgroup of $H$, and so $P$ and $aPa^{-1}$ are conjugate in $H$. It follows that there is $h \in H$ such that $aPa^{-1} = hPh^{-1}$, or equivalently $P = (h^{-1}a)P(h^{-1}a)^{-1}$. This shows that $h^{-1}a \in N_G(P)$. Now, the required factorization is $a = h(h^{-1}a)$. ∎

The following is an application of the Frattini argument. It is convenient to use bar convention for subgroups of $G/H$:

$$\overline{K} := KH/H \quad (\text{for } K \leq G),$$

and in particular, $\overline{G} = G/H$.

**Theorem 2.27** *Let $H$ be a normal subgroup of a finite group $G$ with factor group $\overline{G} := G/H$, and let $P$ be a p-subgroup of $G$. If $(|H|, p) = 1$, then*

$$N_{\overline{G}}(\overline{P}) = \overline{N_G(P)} \quad \text{and} \quad C_{\overline{G}}(\overline{P}) = \overline{C_G(P)}.$$

**Proof** The definition of a factor group gives

$$N_{\overline{G}}(\overline{P}) = \overline{N_G(HP)}.$$

Since $(|H|, p) = 1$, it follows that $P \in \mathrm{Syl}_p(HP)$. In addition, $HP \trianglelefteq N_G(HP)$. Hence, by the Frattini argument, we obtain

$$N_G(HP) = HPN_{N_G(HP)}(P) = HPN_G(P) = HN_G(P).$$

Thus, we conclude that $N_{\overline{G}}(\overline{P}) = \overline{N_G(P)}$.

It is clear that $\overline{C_G(P)} \subseteq C_{\overline{G}}(\overline{P})$. For the converse inclusion, suppose that $Ha \in C_{\overline{G}}(\overline{P})$ is an arbitrary element. Since $C_{\overline{G}}(\overline{P}) < N_{\overline{G}}(\overline{P})$, it follows that there are $h \in H$ and $b \in N_G(P)$ such that $a = hb$. This yields that $Ha = Hb$ and $Hx = b^{-1}(Hx)b = Hb^{-1}xb$, for all $x \in P$. The commutator $b^{-1}xbx^{-1}$ belongs to $H \cap P = \{e\}$. Therefore, we conclude that $b \in C_G(P)$, and so we obtain $Ha = Hb \in \overline{C_G(P)}$. This completes the proof. ∎

**Theorem 2.28** *Let $G = HK$ with $H \trianglelefteq G$, $K \le G$ and $(|H|, p) = 1$. Then, for every $p$-subgroup $P$ of $K$, we have*

$$N_G(P) = \big(H \cap N_G(P)\big)\big(K \cap N_G(P)\big).$$

**Proof** Let $\overline{G} := G/H$ and $L = H \cap K$. By Theorem 2.27, we obtain

$$N_{K/L}\Big(\frac{PL}{L}\Big) = \frac{N_K(P)L}{L}.$$

Since $K/L \cong KH/H \,(= \overline{G})$, it follows that

$$\overline{N_G(P)} = N_{\overline{G}}(\overline{P}) = \overline{N_K(P)}.$$

Therefore, we conclude that $N_K(P) \le N_G(P) \le HN_K(P)$, and the result follows from Dedekind's modular law. ∎

## Exercises

1. Let $P$ be a Sylow $p$-subgroup and $H$ a subgroup of the finite group $G$. Is it true that $P \cap H$ is a Sylow $p$-subgroup of $H$?
2. How many Sylow $p$-subgroups (corresponding to the different primes) do non-abelian groups of orders 21 and 39, respectively, have?
3. What are the orders of all Sylow $p$-subgroups where $G$ has order 18, 24, 54, 72, and 80?
4. Show that a group of order 33 has only one Sylow 3-subgroup.
5. Let $G$ be a non-cyclic group of order 21. How many Sylow 3-subgroups does $G$ have?

6. If $P$ is a normal Sylow $p$-subgroup of a finite group $G$ and $f : G \to G$ is an endomorphism, show that $f(P)$ is a subgroup of $P$.

7. Let $P$ be a normal Sylow $p$-subgroup of $G$. Prove that every inner automorphism of $G$ fixes $P$.

8. Let $H$ and $K$ denote a Sylow 3-subgroup and a Sylow 5-subgroup of a group, respectively. Suppose that $|H| = 3$ and $|K| = 5$. If 3 divides $|N_G(K)|$, show that 5 divides $|N_G(H)|$.

9. If $p^m \big| |G|$ show that $G$ has a subgroup of order $p^m$.

10. If $H$ is a subgroup of order $p^m$ of a finite group $G$, prove that $H$ is contained in some Sylow $p$-subgroup of $G$.

11. If $N$ is a normal subgroup of order $p^m$ of a finite group $G$, show that $N$ is contained in every Sylow $p$-subgroup of $G$.

12. Let $G$ be a finite group and $P$ be a Sylow $p$-subgroup of $G$. Prove that if $a \in N_G(P)$ and $o(a)$ is a power of $p$, then $a \in P$.

13. Let $p$ be a prime and $H$ and $K$ be Sylow $p$-subgroups of a group $G$. Prove that $|N_G(H)| = |N_G(K)|$.

14. If $P$ is a Sylow $p$-subgroup of $G$ and $a, b \in Z(P)$ are conjugate in $G$, prove that they are already conjugate in $N_G(P)$.

15. If $|G| = p^n q$ with $p > q$ primes, prove that $G$ contains a unique normal subgroup of index $q$.

16. Prove that all Sylow $p$-subgroups of a finite group are isomorphic.

17. Let $G$ be a finite group and $P \in \mathrm{Syl}_p(G)$ such that $N_G(P) = P$. Show that $G/G'$ is a $p$-group.

18. Let $G$ be a finite group, $P \in \mathrm{Syl}_p(G)$ and $H \trianglelefteq G$. If $P \trianglelefteq H$, show that $P \trianglelefteq G$.

## 2.3 Sylow $p$-Subgroups of Specific Groups

In this section we study Sylow $p$-subgroups of some special groups. We begin with the following lemma.

**Lemma 2.29** *Let $\ell(k)$ be define by $p^{\ell(k)}|(p^k)!$ but $p^{\ell(k)+1} \nmid (p^k)!$. Then, we have $\ell(k) = 1 + p + \cdots + p^k$.*

**Proof** If $k = 1$, then it is clear that $p|p!$ but $p^2 \nmid p!$. This yields that $\ell(1) = 1$. It is easy to see that the terms in the expression of $(p^k)!$ which can contribute to power of $p$ dividing $(p^k)!$ are only the multiples of $p$, i.e., $p, 2p, \ldots, p^{k-1}p$. So, $\ell(k)$ is the power of $p$ which divides $p(2p)(3p)\ldots(p^{k-1}p) = p^{p^{k-1}}(p^{k-1})!$. This yields that $\ell(k) = p^{k-1} + \ell(k-1)$. Analogously, we obtain $\ell(k-1) = p^{k-2} + \ell(k-2)$, and so on. Therefore, we can write

$$\ell(k) - \ell(k-1) = p^{k-1},$$
$$\ell(k-1) - \ell(k-2) = p^{k-2},$$
$$\vdots$$
$$\ell(2) - \ell(1) = p,$$
$$\ell(1) = 1.$$

By adding the above equalities up, the proof completes.  ∎

**Theorem 2.30** *The symmetric group $S_{p^k}$ has a Sylow $p$-subgroup.*

**Proof** We do the proof by mathematical induction on $k$. If $k = 1$, then we have $\ell(1) = 1$ and $\langle (1\ 2\ \ldots\ n) \rangle$ is a subgroup of order $p$. Now, assume that the result holds for $k - 1$. We prove the theorem for $k$. We separate the integers $1, 2, \ldots, p^k$ into $p$ sets, each with $p^{k-1}$ elements as follows:

$$\{1,\ 2,\ \ldots,\ p^{k-1}\},$$
$$\{p^{k-1}+1,\ p^{k-1}+2,\ \ldots,\ 2p^{k-1}\},$$
$$\vdots$$
$$\{(p-1)p^{k-1}+1,\ \ldots,\ p^k\}.$$

Let $\sigma = \sigma_1 \ldots \sigma_j \ldots \sigma_{p^{k-1}} \in S_{p^k}$, where

$$\sigma_1 = \left(1\ \ p^{k-1}+1\ \ 2p^{k-1}+1\ \ \ldots\ \ (p-1)p^{k-1}+1\right),$$
$$\vdots$$
$$\sigma_j = \left(j\ \ p^{k-1}+j\ \ 2p^{k-1}+j\ \ \ldots\ \ (p-1)p^{k-1}+1+j\right),$$
$$\vdots$$
$$\sigma_{p^{k-1}} = \left(p^{k-1}\ \ 2p^{k-1}\ \ \ldots\ \ (p-1)p^{k-1}\ \ p^k\right).$$

Then, $\sigma$ has the following properties:

(1) $\sigma^p = \mathrm{id}$;
(2) If $\tau \in S_{p^k}$ fixes all $i > p^{k-1}$, then $\sigma^{-1}\tau\sigma$ moves only elements in $\{p^{k-1}+1,\ p^{k-1}+2,\ \ldots,\ 2p^{k-1}\}$. In general, we can check that $\sigma^{-j}\tau\sigma^j$ moves only elements in $\{jp^{k-1}+1,\ jp^{k-1}+2,\ \ldots,\ (j+1)p^{k-1}\}$.

Set $H = \{\tau \in S_{p^k} \mid i\tau = i,\ \text{for } i > p^{k-1}\}$. It is easy to see that $H$ is a subgroup of $S_{p^k}$ and its elements can carry out any permutation on the set $\{1,\ 2,\ \ldots,\ p^{k-1}\}$. This yields that $H \cong S_{p^{k-1}}$. Hence, by induction hypothesis, we conclude that $H$ has a subgroup $P_1$ of order $p^{\ell(k-1)}$. Now, suppose that

$$K = P_1\left(\sigma^{-1}P_1\sigma\right)\left(\sigma^{-2}P_1\sigma^2\right)\ldots\left(\sigma^{-(p-1)}P_1\sigma^{p-1}\right)$$
$$= P_1 P_2 \ldots P_p,$$

where $P_i = \sigma^{-(i-1)}P_1\sigma^{i-1}$, for all $i \geq 1$. Clearly, for each $i$ we have $P_i \cong P_1$, and so its order is $p^{\ell(k-1)}$. In addition, elements in distinct $P_i$'s influence non-overlapping sets of integers, and so commute. Consequently, $K$ is a subgroup of $S_{p^k}$. Since $P_i \cap P_j = \{\text{id}\}$ (for $1 \le i \ne j \le p$), it follows that $|K| = |P_1|^p = p^{p\ell(k-1)}$. Finally, since $\sigma^p = \text{id}$ and $\sigma^{-(i-1)}P_1\sigma^{i-1} = P_i$ (for $i \ge 1$), it follows that $\sigma^{-1}K\sigma = K$. Now, let

$$P = \{\sigma^j\delta \mid \delta \in K, \ 0 \le j \le p - 1\}.$$

Since $\sigma \notin K$ and $\sigma^{-1}K\sigma = K$, it follows that $K$ is a subgroup of $S_{p^k}$ and

$$|P| = p \cdot |K| = p \cdot p^{\ell(k-1)p} = p^{\ell(k-1)p+1}.$$

We have $\ell(k-1) = 1 + p + \cdots + p^{k-2}$, which implies that $p\ell(k-1) + 1 = 1 + p + \cdots + p^{k-1} = \ell(k)$. Since $|P| = p^{\ell(k)}$, it follows that $P$ is a Sylow $p$-subgroup of $S_{p^k}$. ∎

Sylow's Theorems allow us to prove many useful results about finite groups. By using them, we can often conclude a great deal about groups of a particular order if certain hypotheses are satisfied.

**Theorem 2.31** *If $p$ and $q$ are distinct primes with $p < q$, then every group $G$ of order $pq$ has a single subgroup of order $q$ and this subgroup is normal in $G$. Moreover, if $q \not\equiv 1 \pmod{p}$, then $G$ is cyclic.*

***Proof*** We know that $G$ contains a subgroup $Q$ of order $q$. The number of conjugates of $Q$ divides $pq$ and is equal to $1 + kq$ for $k = 0, 1, \ldots$. However, $1 + q$ is already too large to divide the order of the group; hence, $Q$ can only be conjugate to itself. That is, $Q$ must be normal in $G$.

For the second part, consider a Sylow $p$-subgroup $P$ of $G$. Let $n_p$ be the number of Sylow $p$-subgroups. So, $n_p$ divides $[G : P] = q$ and $n_p \equiv 1 \pmod{p}$. Then, $n_p \equiv 1 + kp$ and $1 + kp$ divides $q$. So, $n_p$ is equal to 1 or $q$. But it is given that $q = n_p \not\equiv 1 \pmod{p}$. Hence, $n_p = 1$ and $P$ is a normal subgroup of $G$. Since $|P| = p$, $|Q| = q$ and $p \ne q$, it follows that $P \cap Q = \{e\}$. Then for any $a \in P$ and $b \in Q$, $a^{-1}b^{-1}ab \in P \cap Q$. Hence, $ab = ba$, for all $a \in P$ and $b \in Q$. Consequently, $G = PQ$ and $G$ is an abelian group. Assume that $P = \langle x \rangle$ and $Q = \langle y \rangle$, where $x, y \in G$. We have $\langle xy \rangle = \{x^i y^i \mid i \in \mathbb{N}\}$. Since $(xy)^p = x^p y^p = y^p \ne \{e\}$, it follows that $\langle y^p \rangle = \langle y \rangle \le \langle xy \rangle$. Since $(xy)^q = x^q y^q = x^q \ne \{e\}$, it follows that $\langle x^q \rangle = \langle x \rangle \le \langle xy \rangle$. Hence, we obtain $p \mid |\langle xy \rangle|$ and $q \mid |\langle xy \rangle|$. This implies that $pq \mid |\langle xy \rangle|$. As $|G| \mid |\langle xy \rangle|$, we conclude that $G = \langle xy \rangle$, and so $G$ is cyclic. ∎

**Theorem 2.32** *Let $\mathbb{F}$ be a finite field with $q$ elements, where $q$ is a power of a prime number $p$, and suppose that $\mathcal{P}$ is the subgroup of $GL_n(\mathbb{F})$ consisting of those matrices with all entries below the main diagonal zero, and with the entries on the main diagonal equal to the identity. Then, we have*

*(1) $\mathcal{P} \in \mathrm{Syl}_p\big(GL_n(\mathbb{F})\big)$;*
*(2) $\mathcal{P} \in \mathrm{Syl}_p\big(SL_n(\mathbb{F})\big)$;*
*(3) $\mathcal{P}$ is isomorphic to a Sylow $p$-subgroup of $PGL_n(\mathbb{F})$;*

(4)  $\mathcal{P}$ is isomorphic to a Sylow $p$-subgroup of $PSL_n(\mathbb{F})$;
(5)  The normalizer of $\mathcal{P}$ in $GL_n(\mathbb{F})$ contains all linear transformation $T$ such that

$$v_1 \mapsto \sum_{j=1}^{i} a_{ij} v_j, \quad a_{ij} \in \mathbb{F}, \ a_{ii} \neq 0, \ 1 \leq i, j \leq n,$$

where $\{v_1, \ldots, v_n\}$ is an ordered basis.

**Proof**

(1)  We know that

$$|GL_n(\mathbb{F})| = \prod_{k=0}^{n-1} (q^n - q^k) - q^{n(n-1)/2} m,$$

where $(p, m) = 1$. It remains to note that $|P| = q^{n(n-1)/2}$.
(2)  Since the determinant of any matrix in $\mathcal{P}$ is 1, it follows that $\mathcal{P} \leq SL_n(\mathbb{F})$.
(3)  Consider the canonical map $\pi : GL_n(\mathbb{F}) \to PGL_n(\mathbb{F})$. If $Z = Z(GL_n(\mathbb{F}))$, then we can write

$$\pi(\mathcal{P}) = \{\pi(A) \mid A \in \mathcal{P}\} = \{ZA \mid A \in \mathcal{P}\}$$
$$= \frac{Z\mathcal{P}}{Z} \cong \frac{\mathcal{P}}{Z \cap \mathcal{P}}.$$

Since $Z = \{\lambda I_n \mid \lambda \in \mathbb{F}\}$, it follows that $Z \cap \mathcal{P} = \{I_n\}$. As $|SL_n(\mathbb{F})| = |PGL_n(\mathbb{F})|$, we conclude that $\mathcal{P}$ is isomorphic to a Sylow $p$-subgroup of $PGL_n(\mathbb{F})$.
(4)  Similarly, we consider the canonical map $\pi : SL_n(\mathbb{F}) \to PSL_n(\mathbb{F})$.
(5)  Take $V_i = \langle v_1, \ldots, v_i \rangle$, for all $1 \leq i \leq n$. Clearly, $V_i$'s are subspace of $V$ and they are invariant under $\mathcal{P}$. The chain

$$V_1 \subseteq V_2 \subseteq \cdots \subseteq V_n = V$$

is called a *flag*. We observe that

$$V_1^A \subseteq V_2^A \subseteq \cdots \subseteq V_n^A,$$

for all $A \in \mathcal{P}$. Consequently, each element of $N_{GL_n(\mathbb{F})}(\mathcal{P})$ fixes any subspace $V_i$, for all $i$. Therefore, $N_{GL_n(\mathbb{F})}(\mathcal{P})$ consists of matrices of the form

$$\begin{bmatrix} a_{11} & a_{12} & \cdots & a_{1\,n-1} & a_{1n} \\ 0 & a_{22} & \cdots & a_{2\,n-1} & a_{2n} \\ \vdots & \vdots & \cdots & \vdots & a_{1n} \\ 0 & 0 & \cdots & a_{n-1\,n-1} & a_{n-1\,n} \\ 0 & 0 & \cdots & 0 & a_{nn} \end{bmatrix}$$

$a_{ij} \in \mathbb{F}^*$ and $a_{ij} \neq 0$. This completes the proof.

∎

**Exercises**

1. If $x > 0$ is a real number, define $[x]$ to be $m$, where $m$ is that integer such that $m \leq x < m + 1$. If $p$ is a prime, show that the power of $p$ which exactly divides $n!$ is given by

$$\left[\frac{n}{p}\right] + \left[\frac{n}{p^2}\right] + \cdots + \left[\frac{n}{p^k}\right] + \cdots.$$

2. Let $p$ denote an odd prime.

   (a) Show that the number of Sylow $p$-subgroups in the symmetric group $S_p$ is $(p-2)!$;
   (b) Using the result of (1) and Sylow's Theorems, give a proof of Wilson's Theorem;
   (c) Let $P = \langle (1\ 2\ \ldots\ p) \rangle$, a Sylow $p$-subgroup of $S_p$. Find the order of $N_G(P)$.

3. Use the method for constructing the Sylow $p$-subgroup of $S_{p^k}$ to find generators for

   (a) A Sylow 2-subgroup of $S_8$;
   (b) A Sylow 3-subgroup of $S_9$.

4. If $p$ is a prime number, give explicit generators for a Sylow $p$-subgroup of $S_{p^2}$.
5. Prove that a Sylow 2-subgroup of $A_5$ has exactly 5 conjugates.
6. Prove that a Sylow 2-subgroup of $S_5$ is isomorphic to $D_4$.
7. If $N$ is a normal $p$-subgroup of a finite group $G$, prove that $N \leq P$, for all Sylow $p$-subgroup $P$.
8. Show that a finite abelian group is cyclic if and only if its Sylow $p$-subgroups are cyclic, for all primes $p$.

## 2.4  Worked-Out Problems

**Problem 2.33** If $G$ is a group of order 12, show that $G$ has a normal Sylow 2-subgroup or 3-subgroup.

*Solution* By the Sylow's Third Theorem, $n_2 | 3$, and so we conclude that $n_2 = 1$ or 3. Since $n_3 | 4$ and $n_3 \equiv 1 \pmod 3$, it follows that $n_3 = 1$ or 4. Now, we show that $n_2 = 1$ or $n_3 = 1$. Suppose that $n_3 \neq 1$. Then, we have $n_3 = 4$. Since each Sylow 3-subgroup has order 3, it follows that $G$ has $2n_3 = 8$ elements of order 3. The number of remaining elements is 4. A Sylow 2-subgroup has order 4. Consequently, it fills up the remaining elements, and hence $n_2 = 1$.

∎

**Problem 2.34**

(1) Identify the Sylow 2-subgroups and the Sylow 3-subgroups in $S_3$. How many of each are there?
(2) Exhibit the Sylow 2-subgroups of $S_4$.
(3) Exhibit the Sylow 2-subgroups of $A_5$. How many are there?

*Solution*

(1) We know that $|S_3| = 6$, and so the Sylow 2-subgroups have order 2 and the Sylow 3-subgroups have order 3. Hence, they are cyclic, generated by elements of order 2 and order 3, respectively. Consequently, the Sylow 2-subgroups of $S_3$ are $\langle (1\,2) \rangle$, $\langle (1\,3) \rangle$, $\langle (2\,3) \rangle$, and the unique Sylow 3-subgroup is $\langle (1\,2\,3) \rangle$. Thus, there are 3 Sylow 2-subgroups and there is just one Sylow 3-subgroup.

(2) Since $|S_4| = 24$, it follows that the 2-subgroups have order 8. Suppose that $H$ is any 2-subgroup of $S_4$. Since $K_4$ is a normal 2-subgroup of $S_4$, it follows that $HK_4$ is again a 2-subgroup. Since Sylow 2-subgroups are maximal 2-subgroups, it follows that $K_4$ is contained in every Sylow 2-subgroup. Take $H$ to be a cyclic subgroup of order 4. Then, $H$ is not a subgroup of $K_4$, and so $HK_4$ is a 2-subgroup whose order is strictly greater than 4. So, by Lagrange's Theorem, it just have order 8 and it is a Sylow 2-subgroup of $S_4$. A cyclic subgroup of order 4 in $S_4$ must be generated by a cycle of length 4. There are six cycles of length 4, but they occur in inverse pairs and so there are three distinct cyclic subgroups of order 4. Therefore, we find that three Sylow 2-subgroups, namely

$$\langle K_4, (1\,2\,3\,4) \rangle = \{\text{id}, (1\,2)(3\,4), (1\,3)(2\,4), (1\,4)(2\,3), (1\,2\,3\,4),$$
$$(1\,3), (1\,4\,3\,2), (2\,4)\},$$
$$\langle K_4, (1\,2\,4\,3) \rangle = \{\text{id}, (1\,2)(3\,4), (1\,3)(2\,4), (1\,4)(2\,3), (1\,2\,4\,3),$$
$$(1\,4), (2\,3), (1\,3\,4\,2)\},$$
$$\langle K_4, (1\,3\,2\,4) \rangle = \{\text{id}, (1\,2)(3\,4), (1\,3)(2\,4), (1\,4)(2\,3), (1\,3\,2\,4),$$
$$(1\,4\,2\,3), (1\,2), (3\,4)\}.$$

It is easy to check that there are no others.

(3) Since $|A_5| = 60$, it follows that the Sylow 2-subgroups have order 4. Now, a group of order 4 acting on any set, in this case the set $\{1,\ 2,\ 3,\ 4,\ 5\}$ of which $A_5$ is the group of all even permutations, has orbits of lengths 1, 2 or 4. Since 5 is odd, it follows that a Sylow 2-subgroup must have at least one orbit of length 1, i.e., at least one fixed point. Hence, a Sylow 2-subgroup must be contained in one of the natural subgroups isomorphic to $A_4$. Since in $A_4$ there is a unique subgroup $K_4$ of order 4, it follows that the Sylow 2-subgroups of $A_5$ are the subgroups like

$$\{\text{id}, (1\,2)(3\,4), (1\,3)(2\,4), (1\,4)(2\,3)\}$$

and there are just five of them.

∎

**Problem 2.35** Prove that the multiplicative group of any finite field is cyclic.

*Solution* Let $\mathbb{F}^*$ be the multiplicative group of a finite field $\mathbb{F}$, and suppose that $P$ is a Sylow $p$-subgroup of $\mathbb{F}^*$ with $|P| = p^n$. Then, the orders of elements of $P$ are divisors of $p^n$. Assume that $P$ does not contain an element of order $p^n$. Hence, for every $a \in P$ the equation $a^{p^{n-1}} = e$ holds. However, the equation $a^{p^{n-1}} = e$ has at most $p^{n-1}$ roots in the field $\mathbb{F}$, a contradiction. Thus, we conclude $P$ contains an element of order $p^n$.

Now, let $|\mathbb{F}^*| = p_1^{n_1} \ldots p_k^{n_k}$ be the prime factorization of $|\mathbb{F}^*|$. As we have shown, $\mathbb{F}^*$ contains $k$ elements of orders $p_1^{n_1}, \ldots, p_k^{n_k}$, respectively. Consequently, the product of these elements has the order $|\mathbb{F}^*|$ and hence generates the group $\mathbb{F}^*$.                ■

**Problem 2.36** Let $G$ be a finite group and $P \in \mathrm{Syl}_p(G)$. Prove that for each subgroup $H$ of $G$, there is $a \in G$ such that $H \cap P^a \in \mathrm{Syl}_p(H)$.

*Solution* We can write

$$[G : P] = \sum_{x \in X} [H : H \cap P^a], \qquad (2.4)$$

where $X$ is the set of representatives of distinct double cosets. Since $P$ is a Sylow $p$-subgroup of $G$, it follows that $(p, [G : P]) = 1$. Hence, the right side of (2.4) and $p$ are also relatively prime. Consequently, there is $a \in G$ such that $(p, [H : H \cap P^a]) = 1$. Since $H \cap P^a \leq P^a$, it follows that $H \cap P^a$ is a $p$-subgroup. This shows that $H \cap P^a$ is a Sylow $p$-subgroup of $H$.                ■

**Problem 2.37** Let $G$ be a finite group, $H \trianglelefteq G$ and $P \in \mathrm{Syl}_p(H)$. Prove that there is $Q \in \mathrm{Syl}_p(G)$ such that $Q$ is a subgroup of $N_G(P)$.

*Solution* By Frattini's argument, we have $G = HN_G(P)$. Now, by the second isomorphism theorem we can write

$$\frac{G}{H} = \frac{HN_G(P)}{H} \cong \frac{N_G(P)}{H \cap N_G(P)} = \frac{N_G(P)}{N_H(P)}.$$

Hence, we deduce that $[G : N_G(P)] = [H : N_H(P)]$. But $[H : N_H(P)]$ is the number of Sylow $p$-subgroups of $H$, and it is equal to 1 modulo $p$. Now, since $([G : N_G(P)], p) = 1$, we conclude that $N_G(P)$ contains a Sylow $p$-subgroup of $G$, and this completes the proof.                ■

**Problem 2.38** Let $G$ be a non-trivial finite group and suppose that $P$ is a $p$-subgroup of $Aut(G)$. Prove that there is a Sylow $q$-subgroup $Q$ of $G$ such that $f(Q) = Q$, for all $f \in P$.

*Solution* We consider two cases as follows:

*Case 1*: Let $p \nmid |G|$. Since $G$ is non-trivial, it follows that there is a prime $q$ ($\neq p$) such that $q \mid |G|$. By Sylow's Third Theorem we have $|\mathrm{Syl}_q(G)| \cong 1 (\mathrm{mod}\ q)$ and $|\mathrm{Syl}_q(G)| \mid |G|$. Since $P \leq Aut(G)$, it follows that $|f(Q)| = |Q|$, for all $f \in P$ and $Q \in \mathrm{Syl}_q(G)$. This yields that $f(Q) \in \mathrm{Syl}_q(G)$. Now, we define an action of $P$ on $\mathrm{Syl}_q(G)$ as follows:

$$Q^f := f^{-1}(Q),$$

for all $f \in P$ and $Q \in \mathrm{Syl}_q(G)$. Since $p \nmid |G|$, it follows that $p \nmid |\mathrm{Syl}_q(G)|$. By Theorem 2.5, there is $Q \in \mathrm{Syl}_q(G)$ such that $Q^f = Q$, for all $f \in P$. This gives that $f^{-1}(Q) = Q$, or equivalently $f(Q) = Q$, for all $f \in P$.

*Case 2*: Let $p \,||\, |G|$. We know that $|\mathrm{Syl}_p(G)| \cong 1 \,(\mathrm{mod}\, p)$ and $|\mathrm{Syl}_p(G)| \,||\, |G|$. Again, similar to the above argument, we observe that $P$ acts on $\mathrm{Syl}_p(G)$. Since $p \nmid |\mathrm{Syl}_p(G)|$, it follows that there exists $Q \in \mathrm{Syl}_p(G)$ which is invariant under all elements of $P$. Therefore, $Q$ is a Sylow $p$-subgroup of $G$ and $f(Q) = Q$, for all $f \in P$.  ∎

**Problem 2.39** Let $G$ be a finite group of order $pqr$, where $p$, $q$, and $r$ primes such that $p < q < r$. Show that

(1) $G$ has a normal subgroup;
(2) $G$ has a subgroup of prime index;

*Solution*

(1) Assume that $n_s$ is the number of Sylow $s$-subgroups of $G$, where $s \in \{p,\ q,\ r\}$. For each $s$, by Sylow's Third Theorem, we have $n_s \,||\, |G|$ and $n_s = 1 + ks$, for some integer $k$. We claim that some of Sylow $s$-subgroup of $G$ is normal. In order to prove the claim, suppose that no Sylow $s$-subgroup is normal. So, $n_s \geq 1 + s$, for $s \in \{p,\ q,\ r\}$. Since $n_p$ is among $q, r, qr$, and $q < r$, we get $n_p \geq q$. Similarly, since $n_q$ is among $p, r, pr$, and $p < q, r \leq qr$, we conclude that $n_q \geq r$. Finally, since $n_r$ is among $p, q, pq$, and $p, q < r$, we obtain $n_r = pq$. Since each Sylow $s$-subgroup of $G$ is cyclic of prime order, it follows that each of them has $s - 1$ elements of order $s$. By counting the elements of order $p$, $q$, and $r$, respectively, we obtain

$$q(p - 1) + r(q - 1) + pq(r - 1) \leq pqr,$$

which implies that $-q - r + qr \leq 0$. So, we have $qr \leq q + r \leq 2r$. This yields that $qr \leq 2r$ or $q \leq 2$, and it is a contradiction. This completes the proof of (1).

(2) By (1), there is a normal subgroup $N$ of prime order. Assume that $H$ is a Sylow $s$-subgroup of $G$ with $s \neq |N|$. Then, $|H| = s$ and $HN \leq G$. Since $(|H|, |N|) = 1$, it follows that $|H \cap N| = 1$. This gives that $|HN| = |H|\,|N|$. We know that $|G|$ is the product of three primes, two of them are $|N|$ and $|H|$. Therefore, we conclude that

$$[G : HN] = \frac{|G|}{|HN|} = \frac{|G|}{|H||N|}$$

is the third prime.  ∎

## 2.5  Supplementary Exercises

1. Let $G$ be a finite group and suppose that $f$ is an automorphism of $G$ such that $f^3$ is the identity automorphism. In addition, suppose that $f(a) = a$ implies $a = e$. Prove that for every prime $p$ which divides $|G|$, the Sylow $p$-subgroup is normal in $G$.

2. Let $p$ and $q$ be prime numbers.

   (a) Show that a group $G$ of order $p^2q$ or $p^2q^2$ has at least one normal Sylow subgroup;
   (b) Show that if a group $G$ of order $p^3q$ has no normal Sylow subgroup, then $G$ is isomorphic to the symmetric group $S_4$.

3. Let $G$ be a group of order $p^2q$, where $p$ and $q$ are distinct prime numbers.

   (a) Show that if $G$ has distinct Sylow $p$-subgroups $P_1$ and $P_2$, then $P_1 \cap P_2 \trianglelefteq G$;
   (b) Show that if $G$ has more than one Sylow $p$-subgroup, and if $P_1 \cap P_2 = \{e\}$, for any two distinct Sylow $p$-subgroups $P_1$ and $P_2$, then there are $q(p^2 - 1)$ elements of order power of $p$ in $G$, and hence $G$ has just one Sylow $q$-subgroup.

4. Let $G$ be a group of order $p^2q^2$, where $p$ and $q$ are distinct primes, $q \nmid p^2 - 1$ and $p \nmid q^2 - 1$. Prove that $G$ is abelian. List three pairs of primes that satisfy these conditions.

5. Let $|G| = p^n m$ with $(p, m) = 1$. If $n_p = 1 + p$, prove that

$$\left| \bigcap_{P \in \mathrm{Syl}_p(G)} P \right| = p^{n-1}.$$

6. Prove that $Aut(A_5) \cong S_5$.

7. Show that if a group has $1 + p$ Sylow $p$-subgroups of order $p^n$, then any two of these subgroups have just $p^{n-1}$ elements in common.

8. Show that if a group $G$ has $1 + p$ Sylow $p$-subgroups of order $p^n$, then $G$ contains $p^{n+1}$ elements whose orders are divisors of $p^n$.

9. Let $G$ be a finite group and suppose that $H$ is a subgroup of $G$ such that $H \cap H^a = \{e\}$, for all $a \notin H$. Prove that

   (a) $|H|$ and $[G : H]$ are relatively prime;
   (b) If $[G : H]$ is a power of a prime $p$, then $G$ has a unique Sylow $p$-subgroup;
   (c) If another subgroup $K$ of $G$ also satisfies the condition $K \cap K^a = \{e\}$, for all $a \notin K$, then there is $b \in G$ such that $H \cap K^b \neq \{e\}$ (in this part, $H$ and $K$ are supposed non-trivial).

10. Let $G$ be a finite group, $N \trianglelefteq G$ and $p$ be a prime number. Prove that

    (a) $|\mathrm{Syl}_p(N)| \leq |\mathrm{Syl}_p(G)|$;
    (b) $|\mathrm{Syl}_p(G/N)| \leq |\mathrm{Syl}_p(G)|$.

11. Let $P$ be a Sylow $p$-subgroup of $G$ and let $H$ be any subgroup of $G$ which contains $N_G(P)$. Prove that $[G : H] \equiv 1 \pmod{p}$.

# Chapter 3
# Simple Groups

What are simple groups? Galois called a group simple if its only normal subgroups are the identity subgroup and the group itself.

## 3.1  Simple Groups and Examples

Finite simple groups are important because in a sense they are building blocks of all finite groups, similarly to the way prime numbers are building blocks of the integers.

**Definition 3.1** Let $G$ be a group. We say that $G$ is *simple* if it has no normal subgroups other than $\{e\}$ and $G$.

**Theorem 3.2** *Let $G$ be a non-trivial abelian group. Then, $G$ is simple if and only if $G \cong \mathbb{Z}_p$, for $p$ prime.*

**Proof** Let $p$ be a prime. By Lagrange's theorem, $\mathbb{Z}_p$ does not have any non-trivial subgroup, and so it is simple.

Conversely, suppose that $G$ is a non-trivial abelian simple group. Let $a \in G$ be a non-identity element. Since $G$ is abelian, it follows that the subgroup generated by $a$ is normal in $G$. On the other hand, since $G$ is simple, we conclude that $G = \langle a \rangle$; i.e., $G$ is cyclic. Hence, we must have $G \cong \mathbb{Z}$ or $G \cong \mathbb{Z}_n$, for some integer $n$. But $\mathbb{Z}$ is not simple, because $m\mathbb{Z} \trianglelefteq \mathbb{Z}$, for all integer $m$. This means that $G \cong \mathbb{Z}_n$, for some integer $n$. Now, we show that $n = p$ is a prime number. If $n = rs$ with $1 < r, s < n$, then $\langle a^r \rangle \trianglelefteq G$, and it is impossible. Hence, $n = p$ is a prime. ∎

**Example 3.3** If $G$ is a group of order 42, then $G$ is not simple. Indeed, the number of Sylow 7-subgroups of $G$ is of the form $1 + 7k$ and divides 6. This shows that $k$ must be zero and $G$ has just one Sylow 7-subgroup $H$, and so $H \trianglelefteq G$.

© The Author(s), under exclusive license to Springer Nature Singapore Pte Ltd. 2021
B. Davvaz, *Groups and Symmetry*, https://doi.org/10.1007/978-981-16-6108-2_3

*Example 3.4* Let $G$ be a group of order 48. If $H$ is a Sylow 2-subgroup of $G$, then $|H| = 16$. If $H$ is the only Sylow 2-subgroup of $G$, then $G$ is simple. Otherwise, there will be three such. Assume that $K$ $(\neq H)$ is one of these. If $|H \cap K| = 1, 2$, or 4, then $[H : H \cap K] = 16, 8$, or 4. Now, it is easy to check that there exist $16^2$ or $8 \cdot 16$ or $4 \cdot 16$ distinct elements of the form $hk$, where $h \in H$ and $k \in K$. But none of these is a possibility, because $|G| = 48$. So, $H \cap K$ has index 2 in $H$ and in $K$. Hence, we conclude that $H$ and $K$ are subgroups of $N_G(H \cap K)$. Since $H \neq K$, it follows that $H$ is a proper subgroup of $N_G(H \cap K)$. Consequently, we have $16 \big| |N_G(H \cap K)|$ and $|N_G(H \cap K)| > |G|$. This yields that $|N_G(H \cap K)| = 48$, whence $H \cap K \trianglelefteq G$. This yields that $G$ is not simple.

Let $G$ be a finite group and $H \leq G$. According to Theorem 1.38, if $K$ is the kernel of the permutation representation associated with the action of $G$ on the set of right cosets of $H$, then $K$ is the intersection of all conjugate subgroups of $H$. Moreover, $K$ is the greatest normal subgroup of $G$ that is contained in $H$. The following special case is often useful.

**Corollary 3.5** *Let $G$ be a group, $H \leq G$ and $[G : H] = n$. Then, there is a normal subgroup $K$ of $G$ such that $K \leq H$ and $[G : K] \big| n!$.*

*Proof* The factor group $G/K$ in the above argument is isomorphic to a subgroup of $S_n$. Therefore, it is finite and by Lagrange's theorem, its order divides $n!$. ■

**Theorem 3.6** *If $G$ is a simple group and has a subgroup of index $k$, where $k > 2$, then $|G| \big| k!/2$.*

*Proof* Let $X$ be the set of right cosets of $H$ in $G$. The action of $G$ on $X$ gives rise a homomorphism $f : G \to S_k$, namely the permutation representation associated with the $G$-set $X$. The kernel of $f$ is a normal subgroup of $G$ contained in $H$. Since $G$ is simple, it follows that $\mathrm{Ker} f = \{e\}$, and so $f$ may be taught of as an isomorphism of $G$ to a subgroup of $S_k$. Indeed, we can identify $G$ with $\mathrm{Im} f$ and think of $G$ as actually being a subgroup of $S_k$. Now, $G \cap A_k$ is a normal subgroup of index 1 or 2 in $G$. Since $G$ is simple and its order is greater than 2, it follows that $[G : G \cap A_k]$ cannot be 2. Consequently, we obtain $G \leq A_k$, and by Lagrange's theorem, $|G|$ divides $k!/2$, as required. ■

**Definition 3.7** A maximal normal subgroup of a group $G$ is a normal subgroup $M$ not equal to $G$ such that there is no proper normal subgroup $N$ of $G$ properly containing $M$.

**Theorem 3.8** *Let $G$ be a group and $M$ be a normal subgroup of $G$. Then, $G/M$ is simple if and only if $M$ is a maximal normal subgroup of $G$.*

*Proof* Suppose that $M$ is a proper normal subgroup of $G$ such that $\overline{G} := G/M$ is not simple. Then, $\overline{G}$ has a non-trivial subgroup $\overline{N}$. So, there is a subgroup $N$ of $G$ such that $\overline{N} := N/M$ and $M \subset N$. Hence, we conclude that $N \trianglelefteq G$. Since $M \subset N \subset G$, it follows that $M$ is not a maximal normal subgroup of $G$.

Conversely, if $M$ is not a maximal normal subgroup of $G$, then there is a normal subgroup $N$ of $G$ such that $M \subset N \subset G$. Now, $\overline{N} := N/M$ is a non-trivial normal subgroup of $\overline{G}$. This means that $\overline{G}$ is not simple. ∎

**Exercises**

1. Let $G$ and $H$ be groups and $f : G \to H$ be a homomorphism that does not send every element of $G$ into $e$. If $G$ is simple, show that $f$ must be an injection.
2. Prove that every group of order 12, 28, 56, and 200 must contain a normal Sylow subgroup and hence is not simple.
3. If $G$ is a group of order 132, prove that $G$ is not simple.
4. Show that a Sylow $p$-subgroup of $D_n$ is cyclic and normal for every odd prime $p$.
5. Let $M$ be a maximal subgroup of $G$. Prove that if $M \trianglelefteq G$, then $[G : M]$ is finite and equal to a prime.
6. Let $G$ be a finitely generated group not of order 1. By involving Zorn's lemma prove that $G$ contains a maximal (proper) normal subgroup. Deduce that $G$ has a quotient group which is finitely generated simple group.

## 3.2 Non-simplicity Tests

The following theorem is an easy arithmetic test that comes from combining Sylow's third theorem and the fact that groups of prime power order have non-trivial centers.

**Theorem 3.9 (Sylow Test for Non-simplicity).** *Let $n$ be a positive integer that is not prime, and suppose that $p$ is a prime divisor of $n$. If $1$ is the only divisor of $n$ that is congruent to $1$ modulo $p$, then there does not exist a simple group of order $n$.*

**Proof** If $n$ is a prime power, then a group of order $n$ has a non-trivial center. Consequently, it is not simple. If $n$ is not a prime power, then every Sylow subgroup is proper, and so by Sylow's third theorem, we know that the number of Sylow $p$-subgroups of a group of order $n$ is equal to 1 modulo $p$ and divides $n$. Since 1 is the only such number, the Sylow $p$-subgroup is unique, and so by Corollary 2.21, it is normal. ∎

**Theorem 3.10** *An integer of the form $2n$, where $n$ is an odd number greater than 1, is not the order of a simple group.*

**Proof** Suppose that $G$ is a group of order $2n$, where $n > 1$ is odd. Recall from the proof of Cayley's theorem that the mapping $\rho_a$ is an isomorphism from $G$ to a permutation group on the elements of $G$, where $\rho_a(x) = xa$, for all $x \in G$. Since $|G| = 2n$, it follows that by Cauchy's theorem, there exists an element $a \in G$ of order 2. Consequently, if the permutation $\rho_a$ is written in disjoint cycle form, then each cycle must have length 1 or 2; otherwise, $o(a) \neq 2$. But $\rho_a$ can contain no cycles of length 1, because the cycle $(x)$ would mean $x = \rho_a(x) = xa$, and so $a = e$. Hence, in cycle form, $\rho_a$ consists of exactly $n$ transpositions, where $n$ is odd. Therefore, $\rho_a$ is an odd permutation. This means that the set of even permutations in the image of $G$ is a normal subgroup of index 2. Therefore, we conclude that $G$ is not simple. ∎

**Theorem 3.11** *There is no simple group $G$ with $|G| = p^k m$, where $p \nmid m$, $m > 1$, and $p^k \nmid (m - 1)!$ ($p$ is as usual a prime number).*

**Proof** By Sylow's theorem, such a group $G$ must have a Sylow $p$-subgroup $H$ of order $p^k$ with $[G : H] = m$. Let $X$ be the set of right cosets of $H$ in $G$, and let $G$ act on $X$ by the usual right multiplication. So, we have a homomorphism $f : G \to S_m$ with Kerf $\subseteq H$. Since $G$ is assumed to be simple, it follows that its normal subgroups are $\{e\}$ and $G$. This yields that Kerf $= \{e\}$, and the function $f$ is one to one. This means that $G$ would be a subgroup of $S_m$, and hence the order of $G$ would divide $|S_m| = m!$. But $|G| = p^k m$, and $p^k n | n!$ would mean that $p^k | (m - 1)!$, which is contrary to the assumption. Therefore, no such simple group $G$ exists. ∎

Using Theorem 3.11, it is possible to see that there is no non-abelian simple group of order less than 60. We can assume that the order $n$ of $G$ is not a prime power, since these are never simple unless $n$ is a prime, and all such groups are abelian. Theorem 3.11 then takes care of all orders less than 60 except for 30, 40, and 56, and these can be taken care of by using Sylow's theorems and some simple element counting argument.

**Theorem 3.12** **(Index Theorem)**. *If $G$ is a finite group and $H$ is a proper subgroup of $G$ such that $|G| \nmid [G : H]!$, then $H$ must contain a non-trivial normal subgroup of $G$. In particular, $G$ is not simple.*

**Proof** Let $X$ be the set of all left cosets of $H$ in $G$. Consider the function $\theta : G \to S_X$ defined by $\theta(a)(xH) = axH$. We have Ker$\theta \subseteq H$. Since $H \neq G$, it follows that Ker$\theta \neq G$. On the other hand, if Ker$\theta = \{e\}$, then $G \cong Im\theta$, and so $|G| = |Im\theta||S_X|$. As $|S_X| = |X|! = [G : H]!$, we get $|G||[G : H]!$, which is against the hypothesis. Thus, the kernel $\theta$ must be larger than $\{e\}$. This kernel is the largest normal subgroup of $G$ which is contained in $H$. Therefore, we can conclude that $H$ contains a non-trivial normal subgroup of $G$. ∎

**Example 3.13** In this example, we consider the possibility of a simple group $G$ of order 144. By Sylow's theorem, we know that $G$ has either four Sylow 3-subgroups or 16 Sylow 3-subgroups and at least three 2-subgroups. The index theorem rules out the case where $n_3 = 4$, so we know there exist 16 Sylow 3-subgroups. Now, if every pair of Sylow 3-subgroups had only the identity in common, a straightforward counting argument would produce more than 144 elements. So, let $H$ and $K$ be a pair of Sylow 3-subgroups whose intersection has order 3. Then, $H \cap K$ is a subgroup of both $H$ and $K$, and so $N_G(H \cap K)$ must contain both $H$ and $K$ and therefore the set $HK$ ($HK$ need not be a subgroup). Hence, we have

$$|N_G(H \cap K)| \geq |HK| = \frac{|H| \, |K|}{|H \cap K|} = \frac{9 \cdot 9}{3} = 27.$$

At this stage, we have three arithmetical conditions on $k = |N_G(H \cap K)|$. We know 9 divides $k$; $k$ divides 144; and $k \geq 27$. Clearly, then $k \geq 36$ and so $[G : N_G(H \cap K)] \leq 4$. Now, the index theorem gives us the desired contradiction.

**Exercises**

1. If $G$ is a group of order 30, prove that $G$ has normal Sylow 3- and 5-subgroups.
2. Show that a group $G$ of order 90 is not simple.

## 3.3 The Simplicity of Alternating Groups

The first non-abelian simple groups to be discovered were alternating groups $A_n$, for $n \geq 5$. The simplicity of $A_5$ was known to Galous and is crucial in showing that the general equation of degree 5 is not solvable by radicals.

In order to prove the following theorem, we apply the following remark.

**Remark 3.14** For $n \geq 5$, if a normal subgroup $N$ of $A_n$ contains a cycle of length 3, then $N = A_n$.

**Theorem 3.15** *The alternating group $A_n$ is simple if and only if $n \neq 1, 2,$ or 4.*

**Proof** In the first place $A_4$ is not simple because the permutations $(1\ 2)(3\ 4)$, $(1\ 3)(2\ 4)$, $(1\ 4)(2\ 3)$ form together with id a normal subgroup. Of course, $A_1$ and $A_2$ have order 1. On the other hand, $A_3$ is obviously simple. It remains only to show that $A_n$ is simple if $n \geq 5$. Assume that this is false and there exists a proper non-trivial normal subgroup $N$ of $A_n$.

If $N$ contains a cycle of length 3, then by Remark 3.14, we obtain that $N = A_n$. Hence, we deduce that $N$ cannot contain a cycle of length 3.

Now, suppose that $N$ contains a permutation $\sigma$ whose disjoint cycle decompositions involve a cycle of length at least 4, say

$$\sigma = (a_1\ a_2\ a_3\ a_4\ \ldots)\ldots,$$

in this case, $N$ also contains

$$\tau = (a_1\ a_2\ a_3)^{-1}\sigma(a_1\ a_2\ a_3) = (a_2\ a_3\ a_1\ a_4\ \ldots)\ldots.$$

Hence, $N$ contains $\sigma^{-1}\tau = (a_1\ a_2\ a_4)$. Notice that the other cycles cancel here. This is impossible, and consequently non-identity permutations in $N$ must have cycle decompositions involving cycles of lengths 2 or 3. In addition, such permutations cannot involve just one cycle of length 3; otherwise by squaring we would obtain a cycle of length 3 in $N$.

Next, assume that $N$ contains a permutation $\sigma = (a\ b\ c)(a'\ b'\ c')\ldots$ with disjoint cycles. Then, $N$ contains

$$\tau = (a'\ b'\ c)^{-1}\sigma(a'\ b'\ c) = (a\ b\ a')(c\ c'\ b')\dots,$$

and so $\sigma\tau = (a\ a'\ c\ b\ c')\dots$, which is impossible. Thus, we conclude that each permutation in $N$ is a product of an even number of disjoint transpositions.

If $\sigma = (a\ b)(a'\ b') \in N$, then $N$ contains

$$\tau = (a\ c\ b)^{-1}\sigma(a\ c\ b) = (a\ c)(a'\ b'),$$

for all $c$ unaffected by $\sigma$. This implies that $N$ contains $\sigma\tau = (a\ b\ c)$. So, if id $\neq \sigma \in N$, then

$$\sigma = (a_1\ b_1)(a_2\ b_2)(a_3\ b_3)(a_4\ b_4)\dots,$$

the number of transpositions being at least 4. But then $N$ will also contain

$$\begin{aligned}
\tau &= (a_3\ b_2)(a_2\ b_1)\sigma(a_2\ b_1)(a_3\ b_2)\dots \\
&= (a_1\ a_2)(a_3\ b_1)(b_2\ b_3)(a_4\ b_4)\dots,
\end{aligned}$$

and so $\sigma\tau = (a_1\ a_3\ b_2)(a_2\ b_3\ b_1)$, our final contradiction.  ∎

The examples infinite simple groups may be found by using the following elementary result.

**Corollary 3.16** *If the group $G$ is the union of a chain $\{\{H_i \mid i \in I\}$ of simple groups, then $G$ is simple.*

**Proof** Suppose that $N$ is a non-trivial subgroup of $G$. Then, $N \cap H_i \neq \{e\}$, for some $i \in I$, and so $N \cap H_j \neq \{e\}$, for all $H_i \subseteq H_j$. But $N \cap H_j \trianglelefteq H_j$ and $H_j$ is simple, so we obtain $H_j \leq N$ and $N = G$.  ∎

**Example 3.17** Let $S$ be the group of all *finitary permutations* of $\{1,\ 2,\ 3,\ \dots\}$, i.e., all permutations which move only a finite number of the symbols. We define $S(n)$ to be stabilizer of $\{n+1,\ n+2,\ \dots\}$ in $S$. It is clear that $S(n) \cong S_n$. Let $A(n)$ be the image of $A_n$ under the isomorphism. Then, we have $A(1) = A(2) \subset A(3) \subset \cdots$ and by Corollary 3.16

$$A = \bigcup_{n=5,6,\dots} A(n)$$

is an infinite simple group. This is called the *infinite alternating group*.

**Lemma 3.18** *If $H$ is a subgroup of $S_n$, then either every permutation in $H$ is even or exactly half of the permutations are even.*

**Proof** Let $H$ be a subgroup of $S_n$. If $H$ contains no odd permutations, then $H$ contains only even permutations and we have our desired result. Otherwise, suppose that $\tau \in H$ is an odd permutation and define $f : H \to H$ by $f(\sigma) = \tau\sigma$. We observe that $f$ is a bijective function. Moreover, $f$ sends odd permutations to even permutations and vice versa. This implies that $f$ pairs up the odd permutations in $H$ bijectively with the even permutations in $H$. Hence, there are as many odd permutations as even permutations in $H$.  ∎

**Theorem 3.19 (Embedding Theorem).** *If a finite non-abelian simple group $G$ has a subgroup of index $n$, then $G$ is isomorphic to a subgroup of $A_n$.*

**Proof** Suppose that $H$ is the subgroup of index $n$ and $S_n$ the group of all permutations of the $n$ left cosets of $H$ in $G$. There is a non-trivial homomorphism from $G$ into $S_n$. Since $G$ is simple and the kernel of a homomorphism is a normal subgroup of $G$, we observe that the function from $G$ into $S_n$ is actually one to one, and so $G$ is isomorphic to some subgroup of $S_n$. By Lemma 3.18, any subgroup of $S_n$ consists of even permutations only or half even and half odd. If $G$ were isomorphic to a subgroup of the latter type, the even permutation would be a normal subgroup of index 2, which contradict that the fact $G$ is simple. Consequently, $H$ is isomorphic to a subgroup of $A_n$. ∎

**Example 3.20** Let $G$ be a non-abelian group of order 112. If $G$ is simple, then a Sylow 2-subgroup of $G$ must have index 7. So, by the embedding theorem, $G$ is isomorphic to a subgroup of $S_7$. But 112 does not divide $|A_7|$, a contradiction. Hence, $G$ is not simple.

**Theorem 3.21** *If $G$ is a simple group of order 60, then $G \cong A_5$.*

**Proof** By Sylow's third theorem, the number of Sylow 5-subgroups is 1 or 6. Since $G$ is simple, it follows that there must be 6. Hence, we have already 24 non-identity elements of $G$. The number of Sylow 2-subgroups is either 3, 5 or 15. There cannot be 3, because otherwise $G$ would have a subgroup of index 3, and so there would be a homomorphism of $G$ into $S_3$, with kernel not equal to $G$. Assume that there are 15 Sylow 2-subgroups, each of order 14. If no two intersect in more than $\{e\}$, we pick up 45 non-identity elements of $G$ disjoint from 24 we already have. This is impossible. Thus, if $G$ has 15 Sylow 2-subgroups, then at least one pair intersect non-trivially in a subgroup of order 2. We name this pair of subgroups $H$ and $K$. We observe that if $G = \langle H, K \rangle$, then $H \cap K$ is central and hence normal in $G$, which is impossible. Consequently, $\langle H, K \rangle$, which contains at least $4/2 \cdot 4$ elements, must have index 3 or 5. The former is impossible, as shown above. Thus, the latter holds. Also, note that if $G$ has 5 Sylow 2-subgroups, then we have immediately a subgroup of index 5. Therefore, we have reduced the problem to showing that if $G$ is a simple group of order 60 and with a subgroup of index 5, then $G \cong A_5$. Now, the usual argument shows that $G$ is (isomorphic to) a subgroup of $S_5$. Clearly, $[S_5 : G] = 2$, and so $G \trianglelefteq S_5$. This yields that $G \cap A_5 \trianglelefteq A_5$ and hence $G \cap A_5 = A_5$ or $\{\mathrm{id}\}$. This latter is easily seen to be impossible, whence $G = A_5$ follows. ∎

### Exercises

1. Let $G$ be a group of order 60, and let $P$ be a subgroup of order 5. Show that either $G \cong A_5$ or $P \trianglelefteq G$.
2. Let $G$ be a group of order 120 whose Sylow 5-subgroups are not normal. Show that either $G \cong S_5$ or $G$ has a normal subgroup $N$ of order 2 such that $G/N \cong A_5$.
3. If $S$ is the group of all finitary permutation on the set $\{1, 2, 3, \ldots\}$ and $A$ is the alternating subgroup, prove that $A$ is the only non-trivial normal subgroup of $S$.

## 3.4   The Simplicity of the Projective Special Linear Groups

In this part, we will prove simplicity of $PSL_n(\mathbb{F})$ using a criterion of Iwasawa [43] from 1941 that relates simple quotient groups and doubly transitive group actions.

**Theorem 3.22   (Iwasawa's Criterion)**. *Let $G$ act doubly transitively on a set $X$, and the following statements hold:*

*(1)*  $G' = G$;
*(2)*  *For some $x \in X$, the group $\mathrm{Stab}_G(x)$ has an abelian normal subgroup whose conjugate subgroup generates $G$.*

*Then, the factor group $G/K$ is a simple group, where $K$ is the kernel of action of $G$ on $X$.*

**Proof**  Suppose that $N$ is a normal subgroup of $G$ such that $K \subseteq N \subseteq G$. Let $x \in X$. By Theorem 1.52, $\mathrm{Stab}_G(x)$ is a maximal subgroup of $G$. Since $N \trianglelefteq G$, it follows that $N\mathrm{Stab}_G(x)$ is a subgroup of $G$ such that $\mathrm{Stab}_G(x) \leq N\mathrm{Stab}_G(x) \leq G$. Since $\mathrm{Stab}_G(x)$ is a maximal subgroup of $G$, it follows that $N\mathrm{Stab}_G(x) = \mathrm{Stab}_G(x)$ or $N\mathrm{Stab}_G(x) = G$.

If $N\mathrm{Stab}_G(x) = \mathrm{Stab}_G(x)$, then $N \subseteq \mathrm{Stab}_G(x)$. This means that $N$ fixes $x$. Consequently, $N$ does not act transitively on $X$. This implies that $N$ acts trivially on $X$. Hence, we conclude that $N \subseteq K$, and so $N = K$.

Now, let $N\mathrm{Stab}_G(x) = G$. Suppose that $H$ is the abelian normal subgroup of $\mathrm{Stab}_G(x)$ such that its conjugate subgroups generate $G$. Since $H \trianglelefteq G$, it follows that $NH \trianglelefteq N\mathrm{Stab}_G(x) = G$. We have $aHa^{-1} \subseteq a(NH)a^{-1} = NH$, for all $a \in G$. This yields that $NH$ contains all the conjugate subgroups of $H$. Hence, by hypothesis, we deduce that $NH = G$. Therefore, we obtain

$$\frac{G}{N} = \frac{NH}{N} \cong \frac{H}{N \cap H}.$$

Since $H$ is abelian, it follows that $G/N$ is abelian too. So, we have $G' \subseteq N$. Finally, since $G = G'$, we conclude that $N = G$.  ∎

We know that $SL_n(\mathbb{F})$ is generated by the matrices of the form $E_{ij}(\lambda) = I_n + e_{ij}(\lambda)$, with $i \neq j$, $\lambda \in \mathbb{F}^*$ and $e_{ij}(\lambda)$ is a matrix with $\lambda$ in $ij$th entry and zeroes elsewhere. Now, before we discuss about simplicity of $SL_n(\mathbb{F})$, we give the following lemma.

**Lemma 3.23**  *For any $\lambda \in \mathbb{F}^*$, $E_{ij}(\lambda)$ is conjugate to $E_{12}(1)$.*

**Proof**  Suppose that $T = E_{ij}(\lambda)$. For the standard basis $\{e_1, \ldots, e_n\}$ of $V$, we have

$$T(e_k) = \begin{cases} e_k & \text{if } k \neq j \\ \lambda e_i + e_j & \text{if } k = j. \end{cases}$$

Now, we consider another ordered basis $\{v_1, \ldots, v_n\}$ of $V$ such that $v_1 = \lambda e_i$, $v_2 = e_j$, and $v_3, \ldots, v_n$ in an ordering of other standard basis vectors of $V$ besides $e_i$ and $e_j$. Then, we have

$$T(v_1) = \lambda T(e_i) = \lambda e_i = v_1,$$
$$T(v_2) = T(e_j) = \lambda e_i + e_j = v_1 + v_2,$$
$$T(v_k) = v_k, \text{ for all } k \geq 3.$$

The matrix representation of $T$ related to this basis is $E_{12}(1)$. Therefore, there is $A \in GL_n(\mathbb{F})$ with $A(e_k) = v_k$, for all $k$, and

$$T = AE_{12}(1)A^{-1}.$$

If $\det(A) = 1$, we are done. Otherwise, let $c = 1/\det(A)$. It is not difficult to check that

$$T = BE_{12}(1)B^{-1},$$

where

$$B(e_k) = \begin{cases} v_k & \text{if } k < nj \\ cv_n & \text{if } k = n. \end{cases}$$

We observe that the column of matrix representation of $B$ is the same as the column matrix representation of $A$ expect for the $n$th column, and $\det(B) = c\det(A) = 1$. This gives that $B \in SL_n(\mathbb{F})$, and hence $T$ is conjugate to $E_{12}(1)$ in $SL_n(\mathbb{F})$. ∎

**Theorem 3.24** *Let $V$ be a vector space of dimension $n$ on a field $\mathbb{F}$. Then, the action $SL_n(\mathbb{F})$ on the linear subspaces of $V$ is doubly transitive. Moreover,*

*(1) Kernel of the action is the center of $SL_n(\mathbb{F})$.*

*(2) For some point, the stabilizer has an abelian normal subgroup whose conjugate subgroup generates $SL_n(\mathbb{F})$.*

***Proof*** Suppose that $X = \{\langle v \rangle \mid 0 \neq v \in V\}$. Let $(\langle v \rangle, \langle w \rangle)$ and $(\langle v' \rangle, \langle w' \rangle)$ be two pairs of elements of $X$ such that $\langle v \rangle \neq \langle w \rangle$ and $\langle v' \rangle \neq \langle w' \rangle$. Hence, the vectors $v$ and $w$ are linearly independent as well as the vectors $v'$ and $w'$. Therefore, there is $A \in SL_n(\mathbb{F})$ such that $Av = v'$ and $Aw = w'$. If $\det(A) = 1$, then we are done. If $\det(A) = c \neq 0$, then we define $Bv := Av = v'$, $Bw := c^{-1}w'$, and the effects of $A$ and $B$ on other vectors are same. It is clear that $\det(B) = \det(A)$. Moreover, we have $\langle v \rangle^B = \langle v' \rangle$ and $\langle w \rangle^B = \langle w^B \rangle = \langle c^{-1}w' \rangle = \langle w' \rangle$. Therefore, the action is doubly transitive.

(1) If $A \in SL_n(\mathbb{F})$ lies in the kernel of this action, then $\langle v \rangle^A = \langle v \rangle$, for all nonzero vector $v \in V$. Hence, we obtain $A = aI_n$ with $a^n = 1$.

(2) Now, let $v_0 = \begin{bmatrix} 1 \\ 0 \\ \vdots \\ 0 \end{bmatrix}$ and consider the point $\langle v_0 \rangle \in X$. Then, we have

$$\text{Stab}_G(\langle v_0 \rangle) = \{A \in SL_n(\mathbb{F}) \mid \langle v_0 \rangle^A = \langle v_0 \rangle \} = \left\{ \begin{bmatrix} a & * \\ 0 & B \end{bmatrix} \mid a \in \mathbb{F}^* \right\},$$

where $a = 1/\det(B)$, $B \in GL_{n-1}(\mathbb{F})$ and $*$ is a row vector of length $n - 1$. Now, we define $\theta : \text{Stab}_G(\langle v_0 \rangle) \to GL_{n-1}(\mathbb{F})$ by $\begin{bmatrix} a & * \\ 0 & B \end{bmatrix} \mapsto B$. Clearly, $\theta$ is a homomorphism and its kernel consists of all matrices of the form

$$\begin{bmatrix} a & * \\ 0 & I_{n-1} \end{bmatrix}.$$

Consequently, $\text{Ker}\theta$ is an abelian normal subgroup of $\text{Stab}_G(\langle v_0 \rangle)$. Finally, we explain the subgroups of $SL_n(\mathbb{F})$ that are conjugate to $\text{Ker}\theta$ generate $SL_n(\mathbb{F})$. We know that $SL_n(\mathbb{F})$ generates by matrices of the form $E_{ij}(\lambda)$. Moreover, by Lemma 3.23, any $E_{ij}(\lambda)$ is conjugate to $E_{12}(1)$. Hence, we conclude that the conjugates of $E_{12}(1)$ generate $SL_n(\mathbb{F})$. Since $E_{12}(1) \in \text{Ker}\theta$, it follows that the subgroups of $SL_n(\mathbb{F})$ that are conjugate to $\text{Ker}\theta$ generate $SL_n(\mathbb{F})$.   ∎

**Theorem 3.25** *If either $n > 2$ or $n = 2$ and $|\mathbb{F}| > 3$, then $PSL_n(\mathbb{F})$ is simple.*

**Proof** By considering the derived subgroup of $SL_n(\mathbb{F})$ and Theorem 3.24, we have all conditions of Iwasawa's criterion.   ∎

**Theorem 3.26** *The groups $PSL_2(\mathbb{F}_2)$ and $PSL_2(\mathbb{F}_3)$ are not simple.*

**Proof** Since $PSL_2(\mathbb{F}_2) \cong S_3$ and $PSL_2(\mathbb{F}_3) \cong A_4$, it follows that these groups are not simple.   ∎

**Exercises**

1. What is the Sylow 2-subgroup of $SL_2(\mathbb{F}_3)$?
2. Prove that

   (a) $PSL_2(\mathbb{F}_7) \cong PSL_3(\mathbb{F}_2)$.
   (b) $PSL_4(\mathbb{F}_2) \cong A_8$.

3. Elements of order 2 are *involutions*. Prove that $PSL_2(\mathbb{F})$ has a unique conjugacy class of involutions.
4. For any given finite group $G$, prove that $GL_n(\mathbb{F})$ contains a subgroup isomorphic to $G$.
5. Prove that $PGL_n(\mathbb{C}) = PSL_n(\mathbb{C})$, for each $n$.
6. Prove that $PGL_n(\mathbb{R}) \cong PSL_n(\mathbb{R})$ if and only if $n$ is odd.
7. If $n > 1$, prove that $PGL_n(\mathbb{Q}) \not\cong PSL_n(\mathbb{Q})$.

## 3.5 Worked-Out Problems

**Problem 3.27** Let $G$ be a simple group of order 168. How many elements of order 7 are there in $G$?

*Solution* Since $|G| = 2^3 \cdot 3 \cdot 7$, it follows that every element of order 7 generates a cyclic subgroup of order 7, so let us count the number of such subgroups. By Sylow's third theorem, the number of subgroups of order 7 is $n_7 \equiv 1 \pmod{7}$ and $n_7|168$. The only solutions are 1 and 8. There cannot be only one Sylow 7-subgroup, since it would be normal. So, there are 8 Sylow 7-subgroups. Every two of these subgroups intersects in a subgroup whose order divides 7, and so it must be the trivial subgroup. Each other element in these subgroups must have order 7. Therefore, we conclude that the number of elements of order 7 is 48. ∎

**Problem 3.28** If $G$ is a group of order 231, prove that $|Z(G)| \geq 11$ (in particular, $G$ is not simple).

*Solution* The number $n_{11}$ of Sylow 11-subgroups of $G$ satisfies $n_{11} \equiv 1 \pmod{11}$ and $n_{11}|3^2 \cdot 7 = 63$. The only possibility is $n_{11} = 1$, and so $G$ has a unique normal Sylow 11-subgroup $H$. We claim that $H \subseteq Z(G)$. In order to prove the claim, suppose otherwise. Then, $ab \neq ba$, for some $a \in G$ and $b \in H$, which implies that $b \neq aba^{-1}$. Since $b \in H$ and $H \trianglelefteq G$, it follows that $aba^{-1} \in H$. Since $H$ is cyclic, it follows that $aba^{-1} = b^m$, for some $2 \leq m \leq 10$. Applying the conjugation by $a$ operation repeatedly, and noting that

$$ab^n a^{-1} = (aba^{-1})^n = (b^m)^n = b^{mn},$$

it follows that $a^k b a^{-k} = b^{m^k}$, for any positive integer $k$. In particular, taking $k = o(a)$, we obtain

$$b = ebe = a^{o(a)} b a^{-o(a)} = b^{m^{o(a)}},$$

and since $b$ has order 11, it follows that $m^{o(a)} \equiv 1 \pmod{11}$. In other words, $o(a)$ is divisible by the order of $m$ as an element of $\mathbb{Z}_{11}^*$. But this is a group of order 10, and $m$ is not the identity, so by Lagrange's theorem, the order of $m$ in this group is 2, 5, or 10. By Lagrange's theorem again, none of these can divide the order any element of any element $a \in G$, because $|G| = 3 \cdot 7 \cdot 11$, and this is a contradiction. Therefore, $H \subseteq Z(G)$ and $|Z(G)| \geq |H| = 11$, as desired. ∎

**Problem 3.29** Let $G$ be a finite non-abelian group of order $n$ with the property that $G$ has a subgroup of order $k$ for each positive integer $k$ dividing $n$. Prove that $G$ is not a simple group.

*Solution* Suppose that $|G| = n$ and $p$ is the smallest prime dividing $|G|$. If $G$ is a $p$-group, then $Z(G)$ is a proper and non-trivial subgroup of $G$, and there is nothing more to prove. So, we may assume that $G$ has a composite order. Then, by assumption $G$ has a subgroup $H$ of index $p$ in $G$, i.e., $[G : H] = p$. We know that $G$ acts on the right cosets of $H$ by right multiplication. Hence, there exists a homomorphism $f : G \to S_p$. Consequently, $G/\text{Kerf}$ is isomorphic to a subgroup of $S_p$. Since $p$ is the smallest prime dividing the order of $G$, it follows that $|G/\text{Kerf}|\,|\,p!$. This yields that $|G/\text{Kerf}| = p$. Therefore, $\text{Kerf} \neq \{e\}$; otherwise $\text{Kerf} = \{e\}$ implies that $G$ is abelian and isomorphic to $\mathbb{Z}_p$. But by assumption $G$ is non-abelian.  ∎

**Problem 3.30** If $\mathbb{F}$ is a finite field with $q$ elements, then

(1) $|PGL_n(\mathbb{F})| = \dfrac{|GL_n(F)|}{q-1}$;

(2) $|PSL_n(\mathbb{F})| = \dfrac{|SL_n(F)|}{(n, q-1)}$.

*Solution* (1) $PGL_n(\mathbb{F})$ is the image of $GL_n(\mathbb{F})$ under a homomorphism whose kernel consists of nonzero scalar matrices and so has order $q - 1$.

(2) The center of $SL_n(\mathbb{F})$ is the group of scalar matrices $aI_n$, where $a^n = 1$. The number of solutions in $\mathbb{F}$ of $a^n = 1$ is $(n, q - 1)$; keep in mind Problem 2.35 that the multiplicative group of this field is cyclic of order $q - 1$.  ∎

**Problem 3.31** Prove that $PSL_2(\mathbb{F}_4) \cong PSL_2(\mathbb{F}_5) \cong A_5$.

*Solution* We observe that $|PSL_2(\mathbb{F}_4)| = |PSL_2(\mathbb{F}_5)| = |A_5| = 60$, and these are simple groups. Now, by Theorem 3.21, we know that each group of order 60 is isomorphic to $A_5$.  ∎

## 3.6  Supplementary Exercises

1. Suppose that $G$ is a finite group and $H$ is a subgroup of index 3 in $G$. Prove that $H \trianglelefteq G$, or $G$ contains a subgroup $N$ of index 2 (which is normal).
2. Prove that there are no simple groups of order 396.
3. Show that there are no simple groups of orders 264, 945, or 3864.
4. Suppose that $G$ is a finite simple group and $G$ contains subgroups $H$ and $K$ such that $[G : H]$ and $[G : K]$ are prime. Show that $|H| = |K|$.
5. Suppose that $G$ is an infinite group in which every proper non-trivial subgroup is maximal. Show that $G$ is simple.
6. If $G$ is a non-abelian simple group and $p$ is the smallest prime dividing $|G|$, then either $p^3 \,|\, |G|$ or $12 \,|\, |G|$.
7. Let $n = 2^m k$, where $k$ is odd prime. Prove that the number of Sylow 2-subgroups of $D_n$ is $k$.
   *Hint:* Prove that if $P \in \text{Syl}_p(D_n)$, then $N_{D_n}(P) = P$.
8. Prove that if $|G| = 2^n m$, where $m$ is odd and $G$ has a cyclic Sylow 2-subgroup, then $G$ has a normal subgroup of order $m$.

9. Show that if $N \trianglelefteq GL_2(\mathbb{F})$, where $\mathbb{F}$ has more than three elements, then either $N \leq Z(GL_2(\mathbb{F}))$ or $SL_2(\mathbb{F}) \leq N$.

10. Prove that $PSL_3(\mathbb{F}_4)$ has no elements of order 15, so that $PSL_3(\mathbb{F}_4)$ is not isomorphic with $A_8$.

11. Suppose that $\mathbb{F}_q$ is a finite field with $q = p^n$, where $p$ is a prime. Prove that $PSL_2(\mathbb{F}_q)$ contains a subgroup that is isomorphic to $A_5$ if and only if $p = 5$ or $p^{2n} - 1 \equiv 0 \pmod 5$.

12. Prove that any two simple groups of order 360 are isomorphic, and conclude that $PSL_2(\mathbb{F}_9) \cong A_6$.

    *Hint:* Show that a Sylow 5-subgroup has six conjugates.

13. Prove that the orders of $PSL_4(\mathbb{F}_2)$ and $PSL_3(\mathbb{F}_4)$ are same but they are not isomorphic.

14. For $q \equiv 3 \pmod 8$ and $q \equiv 5 \pmod 8$, prove that the Sylow 2-subgroups of $PSL_2(\mathbb{F}_q)$ are isomorphic to $K_4$.

# Chapter 4
# Product of Groups

In this chapter, we show how one may piece together groups to make larger groups. Also, we show that one can often start with one large group and decompose it into a product of smaller groups. Moreover, we provide a classification of finite abelian groups.

## 4.1 External Direct Product

The purpose of this section is to describe one way in which groups can be built using smaller groups.

**Definition 4.1** The *external direct product* $G_1 \times \cdots \times G_n$ of groups $G_1, \ldots, G_n$ is the set of $n$-tuples $(a_1, \ldots, a_n)$, where $a_i \in G_i$ with the group operation defined componentwise:

$$(a_1, \ldots, a_n)(b_1, \ldots, b_n) = (a_1 b_1, \ldots, a_n b_n),$$

for all $(a_1, \ldots, a_n)$ and $(b_1, \ldots, b_n)$ in $G_1 \times \cdots \times G_n$.

**Theorem 4.2** *If $G_1, \ldots, G_n$ is a finite family of groups, then $G_1 \times \cdots \times G_n$ is a group of order $|G_1| \ldots |G_n|$.*

**Proof** The identity element is $e = (e_1, \ldots, e_n)$, where $e_i$ is the identity element in $G_i$. Moreover, the inverse of each element $(a_1, \ldots, a_n)$ is $(a_1^{-1}, \ldots, a_n^{-1})$. The associativity carries through componentwise. The formula for the order of $G_1 \times \cdots \times G_n$ is clear. ∎

It is clear that the external product of abelian groups is abelian.

© The Author(s), under exclusive license to Springer Nature Singapore Pte Ltd. 2021
B. Davvaz, *Groups and Symmetry*, https://doi.org/10.1007/978-981-16-6108-2_4

**Example 4.3** The group $\mathbb{Z}_2^n$, considered as a set, is just the set of all binary $n$-tuples. The group operation is the "exclusive or" of two binary $n$-tuples. For instance,

$$(01011101) + (01001011) = (00010110).$$

This group is important in coding theory, in cryptography, and in many areas of computer science.

**Theorem 4.4** *The order of an element in an external product of a finite number of finite groups is the least common multiple of the orders of the component elements.*

**Proof** Let $G = G_1 \times \cdots \times G_n$ and $a = (a_1, \ldots, a_n)$ be an arbitrary element of $G$. Suppose that $l$ is the least common multiple of $o(a_1), \ldots, o(a_n)$ and $k = o(a)$. Clearly, we have $a^l = (a_1, \ldots, a_n)^l = (a_1^l, \ldots, a_n^l) = (e_1, \ldots, e_n)$. This yields that $k|l$, and in particular, we get $k \le l$. On the other hand, we have $a^k = e$ if and only if $a_i^k = e_i \in G_i$, for all $1 \le i \le n$, which implies that $o(a_i)|k$. Thus, $k$ is a common multiple of $o(a_i)$, for all $1 \le i \le n$. Since $l$ is the least common multiple of $o(a_i)$, for all $1 \le i \le n$, it follows that $l \le k$. Therefore, we conclude that $l = k$.  ∎

**Example 4.5** In this example, we determine the number of elements of order 7 in $\mathbb{Z}_{49} \times \mathbb{Z}_7$. Let $(a_1, a_2)$ be an element of order 7 in $\mathbb{Z}_{49} \times \mathbb{Z}_7$. By Theorem 4.4, we have $7 = o((a_1, a_2)) = [o(a_1), o(a_2)]$. Now, we consider the following cases:

(1)  If $o(a_1) = 7$ and $o(a_2) = 7$, then there exist 6 choices for $a_1$ and 6 for $a_2$. So, we have 36 choices for $(a_1, a_2)$.
(2)  If $o(a_1) = 7$ and $o(a_2) = 1$, then we obtain a total 6 choices for $(a_1, a_2)$.
(3)  If $o(a_1) = 1$ and $o(a_2) = 7$, then we have a total 6 choices for $(a_1, a_2)$.

Therefore, we deduce that $\mathbb{Z}_{49} \times \mathbb{Z}_7$ has 48 elements of order 7.

**Example 4.6** Although $\mathbb{Z}_2$ is cyclic, $\mathbb{Z}_2 \times \mathbb{Z}_2$ is not cyclic. Since $\mathbb{Z}_2 = \{\bar{0}, \bar{1}\}$, it follows that $\mathbb{Z}_2 \times \mathbb{Z}_2 = \{(\bar{0}, \bar{0}), (\bar{0}, \bar{1}), (\bar{1}, \bar{0}), (\bar{1}, \bar{1})\}$. Note that this is not the same group as $\mathbb{Z}_4$. Both groups have 4 elements, but $\mathbb{Z}_4$ is cyclic of order 4, while in $\mathbb{Z}_2 \times \mathbb{Z}_2$, all elements have order 2, so no element generates the group . Indeed, $\mathbb{Z}_2 \times \mathbb{Z}_2$ is isomorphic to the Klein's four group.

**Theorem 4.7** *Let $m$ and $n$ be two positive integers. Then, $\mathbb{Z}_m \times \mathbb{Z}_n$ is cyclic if and only if $m$ and $n$ are relatively prime.*

**Proof** Suppose that $\mathbb{Z}_m \times \mathbb{Z}_n$ is cyclic and $\mathbb{Z}_m \times \mathbb{Z}_n = \langle (\bar{a}, \bar{b}) \rangle$, for some $(\bar{a}, \bar{b}) \in \mathbb{Z}_m \times \mathbb{Z}_n$. We have $|\mathbb{Z}_m \times \mathbb{Z}_n| = o((\bar{a}, \bar{b})) = mn$. Assume that $l$ is the least common multiple of $m$ and $n$. This yields that $m|l$ and $n|l$. Hence, we have

$$\underbrace{(\bar{a}, \bar{b}) + \cdots + (\bar{a}, \bar{b})}_{l \text{ times}} = (\underbrace{\bar{a} + \cdots + \bar{a}}_{l \text{ times}}, \underbrace{\bar{b} + \cdots + \bar{b}}_{l \text{ times}}) = (\bar{0}, \bar{0}).$$

Consequently, we get $o((\bar{a}, \bar{b})) = mn \le l = [m, n]$. Therefore, we conclude that $(m, n) = 1$.

Conversely, suppose that $m$ and $n$ are relatively prime. Then, we have $[m, n] = mn$. Since $o(\overline{1}) = m$ in $\mathbb{Z}_m$ and $o(\overline{1}) = n$ in $\mathbb{Z}_n$, by Theorem 4.4, it follows that $o((\overline{1}, \overline{1})) = mn$. This shows that $\mathbb{Z}_m \times \mathbb{Z}_n = \langle (\overline{1}, \overline{1}) \rangle$. ∎

This theorem can be extended to a product of more than two factors by similar arguments. We state this as a corollary.

**Corollary 4.8** *The group $\mathbb{Z}_{m_1} \times \cdots \times \mathbb{Z}_{m_n}$ is cyclic and isomorphic to $\mathbb{Z}_{m_1 \ldots m_n}$ if and only if the numbers $m_i$ for $i = 1, \ldots, n$ are such that the greatest common divisor of any two of them is 1.*

**Proof** The result follows immediately from Theorem 4.7 by mathematical induction. ∎

**Example 4.9** The preceding corollary shows that if $n$ is written as a product of powers of distinct prime numbers, as in $n = p_1^{\alpha_1} \times \cdots \times p_k^{\alpha_k}$, then $\mathbb{Z}_n \cong \mathbb{Z}_{p_1^{\alpha_1}} \times \cdots \times \mathbb{Z}_{p_k^{\alpha_k}}$.

We remark that changing the order of the factors in a direct product yields a group isomorphic to the original one. The names of elements have simply been changed via a permutation of the components in the $n$-tuples.

**Remark 4.10** If $C_n$ is an arbitrary cyclic group of order $n$, we know that $C_n \cong \mathbb{Z}_n$; and so, we can say

$$C_{rs} \cong C_r \times C_s \Leftrightarrow (r, s) = 1,$$

Now, we extend Definition 4.1 of external direct product to an arbitrary family of groups. The Cartesian product of a family of groups $\{G_i \mid i \in I\}$ is

$$\prod_{i \in I} G_i = \{f : I \to \bigcup_{i \in I} G_i \mid f(i) \in G_i, \text{ for } i \in I\}.$$

We define a binary operation on $\prod_{i \in I} G_i$ as follows: If $f, g \in \bigcup_{i \in I} G_i$, then

$$fg : I \to \bigcup_{i \in I} G_i$$

is the function given by $i \mapsto f(i)g(i)$. Since $G_i$ is a group, it follows that $f(i)g(i) \in G_i$, for all $i \in I$.

**Theorem 4.11** *If $\{G_i \mid i \in I\}$ is a family of groups, then*

*(1) $\prod_{i \in I} G_i$ is a group;*

*(2) For each $k \in I$, $\pi_k : \prod_{i \in I} G_i \to G_k$ given by $f \mapsto f(k)$ is an epimorphism.*

**Proof** The proof is left as an exercise for the reader. ∎

**Remark 4.12** If we identify $f \in \prod_{i \in I} G_i$ with its image $\{a_i\}_{i \in I}$ (where $a_i = f(i)$ for each $i \in I$) as is done in the case when $I$ is finite, then the binary operation on $\prod_{i \in I} G_i$ is the component multiplication. If $I = \{1, \dots, n\}$, then $\prod_{i \in I} G_i$ is the familiar product $G_1 \times \cdots \times G_n$.

**Theorem 4.13** *Let $\{G_i \mid i \in I\}$ be a family of groups, $H$ be a group and $\{\varphi_i : H \to G_i \mid i \in I$ be a family of homomorphisms. Then, there exists a unique homomorphism $\varphi : H \to \prod_{i \in I} G_i$ such that $\pi_i \circ \varphi = \varphi_i$, for all $i \in i$, and this property determines $\prod_{i \in I} G_i$ uniquely up to isomorphism.*

***Proof*** Define $\varphi : H \to \prod_{i \in I} G_i$ by $\varphi(a) = \{\varphi_i(a) \mid i \in I\}$, for all $a \in H$. Then, $\varphi$ is well defined, and for every $a, b \in I$, we have

$$\varphi(a)\varphi(b) = \{\varphi_i(a)\}_{i \in I}\{\varphi_i(b)\}_{i \in I}$$
$$= \{\varphi_i(ab)\}_{i \in I}$$
$$= \varphi(ab).$$

Hence, $\varphi$ is a homomorphism. Obviously, $\pi_i \circ \varphi(a) = \varphi_i(a)$, for all $a \in H$; i.e., Fig. 4.1 is a commutative diagram. Finally, the uniqueness is evident.    ∎

**Definition 4.14** The *external weak direct product* of a family of groups $\{G_i \mid i \in I\}$ denoted by $\prod_{i \in I}^{w} G_i$ is the set of all $f \in \prod_{i \in I} G_i$ such that $f(i) = e_i$, for all but a finite number of $i \in I$.

If all $G_i$ are abelian, then $\prod_{i \in I}^{w} G_i$ is called the *external direct sum* and is denoted by $\sum_{i \in I} G_i$

**Theorem 4.15** *If $\{G_i \mid i \in I\}$ be a family of groups, then*

**Fig. 4.1** Commutative diagram related to Theorem 4.13

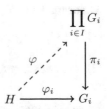

**Fig. 4.2** Commutative diagram related to Theorem 4.16

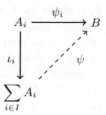

*(1)* $\prod_{i \in I}^{w} G_i$ *is a normal subgroup of* $\prod_{i \in I} G_i$;

*(2)* *For each* $k \in I$, *the function* $\iota_k :: G_k \to \prod_{i \in I} G_i$ *defined by* $\iota_K(a) = \{a_i\}_{i \in I}$, *where* $a_i = e_i$ *for* $i \neq k$ *and* $a_k = a$, *is a monomorphism;*

*(3)* *For each* $i \in I$, $\iota_i(G_i) \trianglelefteq \prod_{i \in I} G_i$.

**Proof** The proof is left as an exercise for the reader. ∎

The functions $\iota_k$ in Theorem 4.15 are called the *canonical injection*.

**Theorem 4.16** *Let* $\{A_i \mid i \in I\}$ *be a family of additive groups. If* $B$ *is an additive group and* $\{\psi_i : A_i \to B \mid i \in I\}$ *is a family of homomorphisms, then there exists a unique homomorphism* $\psi : \sum_{i \in I} A_i \to B$ *such that* $\psi \circ \iota_i = \psi_i$, *for* $i \in I$, *and this property determines* $\sum_{i \in I} A_i$ *uniquely up to isomorphism.*

**Proof** If $\{a_i\}_{i \in I} \in \sum_{i \in I} A_i$ is a nonzero element, then only many of $a_i$ are nonzero, say $a_{i_1}, \ldots, a_{i_r}$. Define $\psi : \sum_{i \in I} A_i \to B$ by $\psi(0) = 0$ and

$$\psi(\{a_i\}_{i \in I}) = \psi_{i_1}(a_{i_1}) + \cdots + \psi_{i_r}(a_{i_r}) = \sum_{i \in I_0} \psi_i(a_i),$$

where $I_0$ is the set $\{i_1, \ldots, i_r\} = \{i \in I \mid a_i \neq 0\}$. Since $B$ is abelian, it is verified that $\psi$ is a homomorphism. Moreover, in Fig. 4.2, $\psi$ is the unique homomorphism that makes a commutative diagram. ∎

**Remark 4.17** Theorem 4.16 is false if the word abelian is omitted (Exercise 11).

Now we present a useful construction associated with direct products.

**Definition 4.18** Let $G$ and $H$ be groups with $A$ a subgroup of $Z(G)$ and $B$ a subgroup of $Z(H)$. Suppose that there exists an isomorphism $\theta : A \to B$. It is easy to check that $K = \{(a, \theta(a)^{-1}) \mid a \in A\}$ is a subgroup of direct product $G \times H$. Since $A$ and $B$ are central, it follows that $K$ is a central subgroup of $G \times H$. The factor group $(G \times H)/K$, denoted by $G \times_\theta H$ is the *central product* of $G$ and $H$ via $\theta$.

**Example 4.19** Let $D_4$ be the dihedral group of order 8 generated by $a$ of order 4 and $b$ of order 2. We know that $Z(D_4) = \{e, a^2\}$. Let $Q_8$ be the quaternion group of order 8 generated by $x$ and $y$, and so its center is $\{e, x^2\}$. Every element of $Q_8$ not in its center has order 4. Suppose that $A = \{e, a^2\}$ and $B = \{e, x^2\}$. We define a function $\theta : A \to B$ by $\theta(e) = e$ and $\theta(a^2) = x^2$. So, $K = \{(e, e), (a^2, x^2)\}$. Since $|G \times H| = 64$, it follows that the central product has 32 elements. Then, we deduce that

(1) The element $(e, e)$ has order 1;
(2) The square of the element $(a^2, e)$ and $(e, x^2)$ is in $K$;
(3) The square of the element $(a, h)$ (for $h$ of order 4) is in $K$;
(4) The square of the element $(a^3, h)$ (for $h$ of order 4) is in $K$;
(5) The square of the element $(g, e)$ (for $g$ of order 2) is in $K$;
(6) The square of the element $(g, x^2)$ (for $g$ of order 2) is in $K$;
(7) All remaining elements of $D_4 \times Q_8$ are of order 4, and none of their squares lies in $K$.

Consequently, there exist 24 elements of the direct product, each of which has an image of order dividing 2 in the central product. This yields that these 24 elements are 12 cosets of $K$ (including the subgroup $K$ itself). So, the central product has an identity and 11 elements of order 2. The remaining 20 elements of the central product all have order 4.

**Exercises**

1. Show that

    (a) If $G \cong A \times B$ and $B \cong C \times D$, then $G \cong A \times C \times D$;
    (b) If $G \cong A \times B \times C$ and $D \cong B \times C$, then $G \cong A \times D$.

2. Let $p$ be a prime number. Prove that $\mathbb{Z}_p \times \mathbb{Z}_p$ has exactly $p + 1$ subgroups of order $p$.
3. Give an example of a proper and non-trivial subgroup of $\mathbb{Z}_4 \times \mathbb{Z}_8$ that cannot be expressed in the form $H_1 \times H_2$, where $H_1 \leq \mathbb{Z}_4$ and $H_2 \leq \mathbb{Z}_8$.
4. Let $H$ and $K$ be finite groups such that $(|H|, |K|) = 1$. Prove that any subgroup of $H \times K$ is of the form $A \times B$, where $A \leq H$ and $B \leq K$.
5. Determine $Aut(\mathbb{Z}_2 \times \mathbb{Z}_2)$?
6. Show that $Aut(\mathbb{Z} \times \mathbb{Z}_4) \cong D_4$.
7. How many isomorphisms are there from $\mathbb{Z}_{12}$ onto $\mathbb{Z}_4 \times \mathbb{Z}_3$?
8. Let $U_n$ be the group of integers modulo $n$ and

$$U_n(k) = \{x \in U_n \mid x \equiv 1 (\mathrm{mod}\ k)\}.$$

If $m$ and $n$ are relatively prime, show that $U_{mn} \cong U_m \times U_n$. Moreover, prove that $U_{mn}(m) \cong U_n$ and $U_{mn}(n) \cong U_m$.

9. Let $p$ and $q$ be odd prime numbers and $m$ and $n$ be positive integers. Explain why $U_{p^n} \times U_{q^n}$ is not cyclic.

10. If $G_1, \ldots, G_n$ are groups, prove that

$$Z(G_1 \times \cdots \times G_n) = Z(G_1) \times \cdots \times Z(G_n).$$

11. Give an example to show that Theorem 4.16 is not true if we delete the word "abelian."

    *Hint*: It suffices to consider the case of two factors $G_1 \times G_2$.

12. Let $G_1$ and $G_2$ be groups, $N_1 \trianglelefteq G_1$ and $N_2 \trianglelefteq G_2$. Show that $N_1 \times N_2$ is a normal subgroup of $G_1 \times G_2$ and $(G_1 \times G_2)/(N_1 \times N_2) \cong G_1/N_1 \times G_2/N_2$.

13. Let $G$ be a group and $N_1, \ldots, N_n$ be normal subgroups of $G$. Prove that $G/\bigcap_{i=1}^{n} N_i$ is isomorphic to a subgroup of $G/N_1 \times \cdots \times G/N_n$.

14. Let $G$ be an abelian group and $H \le G$ such that $G/H$ is an infinite cyclic group. Prove that $G \cong H \times G/H$.

15. A group $G$ is said to be *periodic* if every element of $G$ is of finite order. Show that if $G_1, \ldots, G_n$ are periodic groups, then $G_1 \times \cdots \times G_n$ is also a periodic group.

16. Show that there exist groups $G_1$ and $G_2$ such that each is isomorphic to a subgroup of the other and yet $G_1 \ncong G_2$.

17. Let $A_1, A_2, B_1$ and $B_2$ be groups. Prove that

$$[A_1 \times B_1, A_2 \times B_2] = [A_1, A_2] \times [B_1, B_2].$$

18. If $G_1, \ldots, G_n$ are groups and $G = G_1 \times \cdots \times G_n$, prove that

$$G' = G'_1 \times \cdots \times G'_n.$$

## 4.2  Internal Direct Product

Let $G_1 \times \cdots \times G_n$ be a family of groups and $G = G_1 \times \cdots \times G_n$. We set $\overline{G}_i = \{(e_1, \ldots, e_{i-1}, a, e_{i+1}, \ldots, e_n) | x \in G_i\}$. Then, we have

(1) $\overline{G}_i$ is a subgroup of $G$;
(2) The function $x \to (e_1, \ldots, e_{i-1}, a, e_{i+1}, \ldots, e_n)$ is an isomorphism between $G_i$ and $\overline{G}_i$;
(3) $\overline{G}_i \trianglelefteq G$, for all $1 \le i \le n$;
(4) The $G_i$ commute with each other in the sense that $xy = yx$ if $x \in G_i, y \in G_j$ and $i \neq j$;

Note carefully, however, that the subgroup $\overline{G}_i$ need not commute with itself (the case when $i = j$) unless the group $G_i$ happens to be abelian. The subsets

$$\overline{H}_i = \overline{G}_1 \ldots \overline{G}_{i-1}\overline{G}_{i+1} \ldots \overline{G}_n$$
$$= \{(a_1, \ldots, a_{i-1}, e_i, a_{i+1}, \ldots, a_n) \mid a_j \in G_j, \text{ for } i \neq j\}$$

are also normal subgroups of $G$ and in a group theoretic sense the $\overline{H}_i$ are complementary to the $\overline{G}_i$. We have the following properties:

(a)  $\overline{H}_i \cap \overline{G}_i = \{e\}$;
(b)  $G = \overline{G}_1 \ldots \overline{G}_n$;
(c)  $\overline{H}_i \trianglelefteq G$;
(d)  $G_i \cong G/\overline{H}_i$.

We give the following name to the situation we have just described.

**Definition 4.20** Let $G$ be a group and $H_1, \ldots, H_n$ be subgroups of $G$. We say that $G$ is the *internal direct product* of the subgroups $H_i$ if

(1)  $H_i$ is a normal subgroup of $G$, for all $1 \leq i \leq n$;
(2)  $G = H_1 \ldots H_n$;
(3)  $H_i \cap H_1 \ldots H_{i-1}H_{i+1} \ldots H_n = \{e\}$, for all $1 \leq i \leq n$.

It should be noted that the condition for internal direct product is considerably stronger than $H_i \cap H_j = \{e\}$, for $i \neq j$. This latter condition is not sufficient to ensure we have a direct product.

**Lemma 4.21** *Let $G$ be a group and $H_1, \ldots, H_n$ be subgroups of $G$ satisfying conditions (1) and (2) of Definition 4.20. Then, the following statements are equivalent:*

*(1)  Each $x \in G$ can be written uniquely as a product $x = a_1 \ldots a_n$ with $a_i \in H_i$;*
*(2)  $H_i \cap H_1 \ldots H_{i-1}H_{i+1} \ldots H_n = \{e\}$, for all $1 \leq i \leq n$.*

**Proof**  $(1 \Rightarrow 2)$ Let $x \in H_i \cap H_1 \ldots H_{i-1}H_{i+1} \ldots H_n$ be an arbitrary element. So, $x$ can be decomposed two ways:

$$x = \underbrace{e \ldots e}_{i-1 \text{ times}} a_i \underbrace{e \ldots e}_{n-i \text{ times}} = a_1 \ldots a_{i-1}ea_{i+1} \ldots a_n,$$

where $a_i \in H_i$, for all $1 \leq i \leq n$. Now, unique factorization forces $a_i = e$, for all $1 \leq i \leq n$, and hence $x = e$. This yields that the intersection is trivial.

$(2 \Rightarrow 1)$ Suppose that $x$ has distinct factorization $x = a_1 \ldots a_n = b_1 \ldots b_n$, and let $i$ be the smallest index such that $a_i \neq b_i$. Then, $a_1 = b_1, \ldots, a_{i-1} = b_{i-1}$, and cancelation law yields $a_i a_{i+1} \ldots a_n = b_i b_{i+1} \ldots b_n$. This implies that $b_i^{-1}a_i = b_{i+1} \ldots b_n a_n^{-1} \ldots a_{i+1}^{-1}$, or equivalently

$$b_i^{-1}a_i = (b_{i+1} \ldots b_n)(a_{i+1} \ldots a_n)^{-1} \tag{4.1}$$

Since for each $k$, $H_k \trianglelefteq G$, it follows that $H_{i+1} \ldots H_n$ is a subgroup of $G$. So, the right side of (4.1) belongs to $H_{i+1} \ldots H_n \subseteq H_1 \ldots H_{i-1}H_{i+1} \ldots H_n$, while the left side of (4.1) lies in $H_i$. By (2), we assuming the intersection $H_i \cap H_1 \ldots H_{i-1}H_{i+1} \ldots H_n$ is trivial, so both product reduce to the identity element $e$, and we obtain $a_i = b_i$, and this is a contradiction to the definition of $i$. This completes the proof. ∎

**Theorem 4.22** *If $G$ is the internal direct product of the subgroups $H_1, \ldots, H_n$, then* $G \cong H_1 \times \cdots \times H_n$.

*Proof* We define $\varphi : H_1 \times \cdots \times H_n \to G$ by $(a_1, \ldots, a_n) \mapsto a_1 \ldots a_n$, for every $a_i \in H_i$. In order to show that $\varphi$ is a homomorphism, we first show that elements from distinct $H_i$ commute. Assume that $x \in H_i$ and $y \in H_j$, where $i \neq j$. Since $H_i \trianglelefteq G$, it follows that $x^{-1}(y^{-1}xy) \in H_i$. Since $H_j \trianglelefteq G$, it follows that $(x^{-1}y^{-1}x)y \in H_j$. So, we have $x^{-1}y^{-1}xy \in H_i \cap H_j = \{e\}$. This implies that $x^{-1}y^{-1}xy = e$, and so $xy = yx$.

Now, if $(a_1, \ldots, a_n), (b_1, \ldots, b_n) \in H_1 \times \cdots \times H_n$, then

$$
\begin{aligned}
\varphi\big((a_1, \ldots, a_n) (b_1, \ldots, b_n)\big) &= \varphi\big((a_1b_1, \ldots, a_nb_n)\big) \\
&= a_1b_1 \ldots a_nb_n \\
&= a_1 \ldots a_n b_1 \ldots b_n \\
&= \varphi\big((a_1, \ldots, a_n)\big)\varphi\big((b_1, \ldots, b_n)\big).
\end{aligned}
$$

Thus, $\varphi$ is a homomorphism. Since $G = H_1 \ldots H_n$, it follows that $\varphi$ is onto. Finally, suppose that $(a_1, \ldots, a_n)$ is an arbitrary element of $Ker\varphi$. So, we have $a_1 \ldots a_n = e$. This implies that

$$
\begin{aligned}
a_i &= a_{i-1}^{-1} \ldots a_2^{-1} a_1^{-1} a_n^{-1} \ldots a_{i+1}^{-1} \\
&= a_1^{-1} \ldots a_{i-1}^{-1} a_{i+1}^{-1} \ldots a_n^{-1}.
\end{aligned}
$$

Hence, we obtain $a_i \in H_i \cap H_1 \ldots H_{i-1}H_{i+1} \ldots H_n = \{e\}$. Consequently, we get $a_i = e$, for all $1 \leq i \leq n$. This yields that $Ker\varphi = \{e\}$. Therefore, $\varphi$ is an isomorphism. ∎

Now, we give conditions under which a group $G$ is isomorphic to the weak direct product of a family of its subgroups.

**Definition 4.23** Let $G$ be a group and $\{H_i \mid i \in I\}$ be a family of subgroups of $G$ such that

(1) $H_i$ is a normal subgroup of $G$, for all $i \in I$;
(2) $G = \langle \bigcup_{i \in I} H_i \rangle$;
(3) $H_i \cap \langle \bigcup_{k \neq i} H_k \rangle = \{e\}$, for all $i \in I$.

Then, $G$ is said to be the *internal weak direct product* of the family $\{H_i \mid i \in I\}$ (or *internal direct sum* if $G$ is abelian).

Note that Definition 4.20 is a special case of Definition 4.23.

**Theorem 4.24** *If $G$ be the internal weak direct product of the family $\{H_i \mid i \in I\}$, then* $G \cong \prod_{i \in I}^{w} H_i$.

**Proof** If $\{a_i\}_{i\in I} \in \prod_{i\in I}{}^w H_i$, then $a_i = e$ for all but a finite number of $i \in I$. Let $I_0$ be the finite set $\{i \in I \mid a_i \neq e\}$. Then, $\prod_{i\in I_0} a_i$ is well-defined element of $G$, because if $a \in H_i$ and $b \in H_j$ for $i \neq j$, then $ab = ba$. Define the function $\varphi : \prod_{i\in I}{}^w H_i \to G$ given by $\{a_i\}_{i\in I} \mapsto \prod_{i\in I_0} a_i$ and $\{e\}_{i\in I} \to e$. Then, it is easy to check that $\varphi$ is an epimorphism. Now, if $\{a_i\}_{i\in I} \in Ker\varphi$ is an arbitrary element, then $\prod_{i\in I_0} a_i = e$, and for convenience of notation, we may assume that $I_0 = \{1, \ldots, n\}$. Then, we can write $a_1 a_2 \ldots a_n = e$ with $a_i \in H_i$. Hence, we conclude that $a_1^{-1} = a_2 \ldots a_n \in H_1 \cap \langle \bigcup_{i\neq 1} H_i \rangle$ and consequently $a_1 = e$. Repeating of this argument show that $a_i = e$, for all $i \in I$. Therefore, $\varphi$ is one to one and so an isomorphism.  ∎

### Exercises

1. Let $G$ be a group and $A, B, C$ are subgroups of $G$. If $G = AB = AC$ are the internal direct products, prove that $B \cong C$.
2. Prove that the dihedral group $D_4$ cannot be expressed as an internal direct product of two proper subgroups.
3. Let $H$ and $K$ be subgroups of a group $G$. If $G = HK$ and $x = hk$, where $h \in H$ and $k \in K$ is there any relationship between $o(x), o(h)$ and $o(k)$? What if $G \cong H \times K$?
4. Prove that a group $G$ is the internal weak direct product of a family of its normal subgroups $\{H_i \mid i \in I\}$ if and only if every element $x \in G$ can be uniquely written as a product $x = a_{i_1} \ldots a_{i_n}$ with $i_1, \ldots i_n$ distinct elements of $I$ and $e \neq a_{i_k} \in H_{i_k}$, for all $1 \leq k \leq n$.

## 4.3  Semidirect Product

We describe next an exceedingly useful construction that is a generalization of the direct product of two groups.

**Definition 4.25** Let $K$ and $H$ be groups. If a homomorphism $\varphi : K \to Aut(H)$ is given, then we say that $K$ *acts* on $H$ via $\varphi$ and $K$ is an *operator group* on $H$. The homomorphism $\varphi$ is called an *action* of $K$.

For any $k \in K$, $\varphi(k)$ is an automorphism of $H$. We denote $\varphi(k)(h) := h^k$, for all $h \in H$. The action on $H$ is given if and only if the functions $\varphi(k) : h \mapsto h^k$ are defined for all $k \in K$ and satisfy the following formulas:

$$(uv)^a = u^a v^a,$$
$$u^{ab} = (u^a)^b,$$
$$u^e = u,$$

for all $a, b \in K$ and $u, v \in H$.

**Theorem 4.26** *Let $\varphi$ be an action of a group $K$ on another group $H$. Let $G = \{(k, h) \mid k \in K \text{ and } h \in H\}$. We define a binary operation on $G$ as follows:*

$$(a, h)(b, h') = (ab, \varphi(b)(h)h'),$$

*for all $a, b \in K$ and $h, h' \in H$. Then, $G$ forms a group with respect to this operation.*

*Proof* First, we verify the associative law. Suppose that $a, b, c \in K$ and $h, h', h'' \in H$. Then, we have

$$
\begin{aligned}
\big((a, h)(b, h')\big)(c, h'') &= (ab, h^b h')(c, h'') = \big((ab)c, ((h^b h')^c h'')\big) \\
&= \big(a(bc), ((h^b)^c h'^c)h''\big) = \big(a(bc), h^{bc}(h'^c h'')\big) \\
&= (a, h)(bc, h'^c h'') = (a, h)\big((b, h')(c, h'')\big).
\end{aligned}
$$

Let $e_K$ and $e_H$ denote the identity elements of $K$ and $H$, respectively. By definition $e = (e_K, e_H)$ is the identity and the inverse of $(a, h)$ is given by $\big(a^{-1}, \varphi(a^{-1})(h^{-1})\big)$. Therefore, $G$ forms a group. ∎

**Definition 4.27** The group constructed in Theorem 4.26 is called *semidirect product* of $K$ and $H$ with respect to the action $\varphi$, and we denote this group by $K \ltimes_\varphi H$.

**Theorem 4.28** *Let $K$ and $H$ be groups and $\varphi : K \to Aut(H)$ be a homomorphism. Then, the following are equivalent:.*

*(1) The identity map between $K \ltimes_\varphi H$ and $K \times H$ is an isomorphism;*
*(2) $\varphi : K \to Aut(H)$ is the trivial (identity) homomorphism;*
*(3) $K$ is normal in $K \ltimes_\varphi H$.*

*Proof* $(1 \Rightarrow 2)$: Suppose that $a, b \in K$, $h, h' \in H$ and there exists an isomorphism from $K \ltimes_\varphi H$ to $K \times H$. Then, we have

$$(ab, h^b h') = (a, h)(b, h') = (ab, hh'),$$

where the first equality follows from the group operation in $K \ltimes_\varphi H$ and the second follows from the group operation in $K \times H$. This shows that $h^b h' = hh'$, and so $h^b = h$, for all $h \in H$. Hence, $\varphi(b)$ is the trivial homomorphism.

$(2 \Rightarrow 3)$: Assume that $\varphi$ is trivial. Then, $h^b = h$, for all $b \in K$ and $h \in H$. So, $K \lhd H$. Since $K$ is normal in $K$ and $H$ and $K$ make up all elements of $K \ltimes_\varphi H$, it follows that $K$ is normal in $K \ltimes_\varphi H$.

$(3 \Rightarrow 1)$: Let $K$ be normal in $K \ltimes_\varphi H$. Then, $hb = bh$, for all $b \in K$ and $h \in H$, and the action is trivial, so that

$$(a, h)(b, h') = (ab, h^b h') = (ab, hh'),$$

for all $a, b \in K$ and $h, h' \in H$. ∎

**Theorem 4.29** *Let $K$, $L$, and $H$ be groups. Suppose that $\theta : K \to L$ is an isomorphism and $\psi : L \to Aut(H)$ is a homomorphism. If $\varphi = \psi \circ \theta$, then $K \ltimes_\varphi H \cong L \ltimes_\psi H$.*

**Proof** We define $f : K \ltimes_\varphi H \to L \ltimes_\psi H$ by $f(a, h) = (\theta(a), h)$, for all $a \in K$ and $h \in H$. For every $a, b \in K$ and $h, h' \in H$, we can write

$$
\begin{aligned}
f\big((a, h)(b, h')\big) &= f\big(ab, \varphi(b)(h)h'\big) = \big(\theta(ab), \varphi(b)(h)h'\big) \\
&= \big(\theta(a)\theta(b), \psi \circ \theta(b)(h)h'\big) = \big(\theta(a)\theta(b), \psi(\theta(b))(h)h'\big) \\
&= (\theta(a), h)(\theta(b), h') = f(a, h)f(b, h').
\end{aligned}
$$

This shows that $f$ is a homomorphism.

Next, assume that $(x, h) \in L \ltimes_\psi H$. Since $x \in L$ and $\theta$ is onto, it follows that $\theta(a) = x$, for some $a \in K$. Then, we have $f(a, h) = (x, h)$, and so $f$ is onto. Now, if $f(a, h) = f(b, h')$, then $(\theta(a), h) = (\theta(b), h')$, and hence, $h = h'$ and $\theta(a) = \theta(b)$. Since $\theta$ is one to one, it follows that $a = b$. Hence, we get $(a, h) = (b, h')$. This means that $f$ is one to one. Therefore, $f$ is an isomorphism. ∎

**Corollary 4.30** *Let $K$ and $H$ be groups and $\psi : K \to Aut(H)$ be a homomorphism. Then, for ant $\theta \in Aut(K)$, $\varphi = \psi \circ \theta : K \to Aut(H)$ is a homomorphism so that $K \ltimes_\varphi H \cong K \ltimes_\psi H$.*

**Proof** The result follows from Theorem 4.29 with $L = K$. ∎

Let $H$ be any group. Take $K = Aut(H)$, and let $\varphi$ be the identity map from $K$ to $Aut(H)$ (mapping every element to itself). Then, the semidirect product $Aut(H) \ltimes_\varphi H$ is called the *holomorph* of $H$.

If we set $\overline{K} = \{(a, e_H) \mid a \in K\}$ and $\overline{H} = \{(e_K, h) \mid h \in H\}$, then it is easy to check that $\overline{H} \trianglelefteq G = \overline{KH}$ and $\overline{K} \cap \overline{H} = \{e\}$. This situation is a motivation for following definition.

**Definition 4.31** A group $G$ is said to be an *internal semidirect product* of two subgroups $H$ and $K$ if $H \trianglelefteq G = KH$ and $H \cap K = \{e\}$.

Note that $K$ need not be a normal subgroup of $G$. The definition of an interval semidirect product is not symmetric with respect to $K$ and $H$. So, when it is important to distinguish between them, it should be stand clearly which subgroup is normal in $G$. We will prove that any internal semidirect product is isomorphic to the semidirect product with respect to some action. For this reason, the adjective internal is often omitted.

**Theorem 4.32** *Let $G$ be an internal semidirect product of two subgroups $H$ and $K$ such that $H \trianglelefteq G = KH$ and $H \cap K = \{e\}$. Then, there exists $\varphi : K \to Aut(H)$ with $G \cong K \ltimes_\varphi H$.*

**Proof** We define $\varphi(k)(h) = hkh^{-1}$, for all $h \in H$ and $k \in K$. It is easy to check that $\varphi$ is an action of $K$ on $H$. Each $g \in G$ has a unique expression $g = kh$ with $k \in K$ and $h \in H$. Now, let $f : K \ltimes_\varphi H \to G$ defined by $f(k, h) = kh$, for all $k \in K$ and $h \in H$.

By assumption $G = KH$, so the function $f$ is onto. In addition, for every $a, b \in K$ and $h, h' \in H$, we have $(a, h)(b, h') = (ab, \varphi(b)(h)h')$. Hence, we obtain

$$f((a, h)(b, h')) = ab\varphi(b(h)h'. \tag{4.2}$$

On the other hand, we have

$$f(a, h)f(b, h') = (ah)(bh') = (ab)(b^{-1}hb)h' = (ab)\varphi(b)(h)h'. \tag{4.3}$$

By (4.2) and (4.3), we conclude that the function $f$ is a homomorphism. Finally, we have $Kerf = \{(a, h) \mid f(a, h) = e\} = \{(a, h) \mid k = h^{-1}\}$. Since $H \cap K = \{e\}$, it follows that $Kerf = \{e\}$. This yields that $f$ is an isomorphism. ∎

**Example 4.33** The alternating subgroup $A_n$ is a normal subgroup of $S_n$ and $C_2 = \langle(1\ 2)\rangle$ maps isomorphically onto $S_n/A_n$. Therefore, $S_n \cong C_2 \ltimes A_n$.

**Example 4.34** A cyclic group of order $p^2$, $p$ prime, is not a semidirect product, because it has only one subgroup of order $p$.

**Example 4.35** Let $UT_n(\mathbb{F})$ be the subgroup of upper triangular matrices in $GL_n(\mathbb{F})$. Suppose that $D_n(\mathbb{F})$ is the subgroup of diagonal matrices in $GL_n(\mathbb{F})$ and $IUT_n(\mathbb{F})$ is the subgroup of upper triangular matrices with all their diagonal coefficient equal to 1. Then, $IUT_n(\mathbb{F}) \trianglelefteq UT_n(\mathbb{F})$, $D_n(\mathbb{F}) IUT_n(\mathbb{F}) = UT_n(\mathbb{F})$ and $IUT_n(\mathbb{F}) \cap D_n(\mathbb{F}) = \{I_n\}$. Therefore, we have $UT_n(\mathbb{F}) = D_n(\mathbb{F}) \ltimes IUT_n(\mathbb{F})$.

Note that, when $n \geq 2$, the action of $D_n(\mathbb{F})$ on $IUT_n(\mathbb{F})$ is not trivial. For instance,

$$\begin{bmatrix} a & 0 \\ 0 & b \end{bmatrix}\begin{bmatrix} 1 & c \\ 0 & 1 \end{bmatrix}\begin{bmatrix} a^{-1} & 0 \\ 0 & b^{-1} \end{bmatrix} = \begin{bmatrix} 1 & ac/b \\ 0 & 1 \end{bmatrix},$$

and so $UT_n(\mathbb{F})$ is not the direct product of $D_n(\mathbb{F})$ and $IUT_n(\mathbb{F})$.

**Theorem 4.36** *Let $G$ be a finite group of order $2n$. The following statements are equivalent:*

*(1) $G$ is a didedral group;*
*(2) $G$ is the semidirect product $K \ltimes H$ of two cyclic groups $K = \langle a \rangle$ and $H = \langle h \rangle$ such that $o(a) = 2$, $o(h) = n$ and $h^a = h^{-1}$.*

**Proof** $(1 \Rightarrow 2)$: Suppose that $a, b$ are involutions of $G$ such that $G = \langle a, b \rangle$, and let $K := \langle a \rangle$, $h := ab$ and $H := \langle h \rangle$. Then, we have $h^a = aaba = ba = h^{-1} = babb = h^b$, and so $H \trianglelefteq G$ and $G = KH$. Now, we show that $H \cap K = \{e\}$. Indeed, if $H \cap K \neq \{e\}$, then $a \in H$, and hence $b \in H$. This implies that $h = ab = e$ and $H = \{e\}$, a contradiction.

$(2 \Rightarrow 1)$: Suppose that $b := ah$. Since $b^2 = (aha)h = h^{-1}h = e$, it follows that $b$ is an involution. Consequently, $G = \langle a, b \rangle$ is a dihedral group. ∎

**Exercises**

1.  Show that the quaternion group $Q_8$ may not be decomposed (in a non-trivial way) as a semidirect product.
2.  Show that the symmetric group $S_4$ of degree 4 is isomorphic to a semidirect product of the Klein 4-group $K_4$ by the symmetric group $S_3$.
3.  Show that $GL_2(\mathbb{R})$ is a semidirect product of $SL_2(\mathbb{R})$ and the multiplicative group $\mathbb{R}^*$. Describe an action associated with this semidirect product.
    *Hint:* The action is not unique. Why not?
4.  Describe all semidirect products $\mathbb{Z}_3 \ltimes_\varphi \mathbb{Z}_3$ by finding all possible homomorphisms $\varphi : \mathbb{Z}_3 \to (U_3, \cdot) \cong Aut(\mathbb{Z}_3, +)$.
5.  If $p$ and $q$ are distinct primes, construct all semidirect product of $\mathbb{Z}_p$ by $\mathbb{Z}_q$.
6.  Determine all semidirect products $\mathbb{Z}_7 \ltimes_\varphi \mathbb{Z}_{16}$.
7.  Determine all semidirect products $\mathbb{Z}_2 \ltimes_\varphi \mathbb{Z}_{16}$.

## 4.4    Finite Abelian Groups

In this section, we will solve the classification problem for finite abelian groups. This is done by producing a standard list of finite abelian groups with the property that each finite abelian group is isomorphic to precisely one standard group. The list of standard finite abelian groups with $n$ elements is easy to describe. There is precisely one group with one element. For each integer $n$ greater than 1, and each sequence of positive integers $n_1, n_2, \ldots, n_k$ such that $n_1$ is divisible by $n_2$, $n_2$ is divisible by $n_3$, ..., $n_{k-1}$ is divisible by $n_k$, and $n = n_1 n_2 \ldots n_k$ with $n_k \geq 2$, the standard group corresponding to this sequence is the direct product

$$C_{n_1} \times C_{n_2} \times \cdots \times C_{n_k},$$

where $C_{n_i}$ is the cyclic group of order $n_i$, for all $i = 1, \ldots, n$.

**Theorem 4.37** *Let $G$ be a finite group such that for every prime $p$ dividing $|G|$, the Sylow $p$-subgroup of $G$ is normal. Let $p_1, \ldots, p_k$ be the distinct primes dividing $|G|$, and let $P_i$ be the Sylow $p_i$-subgroup of $G$, for all $1 \leq i \leq k$. Then, $G$ is the internal direct product of $P_1, \ldots, P_k$. In particular, every finite abelian group is the direct product of its Sylow $p$-subgroups.*

**Proof** Since each $P_i$ is a normal subgroup of $G$, it follows that $P_1 P_2 \ldots P_j$, for all $j \leq k$, is a subgroup of $G$. Now, by mathematical induction on $j$, we show that $P_1 P_2 \ldots P_j$ is of order $|P_1||P_2| \ldots |P_j|$. Obviously, it is true for $j = 1$. Assume that it is true for $j - 1$, and we prove it for $j$. We can write

$$|(P_1 P_2 \ldots P_{j-1})P_j| = \frac{|P_1 P_2 \ldots P_{j-1}||P_j|}{|(P_1 P_2 \ldots P_{j-1}) \cap P_j|}.$$

By assumption, we have $|P_1 P_2 \ldots P_{j-1}| = |P_1||P_2| \ldots |P_{j-1}|$. Since $p_j$ does not divide $|P_1 P_2 \ldots P_{j-1}|$, it follows that $|(P_1 P_2 \ldots P_{j-1}) \cap P_j| = 1$, and so

$$|P_1 P_2 \ldots P_j| = |P_1||P_2| \ldots |P_j|.$$

Applying this with $j = k$, shows that $P_1 P_2 \ldots P_k$ is a subgroup of $G$ with $|G|$ elements. Hence, we must have $G = P_1 P_2 \ldots P_k$. Also, we have seen that $(P_1 P_2 \ldots P_{j-1}) \cap P_j = \{e\}$, for all $2 \leq j \leq k$. Therefore, we conclude that $G$ is the internal direct product of $P_1, P_2, \ldots, P_k$.

When $G$ is abelian, the result follows because every subgroup of an abelian group is normal. ∎

Theorem 4.37 reduces the study of finite abelian groups to finite abelian $p$-groups.

**Lemma 4.38** *Let $G$ be a finite abelian $p$-group. If $G$ contains a unique subgroup of order $p$, then $G$ is cyclic.*

**Proof** In order to prove the lemma, we apply mathematical induction on $|G|$. Note that the case $|G| = p$ is obvious. Suppose that $f : G \to G$ is the homomorphism defined by $f(a) = a^p$, for all $a \in G$. By Cauchy's theorem, $G$ has an element of order $p$. Hence, $Kerf$ is the unique subgroup of order $p$, and so is cyclic. If $G = Kerf$, we are done. If $G \neq Kerf$, then $Imf$ is a non-trivial subgroup of $G$. Again, by Cauchy's theorem, $Imf$ has a unique subgroup of order $p$. Since $|Imf| < |G|$, by inductive hypothesis, it follows that $Imf$ is cyclic. Then, by the first isomorphism theorem, we conclude that $G/Kerf$ is cyclic. Let $G/Kerf = \langle xKerf \rangle$, for some $x \in G$. If $g \in G$ is arbitrary, then $gKerf \in G/Kerf$. So, we have $gKerf = (xKerf)^n = x^n Kerf$, for some positive integer $n$. Hence, we can write $g = x^n y$, for some $y \in Kerf$. This shows that $G \subseteq \langle x, Kerf \rangle$. Consequently, we obtain

$$G = \langle x, Kerf \rangle = \langle x \rangle \langle Kerf \rangle. \tag{4.4}$$

By hypothesis, $\langle x \rangle$ has a unique subgroup of order $p$. So, by (4.4), we conclude that $Kerf \subseteq \langle x \rangle$. Therefore, we get $G = \langle x \rangle$. ∎

**Theorem 4.39** *If $G$ is a finite abelian $p$-group and $C$ is a cyclic subgroup of maximal order, then there is a subgroup $B$ of $G$ such that $G$ is an internal direct product of $C$ and $B$.*

**Proof** We proceed by mathematical induction on $|G|$. If $G$ is cyclic, we are done. When $G$ is not cyclic, by Theorem 4.38, $G$ must have more than one subgroup of order $p$, while $C$ has unique cyclic subgroup of order $p$. So, let $H$ be a subgroup of $G$ of order $p$ not contained in $C$. Since $H$ has prime order, it follows that $C \cap H = \{e\}$, which implies that $(CH)/H \cong C$. Since $C$ is a cyclic subgroup of maximal order, it follows that $C = \langle a \rangle$, for some $a \in G$ such that the order of $a$ is greater than the order of other elements of $G$. If $\langle xH \rangle$ is a cyclic subgroup of $G/H$, then the order of $xH$ in $G/H$ divides $o(x)$, which is at most $|C|$. Hence, the order of any cyclic subgroup of

$G/H$ of maximal order is at most $|C|$. Since $(CH)/H \cong C$, it follows that $(CH)/H$ is a cyclic subgroup of $G/H$ of maximal order. Thus, by inductive hypothesis, there is a subgroup $B/H$ of $G/H$ such that

$$\frac{G}{H} = \left(\frac{CH}{H}\right)\left(\frac{B}{H}\right). \tag{4.5}$$

This yields that $G = (CH)B = C(HB) = CB$. Since (4.5) is an internal direct product, it follows that

$$\frac{CH}{H} \cap \frac{B}{H} = \{H\},$$

which implies that $CH \cap B = H$. Therefore, we obtain $C \cap B = C \cap B \cap CH = C \cap H = \{e\}$. This completes the proof. ∎

**Theorem 4.40** *If $G$ is a finite abelian $p$-group, then $G$ is a direct product of cyclic groups.*

**Proof** We proceed by mathematical induction on $|G|$. Let $G$ be a finite abelian $p$-group. If $G$ is cyclic, we are done. Otherwise, suppose that $C$ is a cyclic subgroup of maximal order in $G$. By Theorem 4.39, there is a subgroup $B$ of $G$ such that $G = CB$ and $C \cap B = \{e\}$. Since $|G| = |CB| = |C||B|$ and $p \leq |C|$, it follows that $|B| < |G|$. Now, by the inductive hypothesis, $B$ is a direct product of cyclic groups, and we are done. ∎

More precisely, if $G$ is a finite abelian $p$-group of order $p^n$, then $G$ is a direct product of cyclic groups of orders $p^{\alpha_1}, p^{\alpha_2}, \ldots, p^{\alpha_k}$, where $\alpha_1 \geq \alpha_2 \geq \ldots \geq \alpha_k \geq 1$ and $\alpha_1 + \alpha_2 + \ldots + \alpha_k = n$.

**Corollary 4.41** *A finite abelian group of order $n$ can be written as a direct product*

$$C_{n_1} \times C_{n_2} \times \cdots \times C_{n_k},$$

*where $n_i$ is divisible by $n_j$ for $j > i$, $n_k \geq 2$ and $n_1 n_2 \ldots n_k = n$.*

**Proof** First, by Theorem 4.37, $G$ is an internal direct product of its Sylow subgroups. Then, since each Sylow subgroup is a $p$-group, for a suitable prime, by Theorem 4.40, it can be written as a direct product of cyclic groups. Finally, we use Remark 4.10 in order to complete the proof. ∎

We say a finite abelian $p$-group $G$ is of *type* $(\alpha_1, \alpha_2, \ldots, \alpha_k)$ if

$$G \cong C_{p^{\alpha_1}} \times C_{p^{\alpha_2}} \times \cdots \times C_{p^{\alpha_k}} \text{ and } \alpha_1 \geq \alpha_2 \leq \ldots \geq \alpha_k \geq 1.$$

**Lemma 4.42** *Let $p$ be a prime. For any finite abelian group $G$, the subset $G_p$ of $G$ consisting of elements of order $1$ or $p$ is a subgroup of $G$. Also, the subset $G^p$ of $p$th powers of elements of $G$ is a subgroup of $G$.*

**Proof** It is straightforward. ∎

**Lemma 4.43** *Let G be an abelian p-group of type* $(\alpha_1, \alpha_2, \ldots, \alpha_k)$, *with t the largest integer such that* $\alpha_t > 1$. *Then,* $G_p$ *has order* $p^k$, *and* $G^p$ *has type* $(\alpha_1 - 1, \alpha_2 - 1, \ldots, \alpha_t - 1)$.

**Proof** The claims about the order of $G_p$ and the type of $G^p$ are easily checked since $G^p$ is generated by $x_1^p, x_2^p, \ldots, x_k^p$, where $x_1, x_2, \ldots, x_k$ are generators for $G$. ∎

**Theorem 4.44** *If G is an abelian p-group, then the type of G is uniquely determined. Indeed, if G is of type* $(\alpha_1, \alpha_2, \ldots, \alpha_k)$ *and also of type* $(\beta_1, \beta_2, \ldots, \beta_s)$, *then* $k = s$ *and* $\alpha_i = \beta_i$, *for all* $1 \le i \le k$.

**Proof** We proceed by mathematical induction on $|G|$. By Lemma 4.43, if $G$ has type $(\alpha_1, \alpha_2, \ldots, \alpha_k)$, then $G_p$ has order $p^k$. This implies that $k = s$, and the result is proved if $G = G_p$. Next, we consider the subgroup $G^p$. If $G$ is of type $(\alpha_1, \alpha_2, \ldots, \alpha_k)$ applying Lemma 4.43 shows that $G^p$ is of type $(\alpha_1 - 1, \alpha_2 - 1, \ldots, \alpha_t - 1)$, where $t$ is the largest integer such that $\alpha_t > 1$. Applying induction to $G^p$ gives the result. ∎

Now, we come to the classification theorem for finite abelian groups.

**Theorem 4.45** *(Fundamental Theorem of Finite Abelian Groups). Every finite abelian group G has a unique decomposition in the form*

$$C_{n_1} \times C_{n_2} \times \cdots \times C_{n_k},$$

*where* $n_i$ *is divisible by* $n_j$ *for* $j > i$, $n_k \ge 2$ *and* $|G| = n_1 n_2 \ldots n_k$.

**Proof** Corollary 4.41 gives the existence of such a decomposition for $G$. To prove uniqueness, note that for each prime $p$ dividing $|G|$, the Sylow $p$-subgroup of $G$ is

$$S_p(C_{n_1}) \times S_p(C_{n_2}) \times \cdots \times S_p(C_{n_k}),$$

where $S_p(H)$ denotes the Sylow $p$-subgroups of the abelian group $H$. Now, it is enough we apply Theorem 4.44 to this decomposition. ∎

**Example 4.46** There are exactly 9 abelian groups of order $1000 = 2^3 \cdot 5^3$, namely

$$C_2 \times C_2 \times C_2 \times C_5 \times C_5 \times C_5$$
$$C_2 \times C_2 \times C_2 \times C_5 \times C_{5^2}$$
$$C_2 \times C_2 \times C_2 \times C_{5^3}$$
$$C_2 \times C_{2^2} \times C_5 \times C_5 \times C_5$$
$$C_2 \times C_{2^2} \times C_5 \times C_{5^2}$$
$$C_2 \times C_{2^2} \times C_{5^3}$$
$$C_{2^3} \times C_5 \times C_5 \times C_5$$
$$C_{2^3} \times C_5 \times C_{5^2}$$
$$C_{2^3} \times C_{5^3}.$$

Only the last of these groups is cyclic.

**Exercises**

1. What is the smallest positive integer $n$ such that there are two non-isomorphic groups of order $n$?
2. What is the smallest positive integer $n$ such that there are three non-isomorphic abelian groups of order $n$?
3. What is the smallest positive integer $n$ such that there are exactly four non-isomorphic abelian groups of order $n$?
4. Show that any abelian group of order 546 is cyclic.
5. Show that there are two abelian groups of order 108 that have exactly one subgroup of order 3.
6. Show that there are two abelian groups of order 108 that have exactly four subgroups of order 3.
7. Show that there are two abelian groups of order 108 that have exactly 13 subgroups of order 3.
8. Prove that if a finite abelian group has subgroups of orders $m$ and $n$, then it has a subgroup whose order is the least common multiple of $m$ and $n$.
9. Find all abelian groups (up to isomorphism) of order 360.
10. Show how to get all abelian groups of order $2^3 \cdot 3^4 \cdot 5$.
11. Determine the number of abelian groups (up to isomorphism) of order $(10)^5$.
12. Let $p$ and $q$ be two distinct primes. Find all non-isomorphic abelian groups of order $p^2 q^3$.
13. Prove that an abelian group of order $2^n$ ($n \geq 1$) must have an odd number of elements of order 2.
14. If $Aut(G)$ is abelian, prove that $G$ is cyclic.
15. For any positive integer $n$, and for any prime $p$, prove that the number of non-isomorphic abelian $p$-groups of order $p^n$ is equal to the number of partitions of $n$.
16. Let $p$ be a prime number. Fill in the second row of the table to give the number of abelian groups of order $p^n$, up to isomorphism.

| $n$ | 2 | 3 | 4 | 5 | 6 | 7 | 8 |
|---|---|---|---|---|---|---|---|
| number of groups | | | | | | | |

Let $p, q$ and $r$ be distinct prime numbers. Use the table you created to find the number of abelian groups, up to isomorphism, of the given order:

(a) $p^3 q^4 r^7$;
(b) $(qr)^7$;
(c) $p^3 q^5 r^4$.

## 4.5  Worked-Out Problems

**Problem 4.47**  If $n$ is an odd positive integer and $\mathbb{F}$ is a field, show that $SO_n(\mathbb{F}) \times \mathbb{Z}_2 \cong O_n(\mathbb{F})$.

*Solution*  Let $I_n$ denote the $n \times n$ identity matrix. Both $I_n$ and $-I_n$ commute with every other matrix in $SO_n(\mathbb{F})$, and together they form a subgroup of $SO_n(\mathbb{F})$ of order 2. We define $\varphi : SO_n(\mathbb{F}) \times \{I_n, -I_n\} \to O_n(\mathbb{F})$ by $\varphi((A, U)) = AU$, where $A \in SO_3(\mathbb{F})$ and $U \in \{I_n, -I_n\}$. Then, $\varphi$ is a homomorphism, because

$$\varphi\big((A, U)(B, U)\big) = \varphi\big((AB, UV)\big) = ABUV$$
$$= AUBV = \psi\big((A, U)\big)\varphi\big((B, V)\big),$$

for all $A, B \in SO_n(\mathbb{F})$ and $U, V \in \{I_n, -I_n\}$. Now, if $\varphi((A, U)) = \varphi((B, V))$, then $AU = BV$, giving $\det(AU) = \det(BV)$. But $\det(AU) = \det(A)\det(U) = \det(U)$ because $A \in SO_n(\mathbb{F})$, and similarly $\det(BV) = \det(V)$. Hence, obtain $U = V, A = B$, and we conclude that $\varphi$ is one to one. It only remains to check that $\varphi$ is onto. Given $A \in O_n(\mathbb{F})$, either $A \in SO_n(\mathbb{F})$, in which ease $A = \varphi((A, I_n))$, or $A(-I_n) \in SO_N(\mathbb{F})$ and $A = \varphi(-A, -I_n)$. Finally, we note that $\{I_n, -I_n\}$ is isomorphic to $\mathbb{Z}_2$. This completes the proof.  ∎

**Problem 4.48**  Let $G$ be a group and $G = N_1 \ldots N_n$, where $N_i \trianglelefteq G$ (for $1 \leq i \leq n$). If $|G| = \prod_{i=1}^{n} |N_i|$, prove that $G \cong N_1 \times \cdots \times N_n$.

*Solution*  We set

$$H_i = N_i \cap \Big(\prod_{j \neq i} N_j\Big).$$

We have to show that $H_i = \{e\}$. Indeed, we can write

$$|G| = |N_1 \ldots N_n| = |N_i N_1 \ldots N_{i-1} N_{i+1} \ldots N_n|$$
$$= \frac{|N_i| \, |N_1 \ldots N_{i-1} N_{i+1} \ldots N_n|}{|H_i|}$$
$$\leq \frac{|N_i| \big(|N_1| \ldots |N_{i-1}||N_{i+1}| \ldots |N_n|\big)}{|H_i|}$$
$$= \frac{|N_1| \ldots |N_n|}{|H_i|} = \frac{|G|}{|H_i|}.$$

Therefore, we must have $|H_i| = 1$, and this completes the proof.  ∎

**Problem 4.49**  Let a group $G$ be the internal direct product of subgroups $H_1, \ldots, H_n$ $(n \geq 2)$. Prove that $G/H_i \cong H_1 \ldots H_{i-1} H_{i+1} \ldots H_n$, for all $1 \leq i \leq n$.

*Solution*  Let $1 \leq i \leq n$ be an arbitrary fixed integer. By Lemma 4.21, we know that each $x \in G$ can be written uniquely as a product $x = a_1 \ldots a_n$, for some $a_j \in H_j$ (for $1 \leq j \leq n$). Hence, we define a function $\varphi : G \to H_1 \ldots H_{i-1} H_{i+1} \ldots H_n$ by

$\varphi(x) = a_1 \ldots a_{i-1} a_{i+1} \ldots a_n$. It is easy to check that $\varphi$ is an epimorphism. Now, we determine $Ker\varphi$. We have

$$
\begin{aligned}
Ker\varphi &= \{x \in G \mid \varphi(x) = e\} \\
&= \{a_1 \ldots a_n \mid a_j \in H_j \text{ (for } 1 \leq j \leq n) \text{ and } \varphi(x) = e\} \\
&= \{a_1 \ldots a_n \mid a_j \in H_j \text{ (for } 1 \leq j \leq n) \text{ and } a_1 \ldots a_{i-1} a_{i+1} \ldots a_n = e\} \\
&= \{a_i(a_1 \ldots a_{i-1} a_{i+1} \ldots a_n) \mid a_j \in H_j \text{ (for } 1 \leq j \leq n) \text{ and} \\
&\quad\ a_1 \ldots a_{i-1} a_{i+1} \ldots a_n = e\} \\
&= \{a_i e \mid a_i \in H_i\} = H_i.
\end{aligned}
$$

This shows that $G/H_i \cong H_1 \ldots H_{i-1} H_{i+1} \ldots H_n$, for all $1 \leq i \leq n$. $\blacksquare$

**Problem 4.50** Let $G$ be a finite group and suppose that $a$ $b$ are elements of $G$ such that $ab = ba$ and $(o(a), o(b)) = 1$. Prove that $\langle ab \rangle = \langle a \rangle \times \langle b \rangle$ and $o(ab) = o(a)o(b)$.

*Solution* Let $o(a) = m$ and $o(b) = n$. Note that $H = \langle a, b \rangle$ is an abelian group, where the orders of subgroups $\langle a \rangle$ and $\langle b \rangle$ are relatively prime. Hence, $H \cong \langle a \rangle \times \langle b \rangle$ is a group of order $mn$. Suppose that $x = ab$. The homomorphism $\varphi : \langle x \rangle \to H/\langle a \rangle$ with $x^k \mapsto x^k \langle a \rangle = b^k \langle a \rangle$ is surjective. So, we conclude that $|Im\varphi| = n$ is a divisor of $|\langle x \rangle|$. In a similar way, we deduce that $m$ is a divisor of $|\langle x \rangle|$. Since $(m, n) = 1$, it follows that $o(x) = mn = |H|$. This shows that $H = \langle x \rangle$. $\blacksquare$

**Problem 4.51** Let $G$ be an abelian group and $f : G \to G$ be a homomorphism such that $f \circ f = f$. Prove that $G \cong Kerf \times Imf$.

*Solution* Since $G$ is abelian, it follows that $Kerf \trianglelefteq G$ and $Imf \trianglelefteq G$. Suppose that $x \in G$ is an arbitrary element. Then, we can write

$$
x = xe = xf(e) = xf(x^{-1}x) = xf(x^{-1})f(x).
$$

Since $f\big(xf(x^{-1})\big) = f(x)(f \circ f)(x^{-1}) = f(x)f(x^{-1}) = f(xx^{-1}) = f(e) = e$, it follows that $xf(x^{-1}) \in Kerf$. Also, it is clear that $f(x) \in Imf$. So, we conclude that $G = (Kerf)(Imf)$. Now, let $y \in Kerf \cap Imf$. Then, we have $f(y) = e$ and $y = f(x_0)$, for some $x_0 \in G$. Hence, we obtain $e = f(y) = f\big(f(x_0)\big) = x_0$, which implies that $y = e$. Consequently, $Kerf \cap Imf = \{e\}$, and this completes the proof. $\blacksquare$

**Problem 4.52** Let a group $G$ be the internal direct product of subgroups $H$ and $K$. If there is a non-trivial homomorphism $f : K \to Z(H)$, prove that there is a subgroup $L$ of $G$ such that $L \neq K$ and $G \cong H \times K$.

*Solution* Note that the elements of $H$ and $K$ commute. We define a function $\varphi : K \to G$ by $\varphi(k) = f(k)^{-1}k$, for all $k \in K$. We have to show that $\varphi$ is a homomorphism. Suppose that $k_1, k_2 \in K$ are two arbitrary elements. We have

$$
\begin{aligned}
\varphi(k_1 k_2) &= f(k_1 k_2)^{-1} k_1 k_2 = f(k_1)^{-1} f(k_2)^{-1} k_1 k_2 \\
&= f(k_1)^{-1} k_1 f(k_2)^{-1} k_2 = \varphi(k_1)\varphi(k_2).
\end{aligned}
$$

Hence, $\varphi$ is a homomorphism, and so $\varphi(K)$ is a subgroup of $G$. We set $L = \varphi(K)$. Since $f$ is non-trivial, it follows that there is $k \in K$ such that $f(k) \neq e$. We claim that $L \neq K$. If $L = K$, then $k \in L$. This implies that there is $a \in K$ such that $k = f(a)^{-1}a$, and so $ka^{-1} = f(a)^{-1} \in K \cap H$. Hence, we conclude that $ka^{-1} = e$ and $f(a)^{-1} = e$, which implies that $f(k) = e$, a contradiction.

Now we prove that $G \cong H \times L$. In order to do this, it is enough to show that

(1) $H \trianglelefteq G$ and $L \trianglelefteq G$;
(2) $G = HL$;
(3) $H \cap L = \{e\}$.

(1) We have $H \trianglelefteq G$. So, we show $L \trianglelefteq G$. Let $hk \in HK = G$, where $h \in H$ and $k \in K$, and suppose that $f(z)^{-1}z \in L$, where $z \in K$. Then, we have

$$(hk)^{-1}f(z)^{-1}z(hk) = k^{-1}h^{-1}f(z)^{-1}zhk = k^{-1}f(z)^{-1}zk$$
$$= f(z)^{-1}k^{-1}zk. \tag{4.6}$$

Since $K/Ker f$ is isomorphic to a subgroup of $Z(G)$, it follows that $K/Ker f$ is abelian. So, we have $K' \subseteq Ker f$. Therefore, we conclude that $f(k^{-1}zkz^{-1}) = e$, which implies that $f(k^{-1}zk) = f(z)$. Hence, we can write $f(z)^{-1}k^{-1}zk = \left(f(k^{-1}zk)\right)^{-1}k^{-1}zk$. Since $k, z \in K$, it follows that $k^{-1}zk \in K$, and so

$$f(z)^{-1}k^{-1}zk \in K. \tag{4.7}$$

Now, by (4.6) and (4.7), we conclude that $(hk)^{-1}f(z)^{-1}z(hk) \in L$. This proves that $L \trianglelefteq G$.

(2) Suppose that $hk \in G$ be an arbitrary element, where $h \in H$ and $k \in K$. We can write $hk = hf(k)f(k)^{-1}k$. Since $hf(k) \in H$ and $f(k)^{-1}k \in L$, it follows that $hk \in HL$, and so $G \subseteq HL$. Consequently, we have $G = HL$.

(3) Let $a \in H \cap L$ be arbitrary. Then, there exists $b \in K$ such that $a = f(b)^{-1}b$. Thus, $f(b)a = b \in H \cap K = \{e\}$. This implies that $b = e$ and $f(b)a = e$, and so $a = e$. ∎

## 4.6 Supplementary Exercises

1. Prove that $A_n \times \mathbb{Z}_2$ is not isomorphic to $S_n$, for $n \geq 3$.
2. Why does the construction of Problem 4.47 not lead to an isomorphism from $SO_n(\mathbb{F}) \times \mathbb{Z}_2$ to $O_n(\mathbb{F})$ when $n$ is even? Which elements of $O_n(\mathbb{F})$ commute with every element of $O_n(\mathbb{F})$? Show that $SO_n(\mathbb{F}) \times \mathbb{Z}_2$ is not isomorphic to $O_n(\mathbb{F})$ when $n$ is even.
3. Show that $D_{2n}$ is isomorphic to $D_n \times \mathbb{Z}_2$ when $n$ is odd.
4. Prove that

   (a) $Aut(\mathbb{Z} \times \cdots \times \mathbb{Z}) \cong GL_n(\mathbb{Z})$;

(b) $Aut \underbrace{\left(\mathbb{Z}_{p^m} \times \cdots \times \mathbb{Z}_{p^m}\right)}_{n \text{ times}} \cong GL_n\left(\mathbb{Z}_{p^m}\right)$.

5. Prove that

   (a) $U_4 \cong \mathbb{Z}_2$;
   (b) $U_{2^n} \cong \mathbb{Z}_2 \times \mathbb{Z}_{2^{n-1}}$, for all $n \geq 3$;
   (c) $U_{p^n} \cong \mathbb{Z}_{p^n - p^{n-1}}$, for all $p$ an odd prime.

6. Suppose that $N_1, \ldots, N_n$ are normal subgroups of a group $G$ and define $\varphi : G \to G/N_1 \times \cdots \times G/N_n$ by $\varphi(x) = (xN_1, \ldots, xN_n)$. Prove that

   (a) $Ker\varphi = N_1 \cap \ldots \cap N_n$;
   (b) If each $N_i$ has finite index in $G$ and if $\left(|G/N_i|, |G/N_j|\right) = 1$ for all $i \neq j$, then $\varphi$ is a surjection and

$$[G : N_1 \cap \ldots \cap N_n] = \prod_{i=1}^{n} |G/N_i|.$$

7. Let $G_1$ and $G_2$ be two groups. Prove that there exists a monomorphism from $Aut(G_1) \times Aut(G_2)$ to $Aut(G_1 \times G_2)$. If $(|G_1|, |G_2|) = 1$, then this monomorphism is an isomorphism. Give an example to show that in general this monomorphism need not be an isomorphism.

8. Let $N_1, \ldots, N_n$ be normal subgroups of $G$ such that $|N_i| = m_i$ and $(m_i, m_j) = 1$ for each $i \neq j$. If $|G| = m_1 \ldots m_n$, prove that

   (a) $G \cong N_1 \times \cdots \times N_n$;
   (b) $Aut(G) \cong Aut(N_1) \times \cdots \times Aut(N_n)$.

9. Let $G$ be a group and $G = N_1 \ldots N_n$, where $N_i \trianglelefteq G$ (for $1 \leq i \leq n$). If $(|N_i|, |N_j|) = 1$, for all $i \neq j \in \{1, \ldots, n\}$, prove that $G \cong N_1 \times \cdots \times N_n$.

10. Let $G_1$ and $G_2$ be groups and let $H$ be a subgroup of $G_1 \times G_2$, satisfying: for every $a \in G_1$, there exists $b \in G_2$ such that $(a, b) \in H$ and for every $b \in G_2$, there exists $a \in G_1$ such that $(a, b) \in H$. Putting

$$N_1 = \{a \in G_1 \mid (a, e_2) \in H\} \text{ and } N_2 = \{b \in G_2 \mid (e_1, b) \in H\}.$$

Prove that

   (a) $N_i \trianglelefteq G_i \ (i = 1, 2)$;
   (b) $G_1/N_1 \cong G_2/N_2$.

11. Let $G$ be a finite abelian group such that it contains a subgroup $H_0 \neq \{e\}$ which lies in every subgroup $H \neq \{e\}$. Prove that $G$ must be cyclic. What can you say about $|G|$.

12. Let $G$ be a finite abelian group. Using Exercise 11 show that $G$ is isomorphic to a subgroup of a direct product of a finite number of finite cyclic groups.

13. Prove that the group $\mathbb{Q}$ of rational numbers under addition cannot be written as a direct product of two non-trivial subgroups.

14. Give an example of a finite non-abelian group $G$ which contains a subgroup $H_0 \neq \{e\}$ such that $H_0 \subseteq H$ for all subgroups $H \neq \{e\}$ of $G$.

15. Prove that there does not exist any group with exactly two distinct subgroups of index 2.

16. Let $G$ be a group of order $p^2 q^2$, where $p$ and $q$ are distinct prime numbers and where $q \nmid p^2 - 1$ and $p \nmid q^2 - 1$. Show that $G$ is an external direct product of a group of order $p^2$ and one of order $q^2$. Deduce that $G$ is abelian.

17. Let $V$ be a finite vector space of dimension $n$ over a field $\mathbb{F}$. Prove that as abelian groups $V \cong \mathbb{F}_1 \times \cdots \times \mathbb{F}_n$, where $\mathbb{F}_i \cong \mathbb{F}$, for all $1 \leq i \leq n$.

18. Prove that if $d$ is any divisor of the order of a finite abelian group $G$, then $G$ has a subgroup of order $d$.

19. Use the fundamental theorem of finite abelian groups to prove the fundamental theorem of arithmetic.

# Chapter 5
# Normal Series

To give insights into the structure of a group $G$, we study a normal series of $G$. The idea of normal series of a group and solvability yields invariant of groups (the Jordan–Hölder theorem), showing that simple groups are, in certain sense, building towers of finite groups.

## 5.1 Jordan–Hölder Theorem

In this section, we study composition series. The important thing about composition series is that the composition factors occurring are essentially unique. To recognize when we have a composition series, we need to be able to recognize simple groups.

**Definition 5.1** A sequence of subgroups

$$\{e\} = G_0 \subseteq G_1 \subseteq \cdots \subseteq G_n = G$$

of a group $G$ is called a *normal series* of $G$, if $G_i$ is a normal subgroup of $G_{i+1}$, for $i = 0, \ldots, n - 1$.

The factor groups of this normal series are the groups $G_{i+1}/G_i$ for $i = 0, \ldots, n - 1$; the *length of the normal series* is the number strict inclusions; i.e., the length is the number of non-trivial factor groups.

In general, a normal subgroup of a normal subgroup need not be normal. So the terms in a normal series need not be normal in $G$.

**Example 5.2** The sequence $\{id\} \subseteq A_n \subseteq S_n$ is a normal series for $S_n$.

**Example 5.3** $A_4$ has the normal series $\{id\} \subseteq V \subseteq A_4$ where

© The Author(s), under exclusive license to Springer Nature Singapore Pte Ltd. 2021    105
B. Davvaz, *Groups and Symmetry*, https://doi.org/10.1007/978-981-16-6108-2_5

$$V = \{\text{id}, (1\ 2)(3\ 4), (1\ 3)(2\ 4), (1\ 4)(2\ 3)\}$$

is easily seen to be a normal subgroup to be isomorphic to the Klein 4-group.

**Definition 5.4** A normal series

$$\{e\} = H_0 \subseteq H_1 \subseteq \cdots \subseteq H_m = G$$

is a *refinement of a normal series*

$$\{e\} = G_0 \subseteq G_1 \subseteq \cdots \subseteq G_n = G$$

if $G_0, G_1, \ldots, G_n$ is a subsequence of $H_0, H_1, \ldots, H_m$.

So, a refinement is a normal series containing each of the terms of the original series.

***Example 5.5*** The series $\{0\} \subseteq 72\mathbb{Z} \subseteq 24\mathbb{Z} \subseteq 8\mathbb{Z} \subseteq 4\mathbb{Z} \subseteq \mathbb{Z}$ is a refinement of a series $\{0\} \subseteq 72\mathbb{Z} \subseteq 8\mathbb{Z} \subseteq \mathbb{Z}$.

**Definition 5.6** A *composition series* is a normal series

$$\{e\} = G_0 \subseteq G_1 \subseteq \cdots \subseteq G_n = G,$$

in which, for all $i$, $G_i$ is a maximal normal subgroup of either $G_{i+1}$ or $G_{i+1} = G_i$.

Every refinement of a composition series is also a composition series; it can only repeat some of the original terms.

***Example 5.7*** $\mathbb{Z}_{24}$ has the following composition series:

$$\langle 0 \rangle \subseteq \langle 8 \rangle \subseteq \langle 4 \rangle \subseteq \langle 2 \rangle \subseteq \mathbb{Z}_{24},$$
$$\langle 0 \rangle \subseteq \langle 12 \rangle \subseteq \langle 4 \rangle \subseteq \langle 2 \rangle \subseteq \mathbb{Z}_{24},$$
$$\langle 0 \rangle \subseteq \langle 12 \rangle \subseteq \langle 6 \rangle \subseteq \langle 2 \rangle \subseteq \mathbb{Z}_{24},$$
$$\langle 0 \rangle \subseteq \langle 12 \rangle \subseteq \langle 6 \rangle \subseteq \langle 3 \rangle \subseteq \mathbb{Z}_{24}.$$

**Theorem 5.8** *Every finite group has a composition series.*

***Proof*** Suppose that $|G| = n$. We prove the result by mathematical induction on $n$. If $n = 1$, then the result is trivial. Assume that the result holds for all groups whose order is less than $n$. If $G$ is simple, then $\{e\} = G_0 \subseteq G_1 = G$ is the required composition series. If $G$ is not simple, then $G$ has a maximal subgroup $H$. Since $|H| < n$, by induction hypothesis, it follows that $H$ has a composition series as follows:

$$\{e\} = H_0 \subseteq H_1 \subseteq \cdots \subseteq H_m = H,$$

where $H_{i+1}/H_i$ is simple, for all $i = 0, \ldots, m - 1$. Now, we consider the series

$${e} = H_0 \subseteq H_1 \subseteq \cdots \subseteq H_m = H \subseteq G.$$

Since $H$ is a maximal normal subgroup of $G$, it follows that $G/H$ is simple. Therefore, we conclude that the last series is a composition series of $G$. This completes the proof. ∎

**Theorem 5.9** *If $G$ is an abelian group having a composition series, then $G$ is finite.*

**Proof** We firstly notice that a simple abelian group is a cyclic group of prime order. This follows the fact that any subgroup of an abelian group is normal. Now, let

$${e} = G_0 \subseteq G_1 \subseteq \cdots \subseteq G_n = G$$

be a composition series of $G$, where $G_i \neq G_j$, for all $i \neq j$. Since $G_1/G_0 \cong G_1$ is a simple abelian group, it follows that $|G_1| = p_1$, where $p_1$ is a prime number. Also, $G_2/G_1$ is a simple abelian group, which implies that $|G_2/G_1| = p_2$ for some prime number $p_2$. This yields that $|G_2| = |G_1| \cdot |G_2/G_1| = p_1 p_2$. Proceeding in this manner we conclude that $G$ has $p_1, p_2, \ldots, p_n$ elements, where $p_{i+1} = |G_{i+1}/G_i|$ for $i = 0, 1, \ldots, n - 1$. ∎

**Example 5.10** (**Non-example**). We claim that there is no composition series for $\mathbb{Q}$. Let

$${e} = G_0 \subseteq G_1 \subseteq \cdots \subseteq G_n = \mathbb{Q}$$

be a composition series of $\mathbb{Q}$, where $G_i \neq G_j$, for all $i \neq j$. In order to obtain a contradiction, we construct a subgroup $H$ of $G_1$ which is neither ${0}$ nor $G_1$. Since $\mathbb{Q}$ is abelian, it follows that any subgroup is normal. To do this, choose some nonzero $x \in G_1$ and consider two cases:

*Case 1*: If $x/m \in G_1$ for all nonzero $m \in \mathbb{Z}$, then $G_1$ must be the entire group $\mathbb{Q}$, in which case we let $H = \mathbb{Z}$.

*Case 2*: Suppose that there exists a nonzero $m \in \mathbb{Z}$ with $x/m \notin G_1$. Take $H = {mg | g \in G_1}$. This is a subgroup of $G_1$ because $mg \in G_1$, for all $g \in G_1$. Moreover, we have $H \neq G_1$ because $x \in G_1$ but $x \notin H$.

Therefore, in either case we have constructed a subgroup $H$ such that ${0} \subset H \subset G_1$, and we have the desired contradiction. Hence, our supposition is false and $\mathbb{Q}$ has no composition series.

**Example 5.11** Let $G = \langle a \rangle \cong \mathbb{Z}_{30}$. We consider two composition series of $G$ as follows:
$${e} \subseteq \langle a^{10} \rangle \subseteq \langle a^5 \rangle \subseteq G,$$
$${e} \subseteq \langle a^6 \rangle \subseteq \langle a^2 \rangle \subseteq G.$$

The factor groups of the first normal series are $G/\langle a^5 \rangle \cong \mathbb{Z}_5$, $\langle a^5 \rangle/\langle a^{10} \rangle \cong \mathbb{Z}_2$, and $\langle a^{10} \rangle/{e} \cong \langle a^{10} \rangle \cong \mathbb{Z}_3$. The factor groups of the second normal series are $G/\langle a^2 \rangle \cong \mathbb{Z}_2$, $\langle a^2 \rangle/\langle a^6 \rangle \cong \mathbb{Z}_3$, and $\langle a^6 \rangle \cong \mathbb{Z}_5$. In this case, both composition series have the

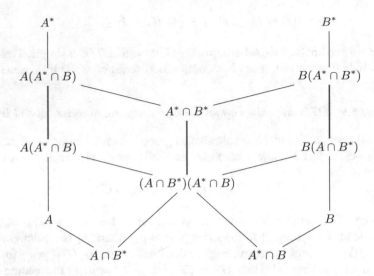

**Fig. 5.1** Butterfly

same length and the factor groups can be paired isomorphically after rearranging them. We give a name to this phenomenon.

**Definition 5.12** Two normal series of a group $G$ are *equivalent* if there is a one-to-one correspondence between their factor groups such that corresponding factor groups are isomorphic.

The next lemma is quite technical. Its value will be immediately apparent in the proof of Theorem 5.14.

**Theorem 5.13 (Zassenhaus' lemma).** *Let $A \trianglelefteq A^*$ and $B \trianglelefteq B^*$ be four subgroups of $G$. Then,*

$$A(A^* \cap B) \trianglelefteq A(A^* \cap B^*),$$
$$B(B^* \cap A) \trianglelefteq B(B^* \cap A^*),$$

*and we have an isomorphism:*

$$\frac{A(A^* \cap B^*)}{A(A^* \cap B)} \cong \frac{B(B^* \cap A^*)}{B(B^* \cap A)}.$$

Figure 5.1 represents the butterfly that goes with its proof.

***Proof*** Since $A \trianglelefteq A^*$, it follows that $A \trianglelefteq A^* \cap B^*$. This implies that $A \cap B^* = A \cap (A^* \cap B^*) \trianglelefteq A^* \cap B^*$. Analogously, we obtain $A^* \cap B \trianglelefteq A^* \cap B^*$. Now, we conclude that $N = (A^* \cap B)(A \cap B^*)$ is a normal subgroup of $A^* \cap B^*$. Now, suppose that $x \in B(B^* \cap A^*)$. Then $x = yx$, for some $y \in B$ and $z \in B^* \cap A^*$. Now, we define $f : B(B^* \cap A^*) \rightarrow (A^* \cap B^*)/N$ by $f(x) = f(yz) = zN$. First, we show that $f$ is well defined. Suppose that $x = yz = y'z'$, where $y, y' \in B$ and $z, z' \in B^* \cap A^*$. Then, we deduce that $z'z^{-1} = y'^{-1}y \in (B^* \cap A^*) \cap B = B \cap A^* \leq N$. It is easy to investigate that $f$ is an onto homomorphism with kernel $B(B^* \cap A)$. Now, by the first isomorphism theorem, we conclude that $B(B^* \cap A) \unlhd B(B^* \cap A^*)$ and

$$\frac{B(B^* \cap A^*)}{B(B^* \cap A)} \cong \frac{B^* \cap A^*}{N}.$$

By moving the symbols $A$ and $B$, we get $A(A^* \cap B) \unlhd A(A^* \cap B^*)$ and an isomorphism of the corresponding factor group to $(B^* \cap A^*)/N$. This shows that the two factor groups in the statement are isomorphic, as desired.   ∎

We can now prove the fundamental theorem on refinements.

**Theorem 5.14 (Schreier refinement theorem).** *Let $G$ be a group. Then, any two normal series of $G$ have equivalent refinements.*

**Proof** Consider two normal series

$$\{e\} = G_0 \subseteq G_1 \subseteq \cdots \subseteq G_n = G \tag{5.1}$$

and

$$\{e\} = H_0 \subseteq H_1 \subseteq \cdots \subseteq H_m = G \tag{5.2}$$

of $G$. We define

$$G_{ij} = G_i(G_{i+1} \cap H_j), 0 \leq i \leq n - 1, 0 \leq j \leq m,$$
$$H_{ij} = H_j(G_i \cap H_{j+1}), 0 \leq i \leq n, 0 \leq j \leq m - 1,$$

Now, by Zassenhaus' lemma, we obtain $G_{ij} \unlhd G_{i\ j+1}$ and $H_{ij} \unlhd H_{i+1\ j}$. Then, the normal series $\{G_{ij} | 0 \leq i \leq n - 1, 0 \leq j \leq m\}$ is a refinement of the normal series (5.1) with $mn$ terms. Also, $\{H_{ij} | 0 \leq i \leq n, 0 \leq j \leq m - 1\}$ is a refinement of the normal series (5.2). Finally, the function pairing $G_{i\ j+1}/G_{ij}$ with $H_{i+1\ j}/H_{ij}$ is a bijection, and Zassenhaus' lemma shows that the corresponding factor groups are isomorphic, i.e.,

$$\frac{G_i(G_{i+1} \cap H_{j+1})}{G_i(G_{i+1} \cap H_j)} \cong \frac{H_j(G_{i+1} \cap H_{j+1})}{H_j(G_i \cap H_{j+1})}.$$

Therefore, we conclude that two refinements are equivalent. This completes the proof.   ∎

A group may have more than one composition series. However, the Jordan–Hölder theorem (named after Camille Jordan and Otto Hölder) states that any two composition series of a given group are equivalent. That is, they have the same composition length and the same composition factors, up to permutation and isomorphism.

**Theorem 5.15 (Jordan–Hölder theorem).** *Every two composition series of a group $G$ are equivalent.*

*Proof* Composition series are normal series, so that every two composition series of $G$ have equivalent refinements. But a composition series is a normal series of maximal length; a refinement of it merely repeats several of its terms, and so its new factor groups have order 1. Therefore, two composition series of $G$ are equivalent.                                                                                    ∎

*Remark 5.16* The Jordan–Hölder theorem does not state the existence of a composition series for a given group.

Hence, once we have determined one composition series for a (say, finite) group, then we have uniquely determined composition factors which can be thought of as the ways of breaking our original group down into simple groups. This is analogous to a "prime factorization" for groups.

If $G$ has a composition series, then the factor groups of this series are called the *composition factors* of $G$.

One should regard the Jordan–Hölder theorem as a unique factorization theorem (fundamental theorem of arithmetic).

**Corollary 5.17** *The primes and their multiplicities occurring in the factorization of an integer $n \geq 2$ are determined by $n$.*

*Proof* Suppose that $n = p_1, p_2, \ldots, p_k$, where the $p_i$s are (not necessarily distinct) primes. If $G = \langle a \rangle$ is a cyclic group of order $n$, then

$$\{e\} \subseteq \langle a^{p_1 \cdots p_{k-1}} \rangle \subseteq \cdots \subseteq \langle a^{p_1 p_2} \rangle \subseteq \langle a^{p_1} \rangle \subseteq \langle a \rangle = G$$

is a normal series. The factor groups have prime orders $p_1, p_1, \ldots, p_k$, respectively. Hence, this is a composition series. Now, the Jordan–Hölder theorem shows that these numbers depend on $n$ alone.                                                                ∎

### Exercises

1. Prove that if $G$ is a group which has a composition series, then any normal subgroup of $G$ and any factor group of $G$ also have composition series with factors isomorphic to composition factors of $G$.

2. Show that a normal series is a composition series if and only if it has maximal length; i.e., every refinement of it has the same length.

3. Show that a normal series is a composition series if and only if its factor groups are either simple or trivial.

4. Show that an abelian group has a composition series if and only if it is finite.

5. Give an example of an infinite group which has a composition series.

6. Let $G$ and $H$ be finite groups. If there are normal series of $G$ and $H$ having the same set of factor groups, prove that $G$ and $H$ have the same composition factors.

7. Let $p$ be a prime number, and let $G$ be a non-trivial $p$-group. Show that $G$ has a normal series
$$\{e\} = G_0 \subseteq G_1 \subseteq \cdots \subseteq G_n = G$$

such that $G_i$ is a normal subgroup of $G$ and $[G : G_i] = pi$ for $i = 0, 1, \ldots, n$.
What are the composition factors of $G$?
8. The dihedral group $D_4$ has seven different composition series. Find all seven.
9. How many different composition series does the quaternion group $Q_8$ have?

## 5.2 Solvable Groups

A solvable group is a group having a normal series such that each normal factor is abelian. The goal of this section is to study the concept of solvable groups.

**Definition 5.18** A group $G$ is said to be *solvable* if it has a normal series $\{e\} = G_0 \subseteq G_1 \subseteq \cdots \subseteq G_n = G$ such that each of its factor groups $G_{i+1}/G_i$ is abelian, for every $i = 0, \ldots, n - 1$.

The above series is referred to as a *solvable series* of $G$.

**Example 5.19** It is clear that from the definition that every abelian group is solvable. So, the concept of solvability may be regarded as a generalization of being abelian.

**Example 5.20** A metabelian group $G$ is a solvable group which has the solvable series $\{e\} \trianglelefteq G' \trianglelefteq G$.

Although there is no complete classification theorem for finite solvable groups, there are several basic properties which are exactly analogous to properties of abelian groups.

**Example 5.21** The groups $S_3$ is solvable groups. It has the series $\{e\} \trianglelefteq A_3 \trianglelefteq S_3$ with factors isomorphic to $\mathbb{Z}_2$ and $\mathbb{Z}_3$.

**Example 5.22** The groups $S_4$ are solvable group. We can take $H_0 = \{id\}$, $H_1 = \{id, (1\,2)(3\,4), (1\,3)(2\,4), (1\,4)(2\,3)\}$, $H_2 = A_4$, and $H_3 = S_4$. Note that $H_3/H_2$ is a group of order 2 that $H_2/H_1$ is a group of order 3, and that $H_1/H_0 \cong H_1$ is a group of order 4, and all of these factor groups are abelian.

**Example 5.23** The dihedral groups $D_n$ contain an element $a$ of order $n$, so $\langle a \rangle$ has index 2, and this shows that it is normal. Thus, $\{e\} \subseteq \langle a \rangle \subseteq D_n$ is a normal series for $D_n$ with both factors cyclic. Thus, $D_n$ is solvable.

**Theorem 5.24** *Any subgroup $H$ of a solvable group $G$ is solvable.*

**Proof** Suppose that $\{e\} = G_0 \subseteq G_1 \subseteq \cdots \subseteq G_n = G$ is a solvable series for $G$. We show that

$$\{e\} = H \cap G_0 \subseteq H \cap G_1 \subseteq \cdots \subseteq H \cap G_n = H$$

is a solvable series for $H$. Since $G_i$ is normal in $G_{i+1}$, where $i = 0, \ldots, n - 1$, we conclude that $N_i = H \cap G_i$ is normal in $N_{i+1} = H \cap G_{i+1}$. Now, we define a mapping $f : N_{i+1} \longrightarrow G_{i+1}/G_i$ by $f(x) = xG_i$, for all $x \in N_{i+1}$. Clearly, $f$ is a homomorphism. Moreover, if $x \in N_{i+1}$, then

$$x \in \mathrm{Ker} f \Leftrightarrow xG_i = G_i \Leftrightarrow x \in G_i \Leftrightarrow x \in H \cap G_i.$$

This yields that $\mathrm{Ker} f = H \cap G_i = N_i$. Hence, by the first isomorphism theorem, $N_{i+1}/N_i \cong f(N_{i+1})$. As $f(N_{i+1})$ is a subgroup of $G_{i+1}/G_i$ and $G_{i+1}/G_i$ is abelian, $f(N_{i+1})$ is also abelian. Consequently, $N_{i+1}/N_i$ is abelian. This proves that $H$ is a solvable group.                                                                            ∎

**Theorem 5.25** *If $N$ is a normal subgroup of a solvable group $G$, then $G/N$ is also solvable.*

**Proof** Suppose that $\{e\} = G_0 \subseteq G_1 \subseteq \cdots \subseteq G_n = G$ is a solvable series for $G$. We consider the following series

$$\{N\} = G_0N/N \subseteq G_1N/N \subseteq \cdots \subseteq G_nN/N = G/N.$$

Let $0 \leq i \leq n - 1$ be an arbitrary element. Suppose that $x \in G_{i+1}N$. Then, $x = gy$ for some $g \in G_{i+1}$ and $y \in N$. Hence,

$$xG_iN = gyG_iN = gyNG_i = gNG_i$$
$$= G_igN = G_igyN = G_iNgy = G_iNx.$$

This proves that $G_iN$ is a normal subgroup of $G_{i+1}N$. Thus,

$$G_{i+1}N/G_iN \cong (G_{i+1}N/N)/(G_iN/N).$$

Now, we define $f : G_{i+1} \to G_{i+1}N/G_iN$ by $f(x) = G_iNx$, for all $x \in G_{i+1}$. Then, $f$ is a homomorphism. As $G_{i+1}N = NG_{i+1}$, for $y \in G_{i+1}N$, we can write $y = zg$, for some $z \in N$ and $g \in G_{i+1}$. Then, $G_iNy = G_iNzg = G_iNg = f(g)$. This shows that $f$ is onto. Since $G_i \subseteq \mathrm{ker} f$, the following function is a homomorphism

$$\overline{f} : G_{i+1}/G_i \to G_{i+1}N/G_iN$$
$$\overline{f}(G_ix) = G_iNx,$$

for all $x \in G_{i+1}$. Clearly, $\overline{f}$ is also onto. Thus, $G_{i+1}N/G_iN$ is a homomorphic image of the abelian group $G_{i+1}/G_i$. So it must be itself abelian. Consequently, each factor group is abelian. This proves that $G/N$ is solvable.                                              ∎

**Theorem 5.26** *Let $N$ be a normal subgroup of a group $G$. If both $N$ and $G/N$ are solvable, then $G$ is solvable.*

**Proof** Since $N$ and $G/N$ are solvable, it follows that there exist the following solvable series for $N$ and $G/N$, respectively,

$$\{e\} = N_0 \subseteq N_1 \subseteq \cdots \subseteq N_m = N,$$
$$\{N\} = G_0/N \subseteq G_1/N \subseteq \cdots \subseteq G_n/N = G/N.$$

Here each $G_i$ is a subgroup of $G$ containing $N$. Since $G_i/N$ is normal in $G_{i+1}/N$, each $G_i$ is normal in $G_{i+1}$. Moreover, $G_0 = N$. Now,

$$\{e\} = N_0 \subseteq N_1 \subseteq \cdots \subseteq N_m = N = G_0 \subseteq G_1 \subseteq \cdots \subseteq G_n = G$$

is a solvable series for $G$. Therefore, we conclude that $G$ is solvable. ∎

**Corollary 5.27** *If two groups $H$ and $K$ are solvable, then $H \times K$ is solvable.*

**Proof** Let $G = H \times K$. Alternatively, we can consider $H \cong (H \times \{e\})$ and $K \cong (\{e\} \times H)$ and clearly, $G = (H \times \{e\})(\{e\} \times K)$. Hence, $G/H \cong K$. Since $K$ is solvable, $G/H$ is solvable. Hence, $G$ is solvable by Theorem 5.26. ∎

**Theorem 5.28** *Any finite p-group is solvable.*

**Proof** Let $G$ be a $p$-group and $|G| = p^k$, for some positive integer $k$. We apply mathematical induction on $k$. If $k = 1$, our group is a cyclic group of prime order. Thus, it is solvable by definition. Assume that the statement holds for all $k \leq n$. We will prove that it holds for $k = n + 1$. Since $G$ is a $p$-group, it follows that $Z(G) \neq \{e\}$. Since $Z(G)$ is abelian, it follows that $Z(G)$ is solvable. Moreover, since $Z(G)$ is non-trivial, it follows that $G/Z(G)$ is a $p$-group and $|G/Z(G)| < p^{n+1}$. So by the inductive assumption, $G/Z(G)|$ is solvable. Finally, using Theorem 5.26, $G$ is also solvable and we are done. ∎

**Lemma 5.29** *Let $N$ be a normal subgroup of $G$ and $G/N$ be abelian. If $x, y \in G$, then $xyx^{-1}y^{-1} \in N$.*

**Proof** It is straightforward. ∎

**Theorem 5.30** *$S_n$ is not solvable for $n \geq 5$.*

**Proof** Suppose that $S_k$ is solvable for some $k \geq 5$ and

$$\{e\} = G_0 \subseteq G_1 \subseteq \cdots \subseteq G_{m-1} \subseteq G_m = S_k$$

is a solvable series for $S_k$. By induction, we show that $G_{m-i}$ contains all of the cycles of length 3, where $0 \leq i \leq m$. Indeed, since $G_0 = \{e\}$, we obtain a contradiction and the proof completes.

Let $(x \ y \ z)$ be a cycle of length 3 in $S_k$. Since $k \geq 5$, there exist $u, v$ distinct from $x, y, z$. Now, suppose that $\sigma = (z \ u \ y)$ and $\delta = (y \ x \ v)$. Since $G_{m-1}$ is a normal subgroup of $S_k$ and $S_k/G_{m-1}$ is abelian, it follows that $\sigma\delta\sigma^{-1}\delta^{-1} \in G_{m-1}$. Therefore,

$$(x \ y \ z) = (z \ u \ y)(y \ x \ v)(z \ y \ u)(y \ v \ x) = \sigma\delta\sigma^{-1}\delta^{-1}$$

is an element of $G_{m-1}$.

Now, let $G_{m-j}$ contain all of the cycles of length 3. Since $(x\ y\ z) \in G_{m-j}$ and $G_{m-j-1}$ is a normal subgroup of $G_{m-j}$, similar to the above discussion, we obtain $(x\ y\ z) \in G_{m-j-1}$.                                                                                                           ∎

**Definition 5.31** Let $G$ be a group. We define the sequence of subgroups $G^{(i)}$ of $G$ inductively by

(1) $G^{(0)} = G$;
(2) $G^{(1)} = G'$, commutator subgroup of $G$;
(3) $G^{(i)} = (G^{(i-1)})'$, commutator subgroup of $G^{(i-1)}$, if $i > 1$.

**Theorem 5.32** *A group $G$ is solvable if and only if $G^{(n)} = \{e\}$ for some $n \geq 1$.*

**Proof** Suppose that $G$ is solvable and $\{e\} = G_0 \subseteq G_1 \subseteq \cdots \subseteq G_n = G$ is a solvable series for $G$. We prove inductively that

$$G^{(k)} \subseteq G_{n-k}, \tag{5.3}$$

for all $0 \leq k \leq n$. If $k = 0$, then $G^{(0)} = G = G_0$. Let $G^{(k)} \subseteq G_{n-k}$ for some $k$. This implies that $G^{(k+1)} = (G^{(k)})' \subseteq (G_{n-k})'$. Since $G_{n-k}/G_{n-k-1}$ is abelian, it follows that $(G_{n-k})' \subseteq G_{n-k-1}$. Consequently, $G^{(k+1)} \subseteq G_{n-k-1}$. Thus, by induction, (5.3) holds. In particular, $G^{(n)} \subseteq G_0$. Therefore, $G^{(n)} = \{e\}$.

Conversely, assume that $G^{(n)} = \{e\}$ for some $n$. Then

$$\{e\} = G^{(n)} \subseteq G^{(n-1)} \subseteq \cdots \subseteq G^{(1)} \subseteq G^{(0)} = G$$

is a normal series for $G$ such that $G^{(j)}/G^{(j+1)} = G^{(j)}/(G^{(j)})'$ is abelian. Thus, $G$ is solvable.                                                                                                                            ∎

In Theorem 5.32, if $n$ is the least such integer, then $G$ is said to be *solvable of length $n$*.

**Example 5.33** Let $\mathrm{IUT}_n(\mathbb{R})$ denote the set of all $n \times n$ matrices of the form

$$A = \begin{bmatrix} 1 & a_{12} & a_{13} & \cdots & a_{1n} \\ 0 & 1 & a_{23} & \cdots & a_{2n} \\ 0 & 0 & 1 & \cdots & a_{3n} \\ \vdots & \vdots & \vdots & \vdots & \vdots \\ 0 & 0 & 0 & \cdots & 1 \end{bmatrix},$$

where $a_{ij}$ are real numbers. It is not difficult to check that $A^{-1}$ exists and that if

$$A^{-1} = \begin{bmatrix} 1 & b_{12} & b_{13} & \cdots & b_{1n} \\ 0 & 1 & b_{23} & \cdots & b_{2n} \\ 0 & 0 & 1 & \cdots & b_{3n} \\ \vdots & \vdots & \vdots & \vdots & \vdots \\ 0 & 0 & 0 & \cdots & 1 \end{bmatrix},$$

then $b_{i\ i+1} = -a_{i\ i+1}$. It follows that the derived group $\mathrm{IUT}_n(\mathbb{R})^{(1)}$ comprises all matrices whose form is that of $A$ except that the elements $a_{12}, a_{23}, \ldots, a_{n-1\ n}$ on the first superdiagonal are all 0. Similarly, we find that $\mathrm{IUT}_n(\mathbb{R})^{(k)}$ comprises matrices in which the first $2^k - 1$ superdiagonals are full of 0s. Then, it follows that easily that the group $\mathrm{IUT}_n(\mathbb{R})$ is solvable of length $-[-\log_2 n]$.

**Exercises**

1. For what values of $n$ is the alternating group $A_n$ solvable?
2. Define a binary operation on the set $G = \{\ldots, z_{-2}, z_{-1}, z_0, z_1, z_2, \ldots\}$ by

$$
a_n a_m = \begin{cases} a_{n+m} & \text{if } n \text{ is even} \\ a_{n\ -m} & \text{if } n \text{ is odd.} \end{cases}
$$

Prove that $G$ is a group with respect to this operation. Is $G$ solvable?
3. Show that every refinement of a solvable series is a solvable series.
4. Show that a solvable group having a composition series must be finite.
5. Define a binary operation on the set of infinite sequences of integers by

$$
\begin{aligned}
&(a_1, a_2, a_3, \ldots, a_n, \ldots)(b_1, b_2, b_3, \ldots, b_n, \ldots) \\
&= \big(a_1 + b_1, (-1)^{b_1} a_2 + b_2, (-1)^{b_1+b_2} a_3 + b_3, \ldots, \\
&\quad (-1)^{b_1+b_2+b_3+\cdots+b_{n-1}} a_n + b_n, \ldots \big).
\end{aligned}
$$

Prove that $G$ is a group with respect to this operation. Is $G$ solvable?
6. If $H$ and $K$ are solvable subgroups of a group $G$ with $H \trianglelefteq G$, show that $HK$ is solvable.
7. Let $G$ be a solvable group. If all Sylow subgroups of $G$ are cyclic, prove that $G'$ is abelian.
8. Prove that the following are equivalent for a finite group $G$:

   (a)  $G$ is solvable.
   (b)  Every non-trivial normal subgroup of $G$ has a non-trivial abelian quotient group.
   (c)  Every non-trivial factor group of $G$ has a non-trivial abelian normal subgroup.

9. Prove that a maximal subgroup $M$ of a finite solvable group $G$ has its index equal to a power of some prime.

## 5.3  Nilpotent Groups

The aim of this section is to introduce the reader to the study of nilpotent groups. We begin with:

**Definition 5.34**  A group $G$ is said to be *nilpotent* if it has a series of subgroups

$$\{e\} = G_0 \subseteq G_1 \subseteq \cdots \subseteq G_n = G$$

such that for every $1 \le i \le n$,

(1)  $G_i$ is a normal subgroup of $G$.
(2)  $G_i/G_{i-1} \le Z(G/G_{i-1})$.

Such series is called a *central series* of $G$. For a nilpotent group, the smallest $n$ such that $G$ has a central series of length $n$ is called the *nilpotency class* of $G$, and $G$ is said to be *nilpotent of class $n$*. Let $H$ and $K$ be subgroups of $G$. We recall that $[H, K] = \langle h^{-1}k^{-1}hk | h \in H, k \in K \rangle$.

**Lemma 5.35** *A series of $G$, say $\{e\} = G_0 \subseteq G_1 \subseteq \cdots \subseteq G_n = G$, is a central series if and only if for each $i = 1, \ldots, n$,*

$$[G_i, G] \le G_{i-1}.$$

**Proof** If the given series is a central series, then for each $i = 1, \ldots, n$, $G_{i-1}$ is a normal subgroup of $G$ and $G_i/G_{i-1} \le Z(G/G_{i-1})$. Then, for any $x \in G_i$ and $y \in G$,

$$(xG_{i-1})(yG_{i-1}) = (yG_{i-1})(xG_{i-1}),$$

that is $xyG_{i-1} = yxG_{i-1}$. Hence, $x^{-1}y^{-1}xy \in G_{i-1}$. This implies that $[G_i, G] \le G_{i-1}$.

Conversely, suppose that for each $i = 1, \ldots, n$, $[G_i, G] \le G_{i-1}$. Let $x \in G_i$ and $y \in G$. Then, $x^{-1}y^{-1}xy \in G_{i-1}$. In particular, since $G_{i-1} \le G_i$, if $x \in G_{i-1}$ then $y^{-1}xy \in G_{i-1}$. Thus, $G_{i-1}$ is a normal subgroup of $G$. Moreover, $xyG_{i-1} = yxG_{i-1}$ for every $x \in G_i$ and $y \in G$, and so $G_i/G_{i-1} \le Z(G/G_{i-1})$. Thus, the series is a central series. ∎

**Example 5.36** Any abelian group is nilpotent.

**Example 5.37** A central series for $Q_8$ is $\{1\} = G_0 \trianglelefteq G_1 = \{1, -1\} \trianglelefteq G_2 = Q_8$. Indeed, we notice that $G_1$ is the center of $Q_8$, and so it is normal in $Q_8$, which shows that this is a normal series. Now, we have

$$\frac{G_1}{G_0} \cong \{1, -1\} \subseteq Z\left(\frac{Q_8}{G_0}\right) \cong \{1, -1\} \text{ and } \frac{G_2}{G_1} \cong \frac{Q_8}{Z(Q_8)}.$$

Since $Q_8/Z(Q_8)$ is of order 4, it is commutative. Hence, we conclude that $Q_8/Z(Q_8) \subseteq Z(Q_8/Z(Q_8))$. Finally, the central series cannot be shortened, since $Q_8$ is not abelian. Consequently, the nilpotency class of $Q_8$ is 2.

**Theorem 5.38** *Any finite p-group is nilpotent.*

**Proof** Suppose that $G$ is a $p$-group and $|G| = p^k$, for some positive integer $k$. We use mathematical induction on $k$. If $k = 1$, then $G$ is abelian, and hence, it is nilpotent. Assume that the statement holds for all $k \le n$. We will prove that it holds for $k =$

$n + 1$. Since $G$ is a $p$-group, it follows that $Z(G) \neq \{e\}$. Since $|G/Z(G)| < p^{n+1}$, by hypothesis we conclude that $G/Z(G)$ is nilpotent. Let

$$\{Z(G)\} = G_0/Z(G) \subseteq G_1/Z(G) \subseteq \cdots \subseteq G_m/Z(G) = G/Z(G)$$

be the central series. Then, the series

$$\{e\} \subseteq Z(G) = G_0 \subseteq G_1 \subseteq \cdots \subseteq G_m = G$$

is a central series for $G$. Indeed, we have $G_i \trianglelefteq G$, for all $0 \leq i \leq m$, and

$$\left[ \frac{G_i}{Z(G)}, \frac{G}{Z(G)} \right] \leq \frac{G_{i-1}}{Z(G)}$$

which implies that $[G_i, G] \leq G_{i-1}$. Therefore, $G$ is nilpotent.  ∎

**Theorem 5.39** *If $G$ is nilpotent, then all subgroups and all quotient groups of $G$ are nilpotent.*

**Proof** Suppose that $G$ is nilpotent and $\{e\} = G_0 \subseteq G_1 \subseteq \cdots \subseteq G_n = G$ is a central series for $G$. Let $H$ be a subgroup of $G$ and $N$ be a normal subgroup of $G$. Then, we have

$$\{e\} = H \cap G_0 \subseteq H \cap G_1 \subseteq \cdots \subseteq H \cap G_n = H \tag{5.4}$$

and

$$\{N\} = G_0 N/N \subseteq G_1 N/N \subseteq \cdots \subseteq G_n N/N = G/N. \tag{5.5}$$

For each $i = 1, \ldots, n$, $[G_i, G] \leq G_{i-1}$. Hence,

$$[G_i \cap H, H] \leq H \cap [G_i, G] \leq H \cap G_{i-1}$$

and

$$[G_i N/N, G/N] = [G_i, G]N/N \leq G_{i-1}N/N.$$

Therefore, by Lemma 5.35, (5.4) and (5.5) are central series, so that $H$ and $G/N$ are nilpotent.  ∎

In the following lemma, we present a non-nilpotent group.

**Lemma 5.40** *$S_3$ is not nilpotent.*

**Proof** Assume that $S_3$ is nilpotent. Then, there exists a central series $\{e\} = G_0 \subseteq G_1 \subseteq \cdots \subseteq G_n = S_3$. We have $G_1/G_0 \leq Z(S_3/G_0)$. So, $G_1 \leq Z(S_3) = \{e\}$ which implies that $G_1 = \{e\}$. Similarly, we obtain $G_2 = \{e\}, \ldots, G_n = S_3 = \{e\}$, that is, a contradiction.  ∎

By the above lemma, we see that Theorem 5.26 is not true for nilpotent groups; i.e., if $N$ is a normal subgroup of a group $G$ and both $N$ and $G/N$ are nilpotent, then $G$ is not nilpotent, in general.

**Lemma 5.41**  *If two groups $H$ and $K$ are nilpotent, then $H \times K$ is nilpotent.*

*Proof*  If $H$ and $K$ are both nilpotent, then there are central series

$$\{e\} = H_0 \subseteq H_1 \subseteq \cdots \subseteq H_m = H,$$
$$\{e\} = K_0 \subseteq K_1 \subseteq \cdots \subseteq K_n = K.$$

By inserting repetition of terms if necessary, we may assume without loss generality that $m = n$. Then, we have

$$(\{e\} \times \{e\}) = (H_0 \times K_0) \subseteq (H_1 \times K_1) \subseteq \cdots \subseteq (H_n \times K_n) = G.$$

For each $i = 1, \ldots, n$,

$$[H_i \times K_i, G] = ([H_i, H] \times [K_i, K]) \leq (H_{i-1} \times K_{i-1}).$$

Now, by Lemma 5.35, we conclude that $G$ is nilpotent.                ∎

**Definition 5.42**  We define subgroups $\Gamma_n(G)$ and $Z_n(G)$ of $G$, respectively, as follows. Let $\Gamma_1(G) = G$ and $Z_0(G) = \{e\}$. Then, for each integer $n > 1$,

$$\Gamma_n(G) = [\Gamma_{n-1}(G), G]$$

and for each integer $n > 0$,

$$Z_n(G)/Z_{n-1}(G) = Z(G/Z_{n-1}(G)).$$

Then,

$$G = \Gamma_1(G) \geq \Gamma_2(G) \geq \Gamma_3(G) \geq \cdots,$$
$$\{e\} = Z_0(G) \leq Z_1(G) \leq Z_2(G) \leq \cdots$$

The first sequence is called the *lower central series* of $G$, and the second sequence is called the *upper central series* of $G$.

It is not difficult to see that the terms of the lower and upper central series are normal subgroups of $G$.

**Theorem 5.43**  *The following three statements are equivalent:*

*(1)  $G$ is nilpotent.*
*(2)  $\Gamma_n(G) = \{e\}$ for some integer $n$.*
*(3)  $Z_n(G) = G$ for some integer $n$.*

**Proof** If $\Gamma_n(G) = \{e\}$ for some $n$, then

$$G = \Gamma_1(G) \geq \Gamma_2(G) \geq \cdots \geq \Gamma_n(G) = \{e\}$$

is a central series of $G$, and so $G$ is nilpotent.

Similarly, if $Z_n(G) = G$ for some $n$, then

$$\{e\} = Z_0(G) \leq Z_1(G) \leq \cdots \leq Z_n(G) = G$$

is a central series of $G$, and so $G$ is nilpotent.

Conversely, assume that $G$ is nilpotent and

$$\{e\} = G_0 \subseteq G_1 \subseteq \cdots \subseteq G_n = G$$

is a central series of $G$. We prove first by induction on $i$ that $G_i \leq Z_i(G)$, for all $0 \leq i \leq n$. This is trivial for $i = 0$. Suppose that $i > 0$ and, inductively, that $G_{i-1} \leq Z_{i-1}(G)$. Then, $G_{i-1}Z_{i-1}(G) = Z_{i-1}(G)$. By hypothesis, $G_i/G_{i-1} \leq Z(G/G_{i-1})$. Therefore,

$$G_i Z_{i-1}(G)/Z_{i-1}(G) \leq Z(G/Z_{i-1}(G)) = Z_i(G)/Z_{i-1}(G).$$

Hence, $G_i \leq Z_i(G)$. Thus, the induction argument goes through. In particular, since $G = G_n \leq Z_n(G)$, we obtain $Z_n(G) = G$.

Now, we prove by induction on $j$ that

$$\Gamma_{j+1}(G) \leq G_{n-j},$$

for all $0 \leq j \leq n$. This is trivial for $j = 0$. Assume that $j > 0$ and, inductively, that $\Gamma_j(G) \leq G_{n-j+1}$. Then, by Lemma 5.35, $[G_{n-j+1}, G] \leq G_{n-j}$. Hence,

$$\Gamma_{j+1}(G) = [\Gamma_j(G), G] \leq [G_{n-j+1}, G] \leq G_{n-j}.$$

Again, the induction argument goes through. In particular, since $\Gamma_{n+1}(G) \leq G_0 = \{e\}$, it follows that $\Gamma_{n+1}(G) = \{e\}$. ∎

**Corollary 5.44** *Let $G$ be a nilpotent group. Then, for any central series of $G$, say $\{e\} = G_0 \subseteq G_1 \subseteq \cdots \subseteq G_n = G$, we have*

$$\Gamma_{n-i+1}(G) \leq G_i \leq Z_i(G),$$

*for all $0 \leq i \leq n$. Moreover, the least integer $r$ such that $\Gamma_{r+1}(G) = \{e\}$ is equal to the least integer $r$ such that $Z_r(G) = G$.*

**Proof** Suppose that $r$ is the least integer such that $Z_r(G) = G$. We show that $r$ is also the least integer such that $\Gamma_{r+1}(G) = \{e\}$. Let $\Gamma_{k+1}(G) = \{e\}$ for some $k < r$. We set $G_i = \Gamma_{k-i+1}(G)$. Then,

$$G_k = \Gamma_1(G), \ldots, G_0 = \Gamma_{k+1}(G) = \{e\}.$$

Hence, $\{e\} = G_0 \subseteq G_1 \subseteq \cdots \subseteq G_k = G$ is a central series. Now, by Theorem 5.43, it follows that $Z_k(G) = G$; this is contrary to the definition of $r$. ■

We conclude that if $G$ is a nilpotent group, then the lower central series of $G$ is its most rapidly descending central series and the upper central series of $G$ is its most rapidly ascending central series.

**Lemma 5.45** *For each non-negative integer $i$, $G^{(i)} \leq \Gamma_{i+1}(G)$.*

*Proof* We prove by induction on $i$. If $i = 0$, then $G^{(0)} = G = \Gamma_1(G)$. Suppose that $i > 0$ and, inductively, that $G^{(i-1)} \leq \Gamma_i(G)$. Then,

$$G^{(i)} = [G^{(i-1)}, G^{(i-1)}] \leq [\Gamma_i(G), G] = \Gamma_{i+1}(G).$$

This completes the proof. ■

**Theorem 5.46** *Each nilpotent group is solvable.*

*Proof* Suppose that $G$ is a nilpotent group. By Corollary 5.44, there exists a nonnegative integer $r$ such that $\Gamma_{r+1} = \{e\}$. By Lemma 5.45, we get $G^{(i)} \leq \Gamma_{i+1}(G) = \{e\}$. Now, by Theorem 5.32, we conclude that $G$ is solvable. ■

Lemma 5.40 shows that the converse of Theorem 5.46 is not true in general.

**Exercises**

1. Show that the family of nilpotent groups is not closed under extensions: There is a non-nilpotent group $G$ with a normal subgroup $N$ such that $N$ and $G/N$ are both nilpotent.
2. If $G$ is a nilpotent group of class 2 and if $a \in G$, prove that a function $f : G \to G$ defined by $f(x) = [a, x]$, for all $x \in G$, is a homomorphism. Conclude, in this case, $C_G(a) \trianglelefteq G$.
3. Let $G$ be a nilpotent group whose order is the product of $k$ prime numbers (either equal or distinct). Prove that the nilpotency class for $G$ does not exceed $k - 1$.
4. Prove that every finite nilpotent group $G$ has a sequence of normal subgroups

$$\{e\} = H_0 \subseteq H_1 \subseteq \cdots \subseteq H_n = G$$

   such that factor group $H_{i+1}/H_i$, for all $i = 0, \ldots, n - 1$, is cyclic.
5. Let $p_1$, $p_2$ and $p_3$ be distinct prime numbers, and let $G$ be a nilpotent group of order $p_1 p_2 p_3$. Prove that $G$ is abelian.
6. f $G$ is a nilpotent group of class $n$, prove that $G/Z(G)$ is nilpotent of class $n - 1$.
7. If $G$ is a nilpotent group and $H$ is a minimal normal subgroup of $G$, prove that $H$ is a subgroup of $Z(G)$.

## 5.4 Worked-Out Problems

**Problem 5.47** Prove that the infinite cyclic group $G$ has no composition series.

*Solution* Let $G = \langle a \rangle$ be the infinite cyclic group. Assume that there is a composition series for $G$, i.e., $\{e\} = H_0 \trianglelefteq H_1 \trianglelefteq \cdots \trianglelefteq H_{n-1} \trianglelefteq H_n = G$. Note that $n \neq 1$; otherwise $\{e\} = H_0 \trianglelefteq G$ yields that $G$ is simple, and clearly $G$ is not simple, because for example $\langle a^2 \rangle$ is a proper normal subgroup of $G$. Hence, $n \geq 2$ and the composition series contains a subgroup $G_{n-1}$ different than $\{e\}$ and $G$. Consequently, $G_{n-1}$ is a non-trivial subgroup of $G$, which means it is of the form $G_{n-1} = \langle a^k \rangle$, for some positive integer $k$. But then, $G_{n-1}$ is an infinite cyclic group, so it cannot be simple, which contradicts the definition of composition series. ∎

**Problem 5.48** Show that the group $GL_n(\mathbb{R})$ has a normal series, but no composition series.

*Solution* The series $\{I_n\} \trianglelefteq Z\big(GL_n(\mathbb{R})\big) \trianglelefteq SL_n(\mathbb{R}) \trianglelefteq GL_n(\mathbb{R})$ is a normal series; however $GL_n(\mathbb{R})$ cannot have a composition series, because it contains a normal subgroup isomorphic to the infinite cyclic group (look at matrices of the form a scalar times the identity matrix, which clearly commute with every matrix in $GL_n(\mathbb{R})$). ∎

**Problem 5.49** If $G$ is a group of order $3 \cdot 2^k$, prove that $G$ is solvable.

*Solution* We apply mathematical induction on $k$. If $k = 1$, then $|G| = 6$. Hence, $G$ has only one Sylow 3-subgroup $H$ which is normal, abelian, of order 3 and the quotient $G/H$ is abelian of order 2. So, it holds for $k = 1$. Suppose that the statement is true for $k \leq n$. We will prove that it holds for $k = n + 1$. By Sylow's theorems, we know that $G$ contains at least on Sylow 2-subgroup of order $2^{k+1}$, say $H$. Since $[G : H] = 3$, it follows that $2^{k+1} \cdot 3$ does not divide $3! = 6$. So, $H$ contains a normal subgroup $K$ of $G$. But $|K| = 2^m$; i.e., it is a 2-group, and so it is solvable. On the other hand, $|G/K| = 3 \cdot 2^{k-m}$, and so by induction, assumption $G/K$ is solvable. Finally, by Theorem 5.26, we conclude that $G$ is solvable. This completes the proof. ∎

**Problem 5.50** If $G$ is a group of order $2^2 \cdot 3^k$, prove that $G$ is solvable.

*Solution* We use mathematical induction on $k$. If $k = 1$, then $|G| = 2^2 \cdot 3$. So, by Problem 5.49, we deduce that $G$ is solvable. Next, assume that the statement is true for $k \leq n$. We will prove that it holds for $k = n + 1$. It is clear that $n + 1 \geq 2$. By Sylow's theorems we know that $G$ contains at least one Sylow 3-subgroup of order $3^{n+1}$, let us say $H$. Since $[G : H] = 2^2 = 4$, it follows that $3^{n+1} \cdot 2^2 = 36 \cdot 3^{n-1}$ does not divide 24. So, $H$ contains a normal subgroup $K$ of $G$. But $|K| = 3^m$; i.e., it is a 3-group, and so it is solvable. On the other hand, $|G/K| = 2^2 \cdot 3^{k-m}$. Now, by induction assumption, we conclude that $G/K$ is solvable. Therefore, by Theorem 5.26, $G$ is solvable. ∎

**Problem 5.51** If $G$ is a nilpotent group and $H$ is a proper subgroup of $G$, prove that $H$ is a proper subgroup of $N_G(H)$.

*Solution* Let $G$ be nilpotent of class $k$, and suppose that $n$ is the maximal integer such that $Z_n(G) \le H$. Since $H$ is a proper subgroup of $G$, it follows that $n \le k - 1$. Assume that $a \in Z_{n+1}(G)$ such that $a \notin H$. By the definition $Z_{n+1}(G)/Z_n(G)$ is the center of $G/Z_n(G)$. In particular, $aZ_n(G)$ commutes with $hZ_n(G)$, for all $h \in H$. Thus, we obtain $aha^{-1}Z_n(G) = hZ_n(G)$. This yields that $aha^{-1} = hz$, for some $z \in Z_n(G)$. Since $Z_n(G)$ is a subgroup of $H$, it follows that $aha^{-1} \in H$. Therefore, we conclude that $a \in N_G(H)$. This shows that $H$ is a proper subgroup of $N_G(H)$. ∎

**Problem 5.52** Prove that a finite group $G$ is nilpotent if and only if $G$ is an internal direct product of its Sylow subgroups, so that $G = P_1 \times \cdots \times P_k$, where $p_1, \ldots, p_k$ are distinct primes dividing $|G|$ and $P_i$ is the Sylow $p_i$-subgroup of $G$, for all $1 \le i \le k$.

*Solution* Let $G$ be nilpotent, and suppose that $p$ is a prime dividing $|G|$. Let $P$ be a Sylow $p$-subgroup of $G$. Since all Sylow $p$-subgroups of $N_G(P)$ are conjugate and $P \trianglelefteq N_G(P)$, it follows that $P$ is the unique Sylow $p$-subgroup of $N_G(P)$. Hence, $P$ is the only subgroup of $N_G(P)$ with $|P|$ elements, and so $P$ is a characteristic subgroup of $N_G(P)$. Consequently, $P$ is a normal subgroup of $N_G(N_G(P))$. This implies that $N_G(P) = N_G(N_G(P))$, and so by Problem 5.51, we conclude that $N_G(P) = G$. Therefore, $P$ is a normal subgroup of $G$. This is true for all primes dividing $|G|$, and so by Theorem 4.37, $G$ is an internal direct product of its Sylow subgroups.

Conversely, let $G$ be the internal direct product of its Sylow subgroups. By definition, each $P_i$ is a normal subgroup of $G$. For any normal subgroup $N$ of $G$, by Theorem 2.25, it follows that $P_i N/N$ is a normal Sylow $p_i$-subgroup of $G/N$. Hence, by Theorem 4.37, $G/N$ is also an internal direct product of its Sylow subgroups. We know that $Z(P_1 \times \cdots \times P_k) = Z(P_1) \times \cdots \times Z(P_k)$. Taking $N$ to be $Z(G)$ and using the fact that $Z(G) \ne \{e\}$, it follows that by induction $G$ is nilpotent. ∎

## 5.5 Supplementary Exercises

1. If $G$ is a finite group having a normal series with factor groups $H_0, H_1, \ldots, H_n$, prove that

$$|G| = \prod_{i=0}^{n} |H_i|.$$

2. Let $G = H_1 \times \cdots \times H_n = K_1 \times \cdots \times K_m$, where each $H_i$ and $K_j$ are simple. Prove that $m = n$ and there is a permutation $\sigma \in S_n$ with $K_{i\sigma} \cong H_i$, for all $i$. *Hint:* Construct composition series for $G$.

3. Prove that every group of order 275 is solvable.

4. Prove that every group of order 100 is solvable.

5. Prove that every group of order 84 is solvable.

6. Prove that every finite group $G$ has a unique maximal normal solvable subgroup $M$; moreover, $G/M$ has no non-trivial normal solvable subgroups.

7. Prove that a finite group is solvable if and only if the factors of a composition series are cyclic groups having prime orders.

8. If $p$ and $q$ are primes with $p < q$, prove that every group of order $pq^n$ is solvable. *Hint:* Use Sylow's theorems.

9. Let $G$ and $H$ be groups such that $G' \cong H'$ and $G/G' \cong H/H'$. Are $G$ and $H$ necessarily isomorphic?

10. Prove that the following are equivalent:

   (a) Any finite group of odd order is solvable.
   (b) Any finite non-abelian simple group has even order.

11. Prove that a finite group $G$ is solvable if and only if $H \neq H'$, for all subgroups $H \neq \{e\}$ of $G$.

12. Prove that if $G$ contains a non-abelian simple subgroup $H$, then $G$ is not solvable.

13. Prove that every group of order less than 60 is solvable.

14. Let $G_i$ be a $p_i$-group (for all $i = 1, \ldots, n$), where all of the primes $p_1, \ldots, p_n$ are distinct. Let $H$ be the set of all sequences $(a_1, \ldots, a_n)$, where $a_i \in G_i$ (for all $i = 1, \ldots, n$). Define an operation on $H$ by $(a_1, \ldots, a_n)(b_1, \ldots, b_n) = (a_1 b_1, \ldots, a_n b_n)$. Prove that $H$ is a nilpotent group and that its nilpotency class is equal to the largest of the nilpotency classes of the $G_i$.

15. Prove that in a finite group $G$ the set of all normal subgroups of $G$ which are nilpotent groups has a universally maximal (under inclusion) element, i.e., a nilpotent normal subgroup which contains all other nilpotent normal subgroups.

16. Prove that the dihedral group of order $2n$ is nilpotent if and only if $n$ is a power of 2.

17. Let $G$ be a finite nilpotent group of order $n$. If $m | n$, prove that $G$ has a subgroup of order $m$.

18. Let $G$ be a finite group. Prove that $G$ is nilpotent if and only if every non-trivial factor group of $G$ has a non-trivial center.

19. Let $G$ be nilpotent but not abelian. Let $H \leq G$ be maximal with respect to being normal in $G$ and abelian. Prove that $H = C_G(H)$.

20. Prove that

   (a) Every group of order $3^2 \cdot 5 \cdot 17$ is abelian.
   (b) Every group of order $3^2 \cdot 5 \cdot 17$ is nilpotent.

21. If $G$ is a nilpotent permutation group of degree $n$, prove that the order of $G$ is at most $2^{n-1}$.

22. Prove that in a finite nilpotent group, each Sylow subgroup is normal.

23. If $G$ is a finite group in which every proper subgroup is nilpotent, prove that $G$ is solvable.

24. If $G$ is a nilpotent transitive permutation group of degree $n$, prove that the order of $G$ divides

$$\prod_{p|n} p^{\frac{p^k-1}{p-1}},$$

where $k$ is the highest power to which $p$ divides $n$ and the product is taken over all primes $p$ which divide $n$.

# Chapter 6
# Free Groups and Presentations

In this chapter, after a short introduction of categories and free objects, we prove that free groups exist in the category of groups. We shall use these to develop a method of describing groups in terms of "generators and relations." Indeed, a group is free if no relation exists between its group generators other than the relationship between an element and its inverse required as one of the defining properties of a group.

## 6.1  Categories and Free Objects

In every day speech, we think of a category as a kind of thing. A category consists of a collection of things, all of which are related in some way. In mathematics, a category can also be construed as a collection of things and a type of relationship between pairs of such things. Indeed, a category is a system of related objects. The objects do not live in isolation: There is some notion of map between objects, binding them together.

**Definition 6.1**  A *category* $\mathcal{C}$ consists the following data:

(1) A set $Ob(\mathcal{C})$ of objects (denoted $A, B, C, \ldots$);
(2) A class of disjoint sets, denoted $Hom(A, B)$, one for each pair of objects; an element $f \in Hom(A, B)$ is called a *morphism* from $A$ to $B$ and is denoted $f : A \to B$;
(3) For all triple $A, B, C \in Ob(\mathcal{C})$, a function

$$Hom(A, B) \times Hom(B, C) \to Hom(A, C)),$$

this function is written $(f, g) \mapsto g \circ f$ and $g \circ f ; A \to C$ is called the *composite* of $f$ and $g$.

© The Author(s), under exclusive license to Springer Nature Singapore Pte Ltd. 2021     125
B. Davvaz, *Groups and Symmetry*, https://doi.org/10.1007/978-981-16-6108-2_6

These data are subject to the following axioms:

(i) Associativity: $h \circ (g \circ f) = (h \circ g) \circ f$, for all $(f, g, h) \in Hom(A, B) \times Hom$
$(B, C) \times Hom(C, D)$.

(ii) Identity: For each $B \in Ob(C)$ there exists a morphism $1_B : B \to B$ such that
for any $f \in Hom(A, B)$ and $g \in Hom(B, C)$,

$$1_B \circ g = g \text{ and } f \circ 1_B = f.$$

**Definition 6.2** A *diagram* in a category $C$ is a directed graph with vertices labelled
by objects, and edges labelled by arrows of $C$. A diagram is commutative if, for every
pair of vertices $A$ and $B$ belonging to it, the labels along any two directed paths from
$A$ to $B$ compose to give the same morphism in $C$.

Each composable pair of morphisms $A \xrightarrow{f} B \xrightarrow{g} C$ in $C$ corresponds to a unique
commutative triangle (see Fig. 6.1),

where $h = g \circ f$ in $C$. The identity axiom for categories is equivalent to the
commutativity of the diagram (Fig. 6.2)

for all morphism $f : A \to B$ in $C$, and the associativity axiom is expressed by the
commutativity of the diagram (Fig. 6.3).

for all composable triples $(h, g, f)$. The red arrow here is the triple composition
$h \circ (g \circ f) = (h \circ g) \circ f$, which may be denoted simply as $h \circ g \circ f$.

**Definition 6.3** In a category $C$ a morphism $f : A \to B$ is called an *isomorphism*
if there exists a morphism $g : B \to A$ such that $g \circ f = 1_A$ and $f \circ g = 1_B$. The
composite of two isomorphism, when defined, is an isomorphism. In this case, we
say that the morphism $f$ is invertible and that $g$ is the inverse of $f$. We may also say
that the objects $A$ and $B$ are isomorphic.

**Lemma 6.4** *Let $C$ be a category and let $\sim$ be the relation on $Ob(C)$ given by saying
$A \sim B$ if $A$ and $B$ are isomorphic. Then, $\sim$ is an equivalence relation.*

**Fig. 6.1** Unique
commutative triangle for
composition of morphisms

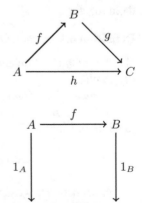

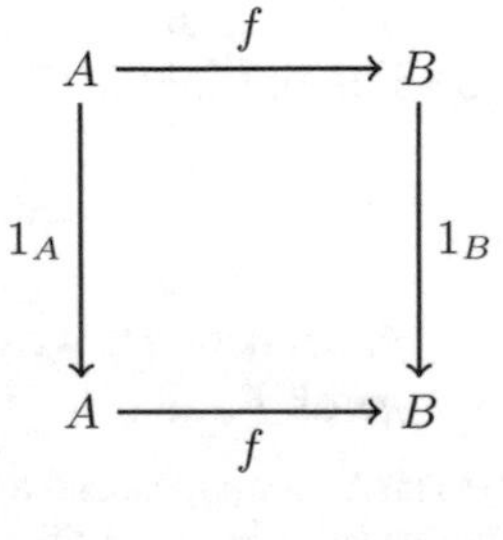

**Fig. 6.2** Identity axiom

**Fig. 6.3** Associativity
axiom

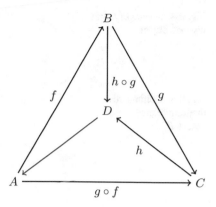

*Proof* It is straightforward.                                                              ■

Almost every known example of a mathematical structure with the appropriate
structure-preserving map yields a category.

**Example 6.5** The category $S$ with objects sets and morphisms the usual functions.
There are variants here: One can consider injective functions instead or again surjective functions. In each case, the category thus constructed is different.

**Example 6.6** The category $G$ with objects groups and morphisms group homomorphisms

**Example 6.7** The category $V$ with objects vector spaces and morphisms linear transformations.

**Example 6.8** The category $R$ with objects rings (with unit) and morphisms ring
homomorphisms.

These examples nicely illustrates how category theory treats the notion of structure
in a uniform manner. Note that a category is characterized by its morphisms, and not
by its objects.

**Example 6.9** A group is essentially the same thing as a category that has only one
object and in which all the morphisms are isomorphisms.

To understand this, first consider a category $C$ with just one object. We denote the
object $A$. Then, $C$ consists of a set, an associative composition function

$$Hom(A, A) \times Hom(A, A) \rightarrow Hom(A, A)$$

and an identity $1_A \in Hom(A, A)$. This would make $Hom(A, A)$ into a group, except
that we have not mentioned inverses. However, to say that every morphism in $C$ is

**Fig. 6.4** A category with
just one object

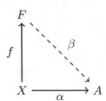

**Fig. 6.5** Commutative
diagram related to free
object $F$

an isomorphism is exactly to say that every element of $Hom(A, A)$ has an inverse
with respect to $\circ$. If we write $G$ for the group $Hom(A, A)$, then the situation is this:

| category $\mathcal{C}$ with single object $A$ | corresponding group $G$ |
|---|---|
| morphisms in $\mathcal{A}$ | elements of $G$ |
| $\circ$ in $\mathcal{C}$ | $\cdot$ in $G$ |
| $1_A$ | $e \in G$ |

The category $\mathcal{C}$ looks something like Fig. 6.4. The arrows represent different morphisms $A \to A$, that is, different elements of the group $G$.

All of the categories in the above examples are *concrete categories*, that is, categories in which objects are sets with additional structure, morphisms are structure-
preserving functions, and the composition law is ordinary composition of functions.

**Definition 6.10** Let $F$ be an object in a concrete category $\mathcal{C}$, $X$ a non-empty set, and
$f : X \to F$ be a function (of sets). Then, $F$ is said to be a *free object* provided that
for any object $A$ of $\mathcal{C}$ and function $\alpha : X \to A$, there corresponds a unique morphism
$\beta : F \to A$ in $\mathcal{C}$ such that $\alpha = \beta \circ f$: this equation expresses the commutativity of
Fig. 6.5.

**Example 6.11** Let $G$ be any group and $x \in G$. Then, the function $\beta : \mathbb{Z} \to G$
defined by $\beta(x) = x^n$ is easily seen to be the unique homomorphism $\mathbb{Z} \to G$ such
that $\beta(1) = x$. Hence, if $X = \{1\}$ and $f : X \to \mathbb{Z}$ is the inclusion function, then $\mathbb{Z}$
is free on $X$ in the category of groups. In other words, in order to determine a unique
homomorphism from $\mathbb{Z}$ to $G$, we need only specify the image of $1 \in \mathbb{Z}$. The additive
group $\mathbb{Q}$ of rational numbers does not have this property. It is easy to see that there
is no non-trivial homomorphism $\mathbb{Q} \to S_3$. Consequently, for any set $X$, function
$f : X \to Q$ and function $\alpha : X \to S_3$ with $f(a) \neq \langle id \rangle$ for some $a \in X$, there is no
homomorphism $\beta : \mathbb{Q} \to S_3$ with $\beta \circ f = \alpha$.

**Exercises**

1. The objects of $\mathcal{R}$ are sets, and a morphism $f : A \to B$ is a relation from $A$ to $B$, i.e., $f \subseteq A \times B$. The identity relation $\{(a, a) \in A \times A \mid a \in A\}$ is the identity morphism on a set $A$. Composition in $\mathcal{R}$ is to be given by:

$$g \circ f = \{(a, c) \in A \times C \mid \exists b ((a, b) \in f \text{ and } (b, c) \in g)\},$$

   for $f \subseteq A \times B$ and $g \subseteq B \times C$. Show that $\mathcal{R}$ is a category.
2. A *pointed set* is a pair $(S, x)$ with $S$ a set and $x \in S$. A morphism of pointed sets $(S, x) \to (S', x')$ is a triple $(f, x, x')$, where $f : S \to S'$ is a function such that $f(x) = x'$. Show that the pointed sets form a category.
3. A *preorder* is a set $P$ equipped with a binary relation $\leq$ that is both reflexive and transitive. Show that any preorder $P$ can be regarded as a category by taking the objects to be the elements of $P$ and taking a unique arrow, $a \to b$ if and only if $a \leq b$.
4. Given categories $\mathcal{A}$ and $\mathcal{B}$, there is a *product category* $\mathcal{A} \times \mathcal{B}$, in which an object of the product category $\mathcal{A} \times \mathcal{B}$ is a pair $(A, B)$, where $A \in \mathrm{Ob}(\mathcal{A})$ and $B \in \mathrm{Ob}(\mathcal{B})$. A morphism $(A, B) \to (A', B')$ in $\mathcal{A} \times \mathcal{B}$ is a pair $(f, g)$, where $f \in Hom(A, A')$ in $\mathcal{A}$ and $g \in Hom(B, B')$ in $\mathcal{B}$. There is only one sensible way to give the definitions of composition and identities in $\mathcal{A} \times \mathcal{B}$; write it down.

## 6.2 Free Groups

We show that free groups exist and a free group is a group which is freely generated by some set. We shall see this observation in continue.

There is nothing in the definition to show that free groups actually exist, a deficiency which will now be remedied.

Let $X$ be a set, and $X^{-1}$ be a set disjoint from $X$ such that $|X| = |X^{-1}|$. We can consider $X^{-1} = \{x^{-1} \mid x \in X\}$.

**Definition 6.12** A *word* on a set $X$ is a sequence $w = (x_1, x_2, \ldots)$, where $x_i \in X \cup X^{-1} \cup 1\}$ for all $i$, such that all $a_i = 1$ from some point on, i.e., there is a non-negative integer $n$ with $a_i = 1$, for all $i > n$. In particular, the constant sequence $(1, 1, 1, \ldots)$ is a word, called the *empty word*, and it is denoted by 1.

Since words contain only a finite number of letters before they become constant, we use more suggestive notation for non-empty words:

$$w = x_1^{\varepsilon_1} x_2^{\varepsilon_2} \ldots x_n^{\varepsilon_n},$$

where $x_i \in X$ and $\varepsilon_i = \pm$ or 0.

**Definition 6.13** Two words are to be considered equal if and only if they have the same elements in corresponding positions.

A word $w$ on $X$ is *reduced* if either $w$ is empty or $w = x_1^{\varepsilon_1} x_2^{\varepsilon_2} \ldots x_n^{\varepsilon_n}$, all $\varepsilon_i = \pm 1$, and $x$ and $x^{-1}$ are never adjacent.

We assume also that the empty word is reduced.

**Definition 6.14** If $w = x_1^{\varepsilon_1} x_2^{\varepsilon_2} \ldots x_n^{\varepsilon_n}$ is a word, then its *inverse* is the word $w^{-1} = x_n^{-\varepsilon_n} \ldots x_2^{-\varepsilon_2} x_1^{-\varepsilon_1}$.

We may define a multiplication on the set of reduced words by juxtaposition, i.e., if $w = x_1^{\varepsilon_1} x_2^{\varepsilon_2} \ldots x_n^{\varepsilon_n}$ and $w' = y_1^{\tau_1} y_2^{\tau_2} \ldots y_m^{\tau_m}$, then

$$ww' = x_1^{\varepsilon_1} x_2^{\varepsilon_2} \ldots x_n^{\varepsilon_n} y_1^{\tau_1} y_2^{\tau_2} \ldots y_m^{\tau_m},$$

with the convention that $w1 = 1w = w$. We observe that this multiplication does not define a binary operation on the set of all reduced words on $X$, because $ww'$ need not be reduced. One can define a new multiplication of reduced words $w$ and $w'$ as the reduced words obtained from $ww'$ after cancelation. More precisely, if $w = x_1^{\varepsilon_1} x_2^{\varepsilon_2} \ldots x_n^{\varepsilon_n}$ and $w' = y_1^{\tau_1} y_2^{\tau_2} \ldots y_m^{\tau_m}$ are non-empty reduced words on $X$ with $n \leq m$, let $k$ be the largest integer ($0 \leq k \leq n$) such that $x_{n-j}^{\varepsilon_{n-j}} = y_{j+1}^{-\tau_{j+1}}$, for $j = 0, 1, \ldots, k-1$. Then, we define

$$\begin{aligned}
ww' &= x_1^{\varepsilon_1} x_2^{\varepsilon_2} \ldots x_n^{\varepsilon_n} y_1^{\tau_1} y_2^{\tau_2} \ldots y_m^{\tau_m} \\
&= \begin{cases}
x_1^{\varepsilon_1} \ldots x_{n-k}^{\varepsilon_{n-k}} y_{k+1}^{\tau_{k+1}} \ldots y_m^{\tau_m} & \text{if } k < n \\
y_{n+1}^{\tau_{n+1}} \ldots y_m^{\tau_m} & \text{if } k = n < m \\
1 & \text{if } k = m = n.
\end{cases}
\end{aligned}$$

If $m > n$, the multiplication is defined analogously. This definition insures that the multiplication of reduced words is a reduced word.

**Theorem 6.15** *If $X$ is a non-empty set and $F$ is the set of all reduced words on $X$, then $F$ is a group under the binary operation defined above and $F = \langle X \rangle$.*

**Proof** We need only verify associativity. This may be done by induction and a tedious examination of cases. Instead, we use the *van der Waerden trick*. For each $x \in X$ and $\varepsilon = \pm 1$, we consider the function $f_{x^\varepsilon} : F \to F$ defined by

$$x_1^{\varepsilon_1} \ldots x_n^{\varepsilon_n} \mapsto \begin{cases}
x^\varepsilon x_1^{\varepsilon_1} \ldots x_n^{\varepsilon_n} & \text{if } x^\varepsilon \neq x_1^{-\varepsilon_1} \\
x_2^{\varepsilon_2} \ldots x_n^{\varepsilon_n} & \text{if } x^\varepsilon = x_1^{-\varepsilon_1} \ (= 1 \text{ if } n = 1).
\end{cases}$$

Since $f_{x^\varepsilon} \circ f_{x^{-\varepsilon}}$ and $f_{x^{-\varepsilon}} \circ f_{x^\varepsilon}$ are both equal to the identity function on $F$, it follows that $f_{x^\varepsilon}$ is a permutation on $F$ with inverse $f_{x^{-\varepsilon}}$. Suppose that $S_F$ is the group of all permutations on $F$ and let $H$ be the subgroup of $S_F$ generated by $\{f_x \mid x \in X\}$. An arbitrary element $g \in H$ (other than the identity) has a factorization

$$g = f_{x_1^{\varepsilon_1}} \circ f_{x_2^{\varepsilon_2}} \circ \ldots \circ f_{x_n^{\varepsilon_n}},$$

where $\varepsilon_i = \pm 1$ and $f_{x^\varepsilon}$ and $f_{x^{-\varepsilon}}$ are never adjacent (or we can cancel). Such a factorization of $g$ is unique, because $g(1) = x_1^{\varepsilon_1} \ldots x_n^{\varepsilon_n}$, and we have already noted that the spelling of a reduced word is unique. Now, the function $\varphi : F \to H$ given by $1 \mapsto id_F$ and $x_1^{\varepsilon_1} \ldots x_n^{\varepsilon_n} \mapsto f_{x_1^{\varepsilon_1}} \circ \ldots \circ f_{x_n^{\varepsilon_n}}$ is clearly a surjection such that $\varphi(ww') = \varphi(w)\varphi(w')$, for all $w, w' \in F$. Since $1 \mapsto x_1^{\varepsilon_1} \ldots x_n^{\varepsilon_n}$ under the function $f_{x_1^{\varepsilon_1}} \circ \ldots \circ f_{x_n^{\varepsilon_n}}$, it follows that $\varphi$ is one to one. The fact that $H$ is a group implies that associativity holds in $F$ and that $\varphi$ is an isomorphism of groups. Finally, it is easy to see that $F = \langle X \rangle$. ∎

**Theorem 6.16** *Given a set X, there exists a free group with basis X.*

**Proof** Let $F$ be the set of all reduced words on $X$. By Theorem 6.15, $F$ is a group under juxtaposition. We show that $F$ is a free group with basis $X$. Let $f : X \to F$ be associated injection function. Assume that $G$ is any group and $\alpha : X \to G$ is any function. We define $\beta : F \to G$ by $\beta(1) = e$, and if $x_1^{\varepsilon_1} \ldots x_n^{\varepsilon_n}$ is a non-empty reduced word on $X$, define

$$\beta\left(x_1^{\varepsilon_1} \ldots x_n^{\varepsilon_n}\right) = \alpha(x_1)^{\varepsilon_1} \ldots \alpha(x_n)^{\varepsilon_n}.$$

Since $G$ is a group and $\varepsilon_i = \pm 1$, the product $\alpha(x_1)^{\varepsilon_1} \ldots \alpha(x_n)^{\varepsilon_n}$ is well defined element of $G$. It is easy to check that $\beta$ is a homomorphism such that $\alpha = \beta \circ f$. Now, if $\gamma : F \to G$ is any homomorphism such that $\alpha = \gamma \circ f$, then

$$\begin{aligned}
\gamma\left(x_1^{\varepsilon_1} \ldots x_n^{\varepsilon_n}\right) &= \gamma(x_1^{\varepsilon_1}) \ldots \gamma(x_n^{\varepsilon_n}) = \gamma(x_1)^{\varepsilon_1} \ldots \gamma(x_n)^{\varepsilon_n} \\
&= \left((\gamma \circ f)(x_1)\right)^{\varepsilon_1} \ldots \left((\gamma \circ f)(x_n)\right)^{\varepsilon_n} \\
&= \alpha(x_1)^{\varepsilon_1} \ldots \alpha(x_n)^{\varepsilon_n} = \beta\left(x_1^{\varepsilon_1} \ldots x_n^{\varepsilon_n}\right).
\end{aligned}$$

This shows that $\beta$ is unique. Therefore, by Definition 6.10, we conclude that $F$ is free with basis $X$. ∎

**Example 6.17** Suppose that the functions $f$ and $g$ of the complex plane are defined by $f(z) = z + 2$ and

$$g(z) = \begin{cases} \dfrac{z}{2z+1} & \text{if } z \neq -\dfrac{1}{2} \\ \dfrac{1}{2} & \text{if } z = -\dfrac{1}{2}. \end{cases}$$

Obviously, $f$ and $g$ are bijective functions. Hence, they generate a group $F$ of permutations of the complex plane. We show that $F$ is free on $\{f, g\}$. A nonzero power of $f$ maps interior of the unit circle to the exterior and a nonzero power of $g$ maps the exterior of the unit circle to interior with 0 removed. The second statement follows from $g(1/z) = 1/(z + 2)$. From this, we observe that no non-trivial reduced word on $\{f, g\}$ can equal 1. This yields that each element of $F$ has a unique expression as a reduced word. Thus, $F$ is free on $\{f, g\}$.

**Example 6.18** Let $\varphi_{a,b,c,d} : \mathbb{C} \to \mathbb{C}$ be a function defined by

$$\varphi_{a,b,c,d} : z \mapsto \frac{az+b}{cz+d},$$

where $ad - bc \neq 0$ and $a, b, c, d \in \mathbb{C}$. The functions presented in Example 6.17 are instance of this kind of complex functions. Such a mapping is known as a *linear fractional transformation*. The function

$$\begin{bmatrix} a & b \\ c & d \end{bmatrix} \mapsto \varphi_{a,b,c,d}$$

is a homomorphism from $GL_n(\mathbb{C})$ to the group of all linear fractional transformations of $\mathbb{C}$ in which

$$A = \begin{bmatrix} 1 & 0 \\ 2 & 1 \end{bmatrix} \quad \text{and} \quad B = \begin{bmatrix} 1 & 2 \\ 0 & 1 \end{bmatrix}$$

map to $f$ and $g$, respectively. Since no non-trivial reduced word on $\{f, \ g\}$ can equal to 1, it follows that the same is true for reduced word on $\{A, \ B\}$. This implies that the group $\langle A, B \rangle$ is free with basis $\{A, \ B\}$.

The extraordinary importance of free groups is underscored by the following result.

**Corollary 6.19** *Every group $G$ is a homomorphic image of a free group.*

**Proof** We construct a set $X = \{x_g \mid g \in G\}$ so that $\alpha : x_g \mapsto g$ is a one-to-one correspondence from $X$ to $G$. If $F$ is a free group with basis $X$, then there is a homomorphism $\beta : F \to G$ extending $\alpha$. Since $\alpha$ is onto, it follows that $\beta$ is an epimorphism. Therefore, we conclude that $G \cong F/Ker\beta$. ∎

The following theorem shows that free groups on sets of equal cardinality are isomorphic.

**Theorem 6.20** *Let $F_1$ and $F_2$ be free groups on $X_1$ and $X_2$, respectively. If $|X_1| = |X_2|$, then $F_1 \cong F_2$.*

**Proof** Suppose that $f_1 : X_1 \to F_1$ and $f_2 : X_2 \to F_2$ are the given injection, and let $\varphi : X_1 \to X_2$ be a one to one correspondence. Since $F_1$ and $F_2$ are free, it follows that there exist commutative diagrams (Fig. 6.6),
where $\beta_1$ and $\beta_2$ are homomorphism. Thus, we have

$$\beta_2 \circ \beta_1 \circ f_1 = \beta_2 \circ f_2 \circ \varphi = f_1 \circ \varphi^{-1} \circ \varphi = f_1,$$

and the diagram (Fig. 6.7) commutes. Since the identity function $id_{F_1}$ makes Fig. 6.7 commutative, it follows that $\beta_2 \circ \beta_1 = id_{F_1}$, by uniqueness. Similarly, we observe that $\beta_1 \circ \beta_2 = id_{F_2}$. This yields that $\beta_1$ is an isomorphism, and consequently $F_1 \cong F_2$. ∎

**Fig. 6.6** Commutative diagrams

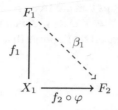

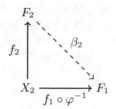

**Fig. 6.7** Commutative diagram

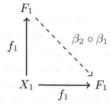

**Fig. 6.8** Projective properties of free groups

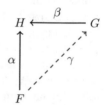

**Theorem 6.21** (Projective Properties of Free Groups). *Let $F$ be a free group and suppose that $G$ and $H$ are two other groups. If $\alpha : F \to H$ is a homomorphism and $\beta : G \to H$ is an epimorphism, then there exists a homomorphism $\gamma : F \to G$ such that $\beta \circ \gamma = \alpha$; in other words, the diagram (Fig. 6.8) is commutative.*

**Proof** Assume that $F$ is a free group with basis $X$. If $x \in X$, then $\alpha(x) \in H = Im\beta$. Since $\beta$ is onto, it follows that there exists $g_x \in G$ such that $\beta(g_x) = \alpha(x)$. Now, we extend the function $x \mapsto g_x$ to a homomorphism $\gamma : F \to G$. Since $\alpha(x) = \beta(g_x) = (\beta \circ \gamma)(x)$, for all $x \in X$, and $F = \langle X \rangle$, it follows that $\beta \circ \gamma = \alpha$. ∎

## Exercises

1. Show that each word can be simplified to only one reduced word.
2. Show that every non-identity element in a free group $F$ has infinite order.
3. Show that the free group on the set $\{x\}$ is an infinite cyclic group, and hence isomorphic to $\mathbb{Z}$.
4. Let $F$ be a free group and $N$ be the subgroup generated by the set $\{x^n \mid x \in F,\ n\ is\ a\ fixed\ integer\}$. Show that $N \trianglelefteq F$.
5. Let $F$ be a free group and let $H \leq F$ have finite index. Show that $H$ intersects every non-trivial subgroup of $F$ non-trivially.
6. Prove that any free group is a free product of infinite cyclic groups.

7. Characterize the abelian groups that are free groups.
8. Let $X$ be the disjoint union $X = Y \cup Z$. If $F$ is a free group with basis $X$ and $N$ is the normal subgroup generated by $Y$, prove that $F/N$ is a free group with basis $\{zN \mid z \in Z\}$.


## 6.3   Generators and Relations

We now present a convenient way to define a group with certain prescribed properties. Simply put, we begin with a set of elements that we want to generate the group, and a set of equations (called relations) that specify the conditions that these generators are to satisfy. Among all such possible groups, we will select one that is as large as possible. This will uniquely determine the group up to isomorphism. To provide motivation for the theory involved, we begin with a concrete example. We have worked with the symmetric group $S_3$ in the following perspective. Assume that $x = (1\,2\,3)$ and $y == (1\,2)$ and write the elements of $S_3$ as $\{e,\ x,\ x^2,\ y,\ xy,\ x^2y\}$. Computing in $S_3$, this way is aided by remembering the following equations: $x^3 = e$, $y^2 = e$, and $yx = x^2y$. Using these equations, any finite length product of $x$, $y$, $x^{-1}$ and $y^{-1}$ can be rewritten as one of the six elements above. What we have secretly done is this: We have built a group presentation for $S_3$. We record this using the following notation: $S_3 = \langle x, y \mid x^3 = e,\ y^2 = e,\ yx = x^2y \rangle$. We say that the right side of the equation above is a presentation for $S_3$. The elements $x$ and $y$ are the generators in this presentation, and the equations $x^3 = e$, $y^2 = e$ and $yx = x^2y$ the relations in this presentation.

**Definition 6.22** Let $X$ be a set and let $Y$ be a set of reduced words on $X$. A group $G$ has *generators* $X$ and *relations* $Y$ if $G \cong F/N$, where $F$ is the free group with basis $X$ and $N$ is the normal subgroup of $F$ generated by $Y$. The ordered pair $\langle X|Y \rangle$ is called a *presentation* of $G$.

A relation $w \in Y$ is often written as $w = e$ to convey its significance in the quotient group $G$ presented.

The following theorem is frequently useful in the discussion of groups with similar presentations.

**Theorem 6.23** (von Dyck's Theorem). *Let $X$ be a set and let $Y$ be a set of reduced words on $X$, and suppose that $G$ is the group defined by the generators $x \in X$ and relations $w = e$, where $w \in Y$. If $H$ is any group such that $H = \langle X \rangle$ and $H$ satisfies the relations $w = e$, for all $w \in Y$, then there is an epimorphism $G \to H$.*

**Proof** If $F$ is the free group on $X$, then the inclusion $X \to H$ induces an epimorphism $\varphi : F \to H$. Since $H$ satisfies the relations $w = e$, for all $w \in Y$, it follows that $Y \subseteq Ker\varphi$. Therefore, we conclude that the normal subgroup $N$ generated by $Y$ in $F$ is contained in $Ker\varphi$. Now, $\varphi$ induces an epimorphism $F/N \to H$.  ∎

**Example 6.24** Suppose that $F$ is the free group with basis $\{x, \ y\}$, and let $N$ be the smallest normal subgroup of $F$ containing the set $\{x^4, \ y^2, \ (xy)^2\}$. We claim that $F/N \cong D_4$. We define $\varphi : F \rightarrow D_4$ by $\varphi(x) = R_0$ (rotation of $\pi/2$) and $\varphi(y) = S_0$ (horizontal reflection). Then, it is easy to see that $\varphi$ is a homomorphism such that $Ker\varphi \subseteq N$. This implies that $F/Ker\varphi \cong D_4$. Suppose that

$$K = \{N, \ xN, \ x^2N, \ x^3N, \ yN, \ xyN, \ x^2yN, \ x^3yN\}.$$

We show that $F/N \cong K$. Each element of $F/N$ can generated by starting $N$ and multiplying on the left by various combinations of $x$'s and $y$'s. Hence, it is enough to show that $K$ is closed under multiplication on the left by $x$ and $y$. It is clear that $K$ is closed under the left multiplication by $x$. For $y$, we must consider eight cases. For instance, we investigate only one of them. The others cases can be done analogously. Consider $y(xN)$. Since $y^2$, $xyxy$, $x^4 \in N$ and $Ny = yN$, it follows that

$$yxN = yxNy^2 = yayNy = x^{-1}(xyxy)Ny = x^{-1}Ny$$
$$= x^{-1}x^4Ny = x^3Ny = x^3yN.$$

After completing the other cases, we observe that $F/N$ has at most eight elements. On the other hand, the factor group $F/Ker\varphi$ has exactly eight elements. Since

$$\frac{F}{Ker\varphi} \cong \frac{F/N}{Ker\varphi/N},$$

it follows that $F/N$ has eight elements too. This yields that $F/N \cong F/Ker\varphi \cong D_4$.

**Example 6.25** Let $G = \langle x, y \mid x^2 = e$ and $y^2 = e \rangle$. This group is called the *infinite dihedral group* $D_\infty$. If $a = xy$, then $G = \langle x, a \rangle$ and $x^{-1}ax = yx = a^{-1}$. Conversely, the relations $x^2 = e$ and $y^2 = e$ are consequences of the relations $x^2 = e$ and $x^{-1}ax = a^{-1}$. Indeed, for given latter we have $y^2 = (x^{-1}a)^2 = x^{-1}axa = e$. Therefore, $G$ also has the presentation

$$\langle x, a \mid x^2 = e \text{ and } x^{-1}ax = a^{-1} \rangle.$$

Also, this group maybe considered as a semidirect product of a cyclic group of order 2 and an infinite cyclic group.

**Definition 6.26** A presentation $G = \langle X|Y \rangle$ is said to be *finitely generated* if $X$ is finite and *finitely related* if $Y$ is finite. If both are finite it is said to be a *finite presentation*.

**Example 6.27** Examples of finitely presented groups include cyclic groups, free groups of finite basis (which have no relations) and finite groups.

**Theorem 6.28** (Neumann's Theorem). *Let $X$ be a set of generators of a finitely presented group $G$. Then, $G$ has a finite presentation of the form*

$$\langle X_0 \mid r_1 = r_2 = \cdots = r_i = e \rangle,$$

*where $X_0 \subseteq X$.*

**Proof** Assume that $G = \langle y_1, \ldots, y_m \mid s_1 = s_2 = \cdots = s_l = e \rangle$ is a presentation of $G$. Let $G = \langle X_0 \rangle$, where $X_0 = \{x_1, \ldots, x_n\} \subseteq X$. Hence, there exist expressions for the $y_i$ in terms of $x_j$'s and $x_j$ in terms of the $y_i$'s, say $y_i = w_i(x)$ and $x_j = v_j(y)$. Consequently, we observe that the following relations hold:

$$s_k\big(w_1(x), \ldots, w_m(x)\big) = e \quad \text{and} \quad x_j = v_j\big(w_1(x), \ldots, w_m(x)\big),$$

where $k = 1, \ldots, l$ and $j = 1, \ldots n$. Now, let $\widehat{G}$ be a group with generators $\widehat{x}_1, \ldots, \widehat{x}_n$ and the above defining relations in $\widehat{x}_1, \ldots, \widehat{x}_n$. By von Dyck's theorem there exists an epimorphism $\widehat{G} \to G$ with $\widehat{x}_i \mapsto x_i$. We set $\widehat{y}_i = w_i(\widehat{x})$. The second set of defining relations shows that $\widehat{G} = \langle \widehat{y}_1, \ldots, \widehat{y}_m \rangle$. As $s_k(\widehat{y}) = e$, by von Dyck's theorem again we conclude that there exists an epimorphism $G \to \widehat{G}$ with $y_i \mapsto \widehat{y}_i$. These epimorphisms are mutually inverse, and consequently they are isomorphisms. Therefore, we conclude that $G$ is generated by $x_1, \ldots, x_n$ subject only to the defining relations in the $x_i$'s listed above.                                               ∎

## Exercises

1. Let $G$ be defined by some set of generators and relations. Show that every factor group of $G$ satisfies the relations defining $G$.
2. Show that the cyclic group of order 6 is the group defined by generators $a, b$ and relations $a^2 = b^3 = a^{-1}b^{-1}ab = e$.
3. Let $G = \langle a, b, c, d \mid ab = c, \ bc = d, \ cd = a \text{ and } da = b \rangle$. Determine $|G|$.
4. Show that $A_4$ has the presentation $\langle x, y \mid x^2 = y^3 = (xy)^3 = e \rangle$.
5. Show that $S_4$ has the presentation $\langle x, y \mid x^4 = y^2 = (xy)^3 = e \rangle$.
6. Prove that the group $G = \langle x, y \mid x^m = e \text{ and } y^n = e \rangle$ is infinite when $m, n \geq 2$.
7. Let $G = \langle x, y \mid x^8 = y^2 = e \text{ and } yxyx^3 = e \rangle$. Show that $|G| \leq 16$. Assuming that $|G| = 16$, find the center of $G$ and the order of $xy$.
8. Let $p$ be a prime number. Prove that $\langle x, y \mid x^p = y^p = (xy)^p = e \rangle$ is infinite if $p > 2$, but that if $p = 2$, it is Klein 4-group.
9. Find a presentation of the group

$$\langle a, b, c \mid abca = c, \ bcab = a \text{ and } cabc = c \rangle$$

in terms of the generators $x = abc$ and $y = ab$.

## 6.4  Coxeter Groups

Although presentations in general can be very difficult, particular kinds of presentations can be very convenient. Our aim will be primarily concerned with the following set of presentations.

**Definition 6.29**  A *Coxeter system* is a pair $(G, X)$ consisting of a group $G$ and a set of generators $X$ for $G$ subject only to relations of the form $(xy)^{f(x,y)} = e$, where

$$f(x, x) = 1, \text{ for all } x;$$
$$f(x, y) \geq 2;$$
$$f(x, y) = f(y, x)$$

(6.1)

When no relation occurs between $x$ and $y$, we set $f(x, y) = \infty$. Consequently, a Coxeter system is defined by a set $X$ and a function $f : X \times X \to \mathbb{N} \cup \{\infty\}$ satisfying (6.1), and the group $G = \langle X | Y \rangle$, where

$$Y = \{(xy)^{f(x,y)} \mid f(x, y) \neq \infty\}.$$

The *Coxeter groups* are those that arise as part of Coxeter system. The cardinality of $X$ is called the *rank* of the Coxeter system.

**Example 6.30**  Up to isomorphism, the only Coxeter system of rank 1 is $)C_2, \{x\})$.

**Example 6.31**  The Coxeter systems of rank 2 are indexed by $f(x, y) \geq 2$.

(1) If $f(x, y)$ is an integer $n$, then the Coxeter system is $(G, \{x, y\})$, where $G = \langle x, y \mid x^2 = y^2 = (xy)^n = e \rangle$, and so $G \cong D_n$;
(2) If $f(x, y) = \infty$, then the Coxeter system is $(G, \{x, y\})$, where $G = \langle x, y \mid x^2 = y^2 = e \rangle$

**Example 6.32**  Let $V = \mathbb{R}$ endowed with the standard positive definite symmetric bilinear form

$$\langle (x_i), (y_i) \rangle = \sum_{i=1}^{n} x_i y_i.$$

We recall that a reflection is a linear transformation $\sigma : V \to V$ sending some nonzero vector $\alpha$ to $-\alpha$ and fixing the points of hyperplane $H_\alpha$ orthogonal to $\alpha$. We write $\sigma_\alpha$ for the reflection defined by $\alpha$. It is given by

$$\sigma_\alpha(v) = v - \frac{2\langle v, \alpha \rangle}{\langle \alpha, \alpha \rangle} \alpha,$$

because this is correct for $v = \alpha$ and for $v \in H_\alpha$, and hence by linearity for all $v \in \langle \alpha \rangle \oplus H_\alpha$. A *finite reflection group* is a finite group generated by reflections. For such a group, it is possible to choose a set $X$ of generating reflections for which $(G, X)$ is a Coxeter system. Thus, the finite reflection groups are all Coxeter groups.

**Example 6.33** We know that the symmetric group $S_n$ is generated by transpositions. Since any transposition is a reflection, it follows that $S_n$ is a finite reflection group, and so a Coxeter group.

**Theorem 6.34** (Structure of Coxeter Groups). *Let $(G, X)$ be the Coxeter system defined by a function $f : X \times X \to \mathbb{N} \cup \{\infty\}$ satisfying (6.1). Then,*

*(1) The natural map $X \to G$ is injection;*
*(2) Each $x \in X$ has order 2 in $G$;*
*(3) For each $x \neq y$ in $X$, $xy$ has order $f(x, y)$ in $G$.*

**Proof** Suppose that $x, y \in X$. Since $(xx)^{f(x,x)} = e$, it follows that $x^2 = e$, and so the order of $x$ is 1 or 2. Also, for any $x, y \in X$, the order $xy$ divides $f(x, y)$, and hence, the theorem says that the elements of $X$ remain distinct in $G$ and that each $x$ and each $xy$ has the largest possible order.

Suppose that $\zeta : X \to \{1, -1\}$ is a function such that $\zeta(x) = -1$, for all $x \in X$, and so it extends to a homomorphism $G \to \{1, -1\}$. As $x$ maps to $-1$, it has order 2. This proves (2). In order to prove the other parts, we consider a vector space $V$ over $\mathbb{R}$ with basis $\{e_x\}_{x \in X}$. We define on $V$ a geometry for which there is distinct reflections $\sigma_x$, for $x \in X$, such that $\sigma_x \sigma_y$ has order $f(x, y)$. We extend the map $x \mapsto \sigma_x$ to a homomorphism $G \to GL(V, \mathbb{R})$. Since $\sigma_x$'s are distinct, it follows that $x$'s must be distinct in $G$. Moreover, since $\sigma_x \sigma_y$ has order $f(x, y)$, it follows that $xy$ in $G$ must have order $f(x, y)$, too. We define a function $\langle \, , \, \rangle : V \times V \to \mathbb{R}$ by

$$\langle e_x, e_y \rangle = \begin{cases} -\cos\left(\dfrac{\pi}{f(x, y)}\right) & \text{if } f(x, y) \neq \infty \\ -1 & \text{otherwise.} \end{cases}$$

It is not difficult to see that $\langle \, , \, \rangle$ is a symmetric bilinear form. Since $\langle e_x, e_y \rangle \neq 0$, it follows that the orthogonal complement of $e_x$ with respect to $\langle \, , \, \rangle$ is a hyperplane $H_x$. This allows us to define a reflection by the rule $\sigma_x(v) = v - 2\langle v, e_x \rangle e_x$, for all $v \in V$. It is clear that $\sigma_x$ is a linear map sending $e_x$ to its negative and fixing the elements of $H_x$, and so $\sigma_x^2 = id_V \in GL(V, \mathbb{R})$. In addition, we have

$$\langle \sigma_x(v), \sigma_y(w) \rangle = \langle v, w \rangle.$$

Obviously, the $\sigma_x$'s are distinct, and so it remains that $\sigma_x \sigma_y$ has order $f(x, y)$. For $x, y \in X$, assume that $V_{(x,y)}$ denote the two-dimensional vector space $\mathbb{R}e_x \oplus \mathbb{R}e_y$. It is clear that $\sigma_x$ and $\sigma_y$ both map $V_{(x,y)}$ into itself. Let $v = ae_x + be_y \in V_{(x,y)}$. If $f(x, y) \neq \infty$, then

$$\langle v, v \rangle = a^2 - 2ab\cos\left(\tfrac{\pi}{f(x,y)}\right) + b^2$$

$$= \left(a - b\cos\left(\tfrac{\pi}{f(x,y)}\right)\right)^2 + b^2 \sin^2\left(\tfrac{\pi}{f(x,y)}\right) > 0,$$

because $\sin\left(\pi/f(x, y)\right) \neq 0$. This means that the restriction of $\langle\ ,\ \rangle$ to $V_{(x,y)}$ is positive definite if $f(x, y) \neq \infty$. If $f(x, y) = \infty$, then $\langle v, v \rangle = a^2 - 2ab + b^2 = (a - b)^2 \geq 0$, and $\langle v, v \rangle = 0$ if $v = e_x + e_y$. Next, we show that the restriction of $\sigma_x\sigma_y$ to $V_{(x,y)}$ has order $f(x, y)$.

When $f(x, y) \neq \infty$, the form $\langle\ ,\ \rangle : V_{(x,y)} \times V_{(x,y)} \to \mathbb{R}$ is positive definite, and so $V_{(x,y)}$ together with $\langle\ ,\ \rangle$ is an Euclidian plane. In addition, $\sigma_x$ and $\sigma_y$ are the reflections in $V_{(x,y)}$ defined by the vectors $e_x$ and $e_y$. Since

$$\langle e_x, e_y \rangle = -\cos\left(\tfrac{\pi}{f(x,y)}\right) = \cos\left(\pi - \tfrac{\pi}{f(x,y)}\right),$$

it follows that the angle between the lines fixed by $e_x$ and $e_y$ is $\pi/f(x, y)$. Now, we observe that $\sigma_x$ and $\sigma_y$ generated a dihedral group $D_{f(x,y)}$ and that $\sigma_x\sigma_y$ has order $f(x, y)$. When $f(x, y) = \infty$, then relative to basis $\{e_x,\ e_y\}$, we have

$$\sigma_x = \begin{bmatrix} -1 & 2 \\ 0 & 1 \end{bmatrix} \text{ and } \sigma_y = \begin{bmatrix} 1 & 0 \\ 2 & -1 \end{bmatrix}.$$

Now, $\sigma_x\sigma_y = \begin{bmatrix} 3 & -2 \\ 2 & -1 \end{bmatrix}$, and so

$$\sigma_x\sigma_y \begin{bmatrix} 1 \\ 1 \end{bmatrix} = \begin{bmatrix} 1 \\ 1 \end{bmatrix} \text{ and } \sigma_x\sigma_y \begin{bmatrix} 1 \\ 0 \end{bmatrix} = \begin{bmatrix} 1 \\ 0 \end{bmatrix} + 2 \begin{bmatrix} 1 \\ 1 \end{bmatrix}.$$

Therefore, we conclude that

$$\left(\sigma_x\sigma_y\right)^m \begin{bmatrix} 1 \\ 0 \end{bmatrix} = \begin{bmatrix} 1 \\ 0 \end{bmatrix} + 2m \begin{bmatrix} 1 \\ 1 \end{bmatrix},$$

which shows that $\sigma_x\sigma_y$ has infinite order. This completes the proof. ∎

**Exercises**

1. By $PGL_2(\mathbb{Z})$ we denote the quotient of the group $GL_2(\mathbb{Z})$ of invertible $2 \times 2$ matrices with integers entries by the central subgroup consisting of the identity element and its negative. Take

$$A = \pm\begin{bmatrix} 0 & 1 \\ 1 & 0 \end{bmatrix},\ B = \pm\begin{bmatrix} -1 & 1 \\ 0 & 1 \end{bmatrix} \text{ and } C = \pm\begin{bmatrix} 1 & 0 \\ 0 & -1 \end{bmatrix}.$$

(a) Prove that $\{A,\ B,\ C\}$ is a generating set for $PGL_2(\mathbb{Z})$;
(b) Using the generating set in (a) to verify that $PGL_2(\mathbb{Z})$ is a quotient of a Coxeter group.

2. Let $F$ be the free group on the set $\{x,\ y\}$ and let $G = C_2$, with generator $a \neq e$. Let $\alpha$ be the homomorphism $F \to G$ such that $\alpha(x) = a = \alpha(y)$. Find a minimal generating set for the kernel of $\alpha$. Is the kernel a free group?

## 6.5    Worked-Out Problems

**Problem 6.35** Let $F$ be a free group on a set $X$. If $|X| > 1$, prove that $F$ has trivial center.

*Solution* Suppose that $F$ has a non-identity element $w$ in its center. Let $w$ be expressed as a reduced word $w = x_1^{\varepsilon_1} \ldots x_n^{\varepsilon_n}$. Assume that $y \in X$ and $y \neq x_1$. Consider $yw = yx_1^{\varepsilon_1} \ldots x_n^{\varepsilon_n}$, where it is a reduced word on $X$. Next, consider $wy = x_1^{\varepsilon_1} \ldots x_n^{\varepsilon_n} y$. If $x_n^{\varepsilon_n} \neq y^{-1}$, then $x_1^{\varepsilon_1} \ldots x_n^{\varepsilon_n} y$ is a reduced word. In this case, since $yw$ begins as a reduced word with $y$ and $wy$ begins with $x_1^{\varepsilon_1} \neq y$, it follows that $yw \neq wy$. Now, let $x_n^{\varepsilon_n} = y^{-1}$. If $n = 1$, then $x_1^{\varepsilon_1} = y^{-1}$ contrary to the choice of $y$. So, suppose that $n > 1$. Then, $wy = x_1^{\varepsilon_1} \ldots x_{n-1}^{\varepsilon_{n-1}}$, and again, it is clear that $wy$ and $yw$ are two different reduced words. Consequently, we have $wy \neq yw$. On the other hand, as $w$ lies in the center, we must have $wy = yw$. This is a contradiction. ∎

**Problem 6.36** Let $F$ be the free group on the set $\{x_1, \ldots, x_n\}$. Show that $F$ has a subgroup of index $m$, for all $1 \leq m \leq n$.

*Solution* Suppose that $C_m$ is the cyclic group of order $m$ generated by an element $a$. Then, there exists an epimorphism $\varphi : F \to C_m$, where $\varphi(x_i) = a$, for all $1 \leq i \leq n$. Now, by the first isomorphism theorem, we have $F/Ker\varphi \cong C_m$. Since $|C_m| = m$, it follows that $|F/Ker\varphi| = m$. This shows that the number of cosets of $Ker\varphi$ in $F$ is $m$. Therefore, $F$ has a subgroup of index $m$. ∎

**Problem 6.37** Let $F$ be the free group with basis $\{a, b, c\}$. Suppose that $N$ be the normal subgroup generated by $c$, i.e., the intersection of all normal subgroups of $F$ containing $c$. Prove that $F/N$ is a free group with basis $\{aN, bN\}$.

*Solution* Let $G$ be a free on $\{x, y\}$, and suppose that $\varphi : F \to G$ is the homomorphism defined by $\varphi(a) = x$ and $\varphi(b) = y$ and $\varphi(c) = e$. Note that as $F$ is free on $\{a, b, c\}$, such a homomorphism $\varphi$ exists. Let $K$ be the kernel of $\varphi$. First, we prove that $K = N$, the normal subgroup generated by $c$. Since $\varphi(c) = e$, it follows that $N$ is contained in $K$. Moreover, assume that $w \in F$ and $w \notin N$. Then, we can write $w = w_1 n_1$, where $w_1$ is a reduced word on $\{a, b\}$ and $n_1 \in N$. So, we have $\varphi(w) = \varphi(w_1 n_1) = \varphi(w_1)\varphi(n_1) = \varphi(w_1)$, and $\varphi(w_1)$ is a reduced word on $\{x, y\}$. Thus, $\varphi(w_1) \neq e$, which implies that $\varphi(w) \neq e$, and hence $w \notin K$. Consequently, if $w \notin N$, then $w \notin K$. This yields that $K$ is contained in $N$. Therefore, as we observed earlier that $N$ is contained in $K$, we conclude that $K = N$.

Since $F$ is generated by $\{a, b, c\}$, it follows that $F/N = \langle aN, bN, cN \rangle = \langle aN, bN \rangle$ because $cN = N$. Now, we prove that $F/N$ is free on $\{aN, bN\}$. Suppose that $(y_1 N)^{\varepsilon_1} \ldots (y_k N)^{\varepsilon_k}$ is a reduced word on $\{aN, bN\}$. Now, $\varphi$ induces an automorphism $\phi : F/K \to G$ defined by $\phi(wK) = \varphi(w)$. Then, we see that $\phi(aK) = x$ and $\phi(bK) = y$. So, we obtain

$$\phi\big((y_1 N)^{\varepsilon_1} \ldots (y_k N)^{\varepsilon_k}\big) = x_1^{\varepsilon_1} \ldots x_k^{\varepsilon_k},$$

where $\phi(y_i N) = x_i \in \{x, y\}$. The word $x_1^{\varepsilon_1} \ldots x_k^{\varepsilon_k}$ is a reduced word on $\{x, y\}$. Since $G$ is free on $\{x, y\}$, it follows that $x_1^{\varepsilon_1} \ldots x_k^{\varepsilon_k} \neq e$. Therefore, we conclude that $(y_1 N)^{\varepsilon_1} \ldots (y_k N)^{\varepsilon_k} \neq N$, and we are done. ∎

**Problem 6.38** Prove that

(1) $SL_2(\mathbb{Z})$ is generated by the matrices

$$A = \begin{bmatrix} 0 & 1 \\ -1 & 0 \end{bmatrix} \text{ and } B = \begin{bmatrix} 0 & -1 \\ 1 & 1 \end{bmatrix}.$$

(2) If $G = \langle x, y \mid x^2 = y^3 = e \rangle$, then $G$ is infinite.

*Solution* (1) It is easy to check that

$$A^2 = -I_2, \ B^2 = \begin{bmatrix} -1 & -1 \\ 1 & 0 \end{bmatrix}, \ B^3 = -I_2 \text{ and } C = AB = \begin{bmatrix} 1 & 1 \\ 0 & 1 \end{bmatrix}.$$

Suppose that $H = \langle A, B \rangle$. Since $C = AB \in H$, it follows that $C^{-1} = \begin{bmatrix} 1 & -1 \\ 0 & 1 \end{bmatrix} \in H$, and so

$$C^n = \begin{bmatrix} 1 & n \\ 0 & 1 \end{bmatrix} \in H,$$

for all $n \in \mathbb{Z}$. Also, we obtain

$$A^2 C^n = \begin{bmatrix} -1 & -n \\ 0 & -1 \end{bmatrix} \in H,$$

for all $n \in \mathbb{Z}$. Obviously, we have $H \subseteq SL_2(\mathbb{Z})$. We claim that $H = SL_2(\mathbb{Z})$. Let $D = \begin{bmatrix} a & b \\ c & d \end{bmatrix}$ be an arbitrary element of $SL_2(\mathbb{Z})$. As $c$ is an integer, we apply mathematical induction on $|c|$. If $c = 0$, then $D = \begin{bmatrix} a & b \\ 0 & d \end{bmatrix}$. Then, we must have $ad = 1$. This implies that

$$D = \begin{bmatrix} 1 & n \\ 0 & 1 \end{bmatrix} \text{ and } D = \begin{bmatrix} -1 & n \\ 0 & -1 \end{bmatrix},$$

where $n \in \mathbb{Z}$. Thus, we conclude that $D \in H$. Now, let $c \neq 0$ and for any matrix $D \in SL_2(\mathbb{Z})$ with condition $|c| < n$ we have $D \in H$ (induction hypothesis). We investigate that the result is true for any matrix $D$ with $|c| = n$. By the division algorithm, we can write $a = cq + r$, where $0 \leq r < |c| = n$. We have

$$A^{-1}C^{-q}D = \begin{bmatrix} 0 & -1 \\ 1 & 0 \end{bmatrix} \begin{bmatrix} 1 & -q \\ 0 & 1 \end{bmatrix} \begin{bmatrix} a & b \\ c & d \end{bmatrix}$$

$$= \begin{bmatrix} -c & -d \\ a - cq & b - dq \end{bmatrix} = \begin{bmatrix} -c & -d \\ r & b - dq \end{bmatrix}.$$

Now, if $r = 0$, then by the above argument $A^{-1}C^{-q}D \in H$, which implies that $D \in H$, and we are done. If $r < n$, then by induction hypothesis, we have $\begin{bmatrix} -c & -d \\ r & b - dq \end{bmatrix} \in H$, or equivalently $A^{-1}C^{-q}D \in H$. Again we conclude that $D \in H$. Therefore, we proved that $SL_2(\mathbb{Z}) \subseteq H$, as desired.

(2) According to previous part, we have $A^2 = -I_2$ and $B^3 = -I_2$. If we set $Z = \{I_2, -I_2\}$, then $Z \trianglelefteq SL_2(\mathbb{Z})$, and in the factor group $PSL_2(\mathbb{Z}) = SL_2(\mathbb{Z})/Z$ we have $(ZA)^2 = Z$ and $(ZB)^3 = Z$. This shows that $PSL_2(\mathbb{Z})$ is generated by $ZA$ and $ZB$. Finally, we assume that $G = \langle x, y \mid x^2 = y^3 = e \rangle$. Then, $PSL_2(\mathbb{Z})$ is a homomorphic image of $G$. In other words, there is a normal subgroup $N$ of $G$ such that $G/N \cong PSL_2(\mathbb{Z})$. Since $PSL_2(\mathbb{Z})$ is infinite, it follows that $G$ is infinite.  ∎

**Problem 6.39** Prove that if $n$ is a positive integer greater than 1, then there exists a presentation of the symmetric group $S_n$ with generators $x_1, x_2, \ldots, x_{n-1}$ and relations

$$x_i^2 = (x_j x_{j+1})^3 = (x_k x_l)^2 = e, \tag{6.2}$$

where $1 \leq i \leq n - 1$, $1 \leq j \leq n - 2$ and $1 \leq l < k - 1 < n - 1$.

*Solution* Suppose that $G$ is a group with the generators $x_1, x_2, \ldots, x_{n-1}$ and the relations listed in (6.2). Let $H = \langle x_1, x_2, \ldots, x_{n-2} \rangle$. By von Dyck's theorem and mathematical induction on $n$, we have $|H| \leq (n - 1)!$. Hence, it suffices to show that $[G : H] \leq n$. Let $H, Hx_{n-1}, Hx_{n-1}x_{n-2}, \ldots, Hx_{n-1}x_{n-2} \ldots x_1$ be $n$ right cosets of $H$ in $G$. We investigate that right multiplication by any $x_j$ permutes these cosets.

If $j < i - 1$, then by (6.2) we obtain $x_i x_j = (x_i x_j)^{-1} = x_j x_i$. Since $x_j \in H$, it follows that

$$(Hx_{n-1} \ldots x_i)x_j = Hx_j x_{n-1} \ldots x_i = Hx_{n-1} \ldots x_i.$$

Now, let $j > i$. As $x_j x_k = x_k x_j$ if $|j - k| > 1$, then

$$(Hx_{n-1} \ldots x_i)x_j = Hx_{n-1} \ldots x_{j+1}(x_j x_{j-1} x_j)x_{j-2} \ldots x_i.$$

Since $(x_{j-1}x_j)^3 = e$, it follows that $x_{j-1}x_j x_{j-1} = x_j x_{j-1} x_j$. So, we conclude that

$$(Hx_{n-1} \ldots x_i)x_j = Hx_{n-1} \ldots x_{j+1}(x_{j-1}x_j x_{j-1})x_{j-2} \ldots x_i$$
$$= Hx_{j-1}x_{n-1} \ldots x_i = Hx_{n-1} \ldots x_i.$$

Finally, we have

$$(Hx_{n-1} \ldots x_i)x_i = Hx_{n-1} \ldots x_{i+1}$$

and

$$(Hx_{n-1} \ldots x_i)x_{i-1} = Hx_{n-1} \ldots x_i x_{i-1}.$$

Since $x_j$'s generate $G$, it follows that every elements of $G$ belongs to one of these cosets and consequently $[G : H] \leq n$.

Now, we consider $n - 1$ adjacent transpositions $\tau_i = (i\ i + 1)$ for $1 \leq i \leq n - 1$. We know that every permutation is a product of transpositions and every transpositions is a product of adjacent transpositions. Indeed, $(i\ j) = (j - 1\ j)(i\ j - 1)(j - 1\ j)$ for $i < j - 1$. In addition, it is easy to check that

$$id = \tau_i^2 = (\tau_j\ \tau_{j+1})^3 = (\tau_k\ \tau_l)^2,$$

for $1 \leq i \leq n - 1, 1 \leq j \leq n - 2$ and $1 \leq l < k - 1 < n - 1$. So, by von Dyck's Theorem, there exists an epimorphism $\varphi : G \to S_n$ such that $\varphi(x_i) = \tau_i$. Since $[G : Ker\varphi] = n!$ and $|G| \leq n!$, it follows that $Ker\varphi = \{e\}$ and hence $\varphi$ is an isomorphism. ∎

## 6.6 Supplementary Exercises

1. Let $z$ be a complex number such that $|z| \geq 2$. Prove that

$$\begin{bmatrix} 1 & 0 \\ z & 1 \end{bmatrix} \text{ and } \begin{bmatrix} 1 & z \\ 0 & 1 \end{bmatrix}$$

   generate a free group.
2. If $N$ is a normal subgroup of $G$ and $G/N$ is free, prove that there exist a subgroup $H$ such that $G = HN$ and $H \cap = \{e\}$.
   Hint: Use projective property.
3. If $N$ is a normal subgroup of $G$ and $G/N$ is free, prove that $G$ is a semidirect product of $N$ by $G/N$.
4. Let $F$ be a free group on $X$. If $|X| \geq 2$, prove that $F$ has an automorphism $\varphi$ with $\varphi(\varphi(a)) = a$, for all $a \in F$ and with no fixed points.
5. If $F_i$ is free on $X_i, i = 1, 2$ and $F_1 \cong F_2$, prove that $|X_1| = |X_2|$.
   Hint: Consider $Hom(F_i, \mathbb{Z}_2)$ and view it as a vector space over $\mathbb{Z}_2$.
6. Prove that every free group is torsion-free.
7. Let $F$ be the free group on the set $X$, and let $Y \subset X$. If $N$ is the smallest normal subgroup of $F$ containing $Y$, prove that $F/N$ is a free group.
8. Let $F$ be the free group on $\{x,\ y\}$. Prove that $F$ has a subgroup $H$ for which $xHx^{-1} \subset H$.
9. Let $G = \langle x, y \mid x^{2n} = e,\ x^n = y^2 \text{ and } y^{-1}xy = x^{-1} \rangle$. Show that $Z(G) = \{e, x^n\}$. Assuming that $|G| = 4n$, show that $G/Z(G)$ is isomorphic to $D_n$. (The group $G$ is called the *dicyclic group* of order $4n$.)

10. Let $G$ be an abelian group with generators $x_1, x_2, \ldots, x_n$ and defining relations consisting of $[x_i, x_j] = e$, for $i < j \leq n$, and $r$ further relations. If $r < n$, prove that $G$ is infinite.

11. Suppose that $G$ is a group with $n$ generators and $r$ relations, where $r < n$. Prove that $G$ is infinite.

12. Let $G$ be a group with presentation $\langle X | Y \rangle$, and $N$ be a normal subgroup of $G$. Suppose that $A \subset N$ such that $\langle A \rangle = N$. We write each $a \in A$ as a word $w_a$ on $X$. If $Y'$ consists of all the relations $w_a = e$, for $a \in A$, prove that $\langle X | Y \cup Y' \rangle$ is a presentation for $G/N$.

13. Prove that any subgroup of a free group is free.

14. Let $N$ be a normal subgroup of a group $G$. If $N$ and $G/N$ are finitely presented groups. Prove that $G$ is finitely presented.

# Chapter 7
# Symmetry Groups of Geometric Objects

In this chapter we will study interactions between geometry and group theory. The notion of isometry is a general notion commonly accepted in mathematics. The word isometry means "preserving distances". An isometry is a transformation that preserves distance.

## 7.1 Isometries of Euclidean Space

We begin with the characterization of isometries of a finite dimensional Euclidean space.

**Definition 7.1** An *isometry* of $\mathbb{R}^n$ is a function $T: \mathbb{R}^n \to \mathbb{R}^n$ which preserves the distance, i.e., for every $X$ and $Y$ in $\mathbb{R}^n$, we have

$$\|T(X) - T(Y)\| = \|X - Y\|.$$

*Example 7.2* Each translation in $\mathbb{R}^n$ is an isometry.

**Theorem 7.3** *Let $T: \mathbb{R}^n \to \mathbb{R}^n$ be an isometry. Then, $T$ can be written uniquely as composition $T = T' \circ F$, where $T'$ is a translation and $F$ is an isometry fixing the origin.*

**Proof** If $T = T' \circ F$, where $T'$ is a translation by a column vector $Y$ and $F$ is an isometry such that $F(0) = 0$, then

$$T(X) = T' \circ F(X) = T'\big(F(X)\big) = F(X) + Y. \tag{7.1}$$

If we consider $X = 0$, then $T(0) = Y$. This shows that $Y$ is specified by $T$. Now, using (7.1), we get $F(X) = T(X) - Y = T(X) - T(0)$. This means that $F$ is nominated by $F$. Consequently, by using the above argument, we define

$$T'(X) = X + T(0),$$
$$F(X) = T(X) - T(0),$$

for every column vector $X$ in $\mathbb{R}^n$. This completes the proof.                                     ∎

**Theorem 7.4** *If* $T : \mathbb{R}^n \to \mathbb{R}^n$ *is a function, then the following conditions are equivalent:*

*(1)  $T$ is an isometry such that $T(0) = 0$;*
*(2)  $T$ is a linear transformation, and the matrix $A$ such that $T(X) = AX$, for all column vector $X$ in $\mathbb{R}^n$ satisfy $A^t A = I_n$.*

**Proof** $(1 \Rightarrow 2)$ By assumption, we have $\|T(X) - T(Y)\| = \|X - Y\|$, for all $X$ and $Y$ in $\mathbb{R}^n$. Since $T(0) = 0$, it follows that $\|T(X)\| = \|X\|$ and $\|T(Y)\| = \|Y\|$. Expanding this out in terms of inner product, we obtain $\langle T(X), T(Y) \rangle = \langle X, Y \rangle$, and this shows that $T$ preserves the inner product. Now, let $\{E_1, \ldots, E_n\}$ be the standard basis for $\mathbb{R}^n$ We know that $\{T(E_1), \ldots, T(E_n)\}$ must also form an orthonormal basis for $\mathbb{R}^n$. Assume that $Y_i = T(E_i)$, for all $1 \le i \le n$, and let $X = c_1 E_1 + \cdots + c_n E_n$ be an arbitrary column vector in $\mathbb{R}^n$, where $c_1, \ldots, c_n$ are scalars. Obviously, we have $c_i = \langle X, E_i \rangle$. So, we conclude that $\langle T(X), Y_i \rangle = \langle T(X), T(E_i) \rangle = c_i$. Since $\{Y_1, \ldots, Y_n\}$ is an orthonormal basis, it follows that

$$T(X) = \sum_{i=1}^{n} \langle T(X), Y_i \rangle Y_i = \sum_{i=1}^{n} c_i Y_i = \sum_{i=1}^{n} c_i T(E_i).$$

This yields that $T$ is a linear transformation. Now, let $A$ be the associated matrix of $T$ relative to the standard basis $\{E_1, \ldots, E_n\}$. We can write $T(X) = AX$, for all column vector $X$ in $\mathbb{R}^n$. We claim that $A^t A = I_n$. Since $\langle T(X), T(Y) \rangle = \langle X, Y \rangle$, it follows that $\langle AX, AY \rangle = \langle X, Y \rangle$. On the other hand, we observe that $\langle AX, AY \rangle = \langle A^t AX, Y \rangle$. So, we conclude that $\langle A^t AX, AY \rangle = \langle X, Y \rangle$, for all column vectors $X$ and $Y$ in $\mathbb{R}^n$. Hence, for each distinct integers $1 \le i, j \le n$, we have

$$\langle A^t A E_i, E_j \rangle = \langle E_i, E_j \rangle = 0 = \text{the } ij \text{ entry of } A^t A;$$
$$\langle A^t A E_i, E_i \rangle = \langle E_i, E_i \rangle = 1 = \text{the } ii \text{ entry of } A^t A.$$

Consequently, we obtain $A^t A = I_n$.

$(2 \Rightarrow 1)$ Suppose that $T$ is a linear transformation and $T(X) = AX$, for all column vector $X$ in $\mathbb{R}^n$, where $A^t A = I_n$. Clearly, we have $T(0) = 0$. First, note that

$$\langle T(X), T(Y) \rangle = \langle AX, AY \rangle = \langle A^t AX, Y \rangle \langle X, Y \rangle. \tag{7.2}$$

By using (7.2), we can write

$$\begin{aligned}
\|T(X) - T(Y)\|^2 &= \langle T(X) - T(Y), T(X) - T(Y)\rangle \\
&= \langle T(X), T(X)\rangle - 2\langle T(X), T(Y)\rangle + \langle T(Y), T(Y)\rangle \\
&= \langle X, X\rangle - 2\langle X, Y\rangle + \langle Y, Y\rangle \\
&= \langle X - Y, X - Y\rangle = \|X - Y\|^2.
\end{aligned}$$

Therefore, we obtain $\|T(X) - T(Y)\| = \|X - Y\|$. This shows that $T$ is an isometry. Moreover, it is clear that $T(0) = 0$. ∎

**Theorem 7.5** *The set of all isometries of $\mathbb{R}^n$ forms a group under composition.*

This group is called *Euclidean group* and is denoted by $E_n$.

**Proof** First, we show that the composition of two isometries of $\mathbb{R}^n$ is also an isometry of $\mathbb{R}^n$. Let $T_1$ and $T_2$ be two isometries of $\mathbb{R}^n$, and let $T = T_1 \circ T_2$. Then, for every column vectors $X$ and $Y$ in $\mathbb{R}^n$, we have

$$\begin{aligned}
\|T(X) - T(Y)\| &= \|T_1 \circ T_2(X) - T_1 \circ T_2(Y)\| \\
&= \|T_1(T_2(X)) - T_1(T_2(Y))\| \\
&= \|T_2(X) - T_2(Y)\| \\
&= \|X - Y\|.
\end{aligned}$$

This shows that $T_1 \circ T_2 \in E_n$. Clearly, the associative low holds and the identity function lies in $E_n$.

Now, suppose that $T : \mathbb{R}^n \to \mathbb{R}^n$ is an arbitrary isometry. By Theorem 7.3, we can write $T = F + T(0)$, where $F$ is an isometry of $\mathbb{R}^n$ such that $F(0) = 0$. Then, by Theorem 7.4, we find an invertible matrix $A$ such that $F(X) = AX$, for all column vector $X$ in $\mathbb{R}^n$. So, we conclude that $T(X) = AX + T(0)$. Now, we observe that $T^{-1}$ exists and its rule is $T^{-1}(X) = A^{-1}(X - T(0))$. Finally, we must show that $T^{-1}$ is an isometry of $\mathbb{R}^n$. Let $X$ and $Y$ be two arbitrary column vectors in $\mathbb{R}^n$. Since $T$ is an isometry, it follows that $\|T(T^{-1}(X)) - T(T^{-1}(Y))\| = \|T^{-1}(X) - T^{-1}(Y)\|$. This implies that $\|X - Y\| = \|T^{-1}(X) - T^{-1}(Y)\|$. Hence, we conclude that $T^{-1}$ is an isometry, and so $T^{-1} \in E_n$. ∎

*Remark 7.6* We say a function $T: \mathbb{R}^n \to \mathbb{R}^n$ has a *limit* $L$ at $A$, and we write $\lim_{X \to A} T(X) = L$ if for all $\epsilon > 0$, we can find $\delta > 0$ such that

$$\|X - A\| < \delta \Rightarrow \|T(X) - L\| < \epsilon.$$

We say that $T: \mathbb{R}^n \to \mathbb{R}^n$ is *continuous* at $A$ if $\lim_{X \to A} T(X) = f(A)$. We say that $T$ is continuous (everywhere) if it is continuous at every point of the domain $\mathbb{R}^n$.

**Theorem 7.7** *Each isometry of $\mathbb{R}^n$ is a continuous function.*

**Proof** It is enough we consider $\delta = \epsilon$. ∎

**Exercises**

1. Write down some isometries that are linear and some that are not.
2. Prove that there are exactly three possibilities for the number of distinct powers of an isometry $T$ of $\mathbb{R}$.
3. Find a counterexample to show that a distance-preserving map need not be surjective.
4. Calculate the commutators $[u, v]$ of various pairs of elements $u, v \in G = E_n$. Find a general form for these commutators and so identify the subgroup $G'$ of $G$ which they generate.

## 7.2  Isometries of $\mathbb{R}^3$

Now, we concentrate on isometries of $\mathbb{R}^3$. We begin with some examples.

*Example 7.8*  Each translation in $\mathbb{R}^3$ is an isometry.

*Example 7.9*  Let $N$ be a unit vector in $\mathbb{R}^3$, and suppose that $\mathcal{P} = \{X \in \mathbb{R}^3 \mid \langle X, N \rangle = 0\}$ is the plane through the origin with normal direction $N$. We define a reflection $T$ across $\mathcal{P}$ from $\mathbb{R}^3$ onto $\mathbb{R}^3$ by $T(X) = X - 2\langle X, N \rangle N$, for all $X \in \mathbb{R}^3$. Then, $T$ is an isometry. Indeed, for every $X, Y \in \mathbb{R}^3$, we have

$$T(X) - T(Y) = \big(X - 2\langle X, N \rangle N\big) - \big(Y - 2\langle Y, N \rangle N\big)$$
$$= (X - Y) - 2\langle X - Y, N \rangle N.$$

So, we can write

$$\|T(X) - T(Y)\|^2 = \|(X - Y) - 2\langle X - Y, N \rangle N\|^2$$
$$= \|X - Y\|^2 - 4\langle X - Y, N \rangle \langle X - Y, N \rangle + 4\langle X - Y, N \rangle^2 \|N\|^2$$
$$= \|X - Y\|^2.$$

This shows that $\|T(X) - T(Y)\| = \|X - Y\|$.

*Example 7.10*  Let $T : \mathbb{R}^3 \to \mathbb{R}^3$ be a rotation in $\mathbb{R}^3$ with an angle $\theta$ defined by

$$T\left(\begin{bmatrix} x \\ y \\ z \end{bmatrix}\right) = \begin{bmatrix} x \cos\theta - y \sin\theta \\ x \sin\theta + y \cos\theta \\ z \end{bmatrix},$$

for all $x, y, z \in \mathbb{R}$. Then, a simple calculation shows that $T$ is an isometry.

**Definition 7.11**  *Collinearity of a set of points* is the property of their lying on a single line. A set of points with this property is said to be *collinear*. Two points are trivially collinear since two points determine a line.

**Theorem 7.12**  *If $T : \mathbb{R}^3 \to \mathbb{R}^3$ is an isometry, then*

*(1)  $T$ preserves collinearity;*
*(2)  $T$ maps any straight line to a straight line.*

**Proof**  (1) We remark that a position vector in $\mathbb{R}^3$ can be represent a point in space. Suppose that $p$, $q$, and $r$ are three distinct points in $\mathbb{R}^3$ with position vectors $P$, $Q$, and $R$, respectively. We use triangle inequality for $P$, $Q$, and $R$ in $\mathbb{R}^3$,

$$\|P - Q\| + \|Q - R\| \le \|P - R\|,$$

with equality if and only if $p$, $q$, and $r$ are collinear with $q$ between $p$ and $r$. Now, if $p$, $q$, and $r$ lie on a line, then their position vectors satisfy triangle equality. Since $T$ is an isometry, it follows that the position vectors of the points $T(p)$, $T(q)$, and $T(r)$ satisfy triangle equality. Therefore, we conclude that they are also collinear.

(2) Suppose that $L$ is a line passing through points $p$ and $q$ in $\mathbb{R}^3$. Then, the image of $L$ is $T(L) = \{T(a) \mid a \in L\}$. By (1), any point $b = T(a)$ in $T(L)$ is collinear with $T(p)$ and $T(q)$. On the other hand, if $b$ is an arbitrary point in the line passing through $T(p)$ and $T(q)$, then $b = T(a)$, for some $a \in L$.  ∎

**Theorem 7.13**  *Let $E_1$, $E_2$ and $E_3$ be the standard basis for $\mathbb{R}^3$. If $T$ is an isometry of $\mathbb{R}^3$ such that $T(0) = 0$, $T(E_1) = E_1$, $T(E_2) = E_2$, and $T(E_3) = E_3$, then $T$ is the identity function.*

**Proof**  Suppose that $X = [x \; y \; z]^t$ is a column vector in $\mathbb{R}^3$ and $T(X) = [a \; b \; c]^t$. We have $\langle X, E_i \rangle = \langle T(X), T(E_i) \rangle = \langle T(X), E_i \rangle$, for $i = 1, 2, 3$. Hence, we conclude that $\langle T(X) - X, E_i \rangle = 0$. Therefore, we get $a - x = 0$, $b - y = 0$, and $c - z = 0$, or $a = x$, $b = y$, and $c = z$. This completes the proof.  ∎

**Theorem 7.14**  *If $A$ and $B$ are two column vectors in $\mathbb{R}^3$ such that $\|A\| = \|B\|$, then there exists a reflection $R$ across a plane $\mathcal{P}$ through the origin such that $R(A) = B$ and $R(B) = A$.*

**Proof**  If $A = B$, then $\mathcal{P}$ is any plane through the origin and $A$. In this case, the reflection $R$ across the plane $\mathcal{P}$ satisfies $R(A) = A = B$ because $R$ fixes any point of $\mathcal{P}$.

If $A \ne B$, then $\|A - B\| \ne 0$. So, $N = (A - B)/\|A - B\|$ is a unit vector. Let $\mathcal{P} = \{X \in \mathbb{R}^3 \mid \langle X, N \rangle = 0\}$ be the plane through the origin with normal vector $N$. Now, we define $R : \mathbb{R}^3 \to \mathbb{R}^3$ by $R(X) = X - 2\langle X, N \rangle N$. Since

$$\|A - B\|^2 = \|A\|^2 - 2\langle A, B \rangle + \|B\|^2 = 2\|A\|^2 - 2\langle A, B \rangle,$$

it follows that

$$R(A) = A - 2\langle A, N \rangle N = A - 2\Big\langle A, \frac{A - B}{\|A - B\|} \Big\rangle \frac{A - B}{\|A - B\|}$$
$$= A - \frac{2\|A\|^2 - 2\langle A, B \rangle}{\|A - B\|^2}(A - B) = A - (A - B) = B.$$

On the other hand, since $R^2$ is the identity function, we conclude that $R(B) = A$, too. ∎

**Theorem 7.15** *Every isometry of $\mathbb{R}^3$ is a composition of at most 4 reflections.*

**Proof** We consider the following steps:

(1) If $T(0) = 0$, then proceed to step (2) by taking $T_1 = T$ preserving the origin. In otherwise, we may assume that $T(0) \neq 0$. Let $M = T(0)/2$, the midpoint of 0 and $T(0)$, and suppose that $\mathcal{P}$ is the plane passing through $M$ with unit normal vector $N = M/\|M\| = T(0)/\|T(0)\|$. This implies that

$$T(0) = \|T(0)\|N. \tag{7.3}$$

We consider the plane $\mathcal{P} = \{X \in \mathbb{R}^3 \mid \langle X, N \rangle = \langle M, N \rangle\}$. Clearly, we obtain

$$\langle M, N \rangle = \left\langle \frac{T(0)}{2}, \frac{T(0)}{\|T(0)\|} \right\rangle = \frac{\|T(0)\|^2}{2\|T(0)\|} = \frac{\|T(0)\|}{2}.$$

Now, let $R_1$ be the reflection in $\mathbb{R}^3$ across the plane $\mathcal{P}$. Then, we have

$$R_1 \circ T(0) = R_1\big(T(0)\big) = T(0) + 2\left(\frac{\|T(0)\|}{2} - \langle T(0), N \rangle\right)N.$$

By using (7.3), we can write

$$R_1 \circ T(0) = T(0) + T(0) - 2\langle T(0), N \rangle N = 2T(0) - 2T(0) = 0.$$

Since $R_1$ and $T$ are isometries, it follows that $T_1 = R_1 \circ T$ is an isometry of $\mathbb{R}^3$ preserving the origin.

(2) If $T_1(E_1) = E_1$, then proceed to step (3) with $T_2 = T_1$ preserving the origin and $E_1$. In otherwise, we may assume that $T_1(E_1) \neq E_1$. Since $T_1$ is an isometry, it follows that $\|T_1(E_1)\| = \|E_1\|$. By Theorem 7.14, there exists a reflection $R_2$ in $\mathbb{R}^3$ across a plane through the origin interchanging $T_1(E_1)$ and $E_1$. Hence, we have $R_2\big(T_1(E_1)\big) = E_1$ and $R_2(0) = 0$. We define $T_2 = R_2 \circ T_1$. Then, $T_2$ is an isometry of $\mathbb{R}^3$ satisfying

$$T_2(0) = R_2 \circ T_1(0) = R_2\big(T_1(0)\big) = R_2(0) = 0,$$
$$T_2(E_1) = R_2 \circ T_1(E_1) = R_2\big(T_1(E_1)\big) = E_1.$$

This shows that $T_2$ preserving the origin and $E_1$.

(3) If $T_2(E_2) = E_2$, then proceed to step (4) with $T_3 = T_2$ preserving the origin, $E_1$ and $E_2$. In otherwise, we may assume that $T_2(E_2) \neq E_2$. Since $T_2$ is an isometry such that $T_2(0) = 0$, it follows that

$$\langle T_2(E_2) - E_2, E_1 \rangle = \langle T_2(E_2), T_2(E_1) \rangle - \langle E_2, E_1 \rangle = 0.$$

Now, we set

$$N = \frac{T_2(E_2) - E_2}{\|T_2(E_2) - E_2\|},$$

and suppose that $\mathcal{P}'$ is the plane through the origin with unit normal vector $N$. By Theorem 7.14, we conclude that there is a reflection $R_3$ in $\mathbb{R}^3$ across the plane $\mathcal{P}'$ interchanging $E_2$ and $T_2(E_2)$. Since $\langle E_1, N \rangle = 0$, it follows that $0$ and $E_1$ lie in $\mathcal{P}'$. Hence, $R_3$ preserves both $0$ and $E_1$. So, we get

$$\begin{aligned}
R_3 \circ T_2(0) &= R_3\big(T_2(0)\big) = R_3(0) = 0, \\
R_3 \circ T_2(E_1) &= R_3\big(T_2(E_1)\big) = E_1, \\
R_3 \circ T_2(E_2) &= R_3\big(T_2(E_2)\big) = E_2.
\end{aligned}$$

Therefore, we conclude that the isometry $T_3 = R_3 \circ T_2$ preserves the origin and both $E_1$ and $E_2$.

(4) If $T_3(E_3) = E_3$, then take $T_4 = T_3$ preserving the origin, $E_1$, $E_2$, and $E_3$. In otherwise, we may assume that $T_3(E_3) \neq E_3$. Then, we obtain

$$\begin{aligned}
\langle T_3(E_3), E_1 \rangle &= \langle T_3(E_3), T_3(E_1) \rangle = \langle E_3, E_1 \rangle = 0, \\
\langle T_3(E_3), E_2 \rangle &= \langle T_3(E_3), T_3(E_2) \rangle = \langle E_3, E_2 \rangle = 0.
\end{aligned}$$

Consequently, $T_3(E_3)$ is parallel to $E_3$. Since $\|T_3(E_3)\| = \|E_3\| = 1$, it follows that $T_3(E_3) = \pm E_3$. Since $T_3(E_3) \neq E_3$, it follows that $T_3(E_3) = -E_3$. Now, we define $R_4$ as the reflection in $\mathbb{R}^3$ across the plane through the origin with the unit normal vector $E_3$. Thus, we have

$$\begin{aligned}
R_4 \circ T_3(0) &= R_4\big(T_3(0)\big) = R_4(0) = 0, \\
R_4 \circ T_3(E_1) &= R_4\big(T_3(E_1)\big) = R_4(E_1) = E_1 - 2\langle E_1, E_3 \rangle E_3 = E_1.
\end{aligned}$$

Moreover, we observe that

$$\begin{aligned}
R_4 \circ T_3(E_2) &= R_4(E_2) = E_2 - 2\langle E_2, E_3 \rangle E_3 = E_2, \\
R_4 \circ T_3(E_3) &= R_4\big(T_3(E_3)\big) = R_4(-E_3) = E_3 - 2\langle E_3, E_3 \rangle E_3 = -E_3.
\end{aligned}$$

Finally, if we take $T_4 = R_4$, then $T_4$ preserves the origin, $E_1$, $E_2$, and $E_3$. So, by Theorem 7.13, $T_4$ is the identity function. This yields that $R_4 \circ R_3 \circ R_2 \circ R_1 \circ T = id$, where $R_i$ is either reflection or the identity function. So, we can write $T = (R_4 \circ R_3 \circ R_2 \circ R_1)^{-1} = R_1^{-1} \circ R_2^{-1} \circ R_3^{-1} \circ R_4^{-1} = R_1 \circ R_2 \circ R_3 \circ R_4$. Therefore, we deduce that $T$ is a composition of at most 4 reflections. ∎

**Corollary 7.16** *If $T : \mathbb{R}^3 \to \mathbb{R}^3$ is an isometry such that $T(0) = 0$, then $T$ can be expressed as a a composition of at most 3 reflections, each of them fixing the origin.*

**Proof** The proof follows from Theorem 7.15, by skipping step (1). ∎

**Exercises**

1. Prove that a reflection $R$ across a plane $\mathcal{P}$ is a linear transformation if and only if $0 \in \mathcal{P}$.
2. Prove that any translation is a composition of two reflections across two parallel planes.
3. For each of the following matrices, show that it represents an isometry of $\mathbb{R}^3$ which fixes the origin and decide whether it represents a rotation about a line or a reflection in a plane. In each case, find the line of rotation or plane of refection:

$$\begin{bmatrix} 1/\sqrt{3} & 1/\sqrt{3} & 1/\sqrt{3} \\ 1/\sqrt{6} & 1/\sqrt{6} & -2/\sqrt{6} \\ 1/\sqrt{2} & -1/\sqrt{2} & 0 \end{bmatrix} \text{ and } \begin{bmatrix} 1/\sqrt{3} & 1/\sqrt{3} & -1/\sqrt{3} \\ 1/\sqrt{6} & 1/\sqrt{6} & 2/\sqrt{6} \\ 1/\sqrt{2} & -1/\sqrt{2} & 0 \end{bmatrix}.$$

4. Prove that any rotation of $\mathbb{R}^3$ can be written as the product of two reflections. Deduce that the product of two rotations of $\mathbb{R}^33$ with intersecting axes is again a rotation. What if the axes are parallel?

## 7.3   Isometries of $\mathbb{R}^2$

Since plane symmetry groups basically describe two-dimensional images, it is necessary to make sense of how such images can have isometries.

**Definition 7.17** A *planer image* is a function $\psi : \mathbb{R}^2 \to \{c_1, \ldots, c_n\}$, where $c_1, \ldots,$ $c_n$ are colors. An *isometry of a planer image* $\psi$ is an isometry $T : \mathbb{R}^2 \to \mathbb{R}^2$ such that $\psi(T(X)) = T(X)$, for all $X \in \mathbb{R}^2$.

There are three basic types of isometries of the plane, translations, refections, rotations. A fourth type, glide reflections, is built up as compositions of refections and translations (see Exercise 4). Indeed, the composition of a refection $R$ across a line $\mathcal{L}$ and a translation $T$ transit parallel to the line of refection $\mathcal{L}$ (in either possible order) is glide reflection. Translations and rotations preserve orientation, and reflections and glide reflection clearly reverse it.

**Definition 7.18** We say two translations are independent if their respective glide vectors are linearly independent.

**Definition 7.19** A *pattern* is a planar image that is invariant under two linearly independent and isometric translations in the plane. Furthermore, in both directions there is a translation of minimal length preserving the pattern.

**Definition 7.20** A *plane symmetry group* is a group of isometries that acts on a two-dimensional repeating pattern.

The notion of *congruent figures* is introduced in terms of moves: Two figures are called *congruent* if there exists a move mapping one of them to the other one.

**Theorem 7.21** *Any isometry maps a segment to a congruent segment.*

**Proof** Let $T : \mathbb{R}^2 \to \mathbb{R}^2$ be an isometry and $AB$ is a segment. We show that $T$ maps $AB$ to segment $T(A)T(B)$ congruent to $AB$. Points lie in the segment $AB$ are determined as those points $X$ for which the triangle inequality $AB \leq AX + XB$ turns to identity. More precisely, $X$ belongs to the segment $AB$ if and only if $AB = AX + XB$. Since $T$ is an isometry, it follows that $T(A)T(B) = AB, T(A)T(X) = AX$, and $T(X)T(B) + XB$, for any point $X$. Therefore, we have $T(A)T(B) = T(A)T(X) + T(X)T(B)$ if and only if $AB = AX + XB$. Consequently, $T(X)$ lies in the segment $T(A)T(B)$ if and only if $X$ lies to the segment $AB$.      ∎

A *collineation* is a transformation that maps lines to lines.

**Theorem 7.22** *Isometries are collineations.*

**Proof** Suppose that $T \colon \mathbb{R}^2 \to \mathbb{R}^2$ is an isometry and $\mathcal{L}$ is a line in $\mathbb{R}^2$. Then, there exist two distinct points $A$ and $B$ such that

$$\mathcal{L} = \{P \in \mathbb{R}^2 \mid \|A - P)\| = \|B - P\|\}.$$

Since $T$ is one to one, it follows that $T(A)$ and $T(B)$ are distinct. Let

$$\mathcal{L}' = \{Q \in \mathbb{R}^2 \mid \|T(A) - Q)\| = \|T(B) - Q\|\}.$$

We prove that $\mathcal{L}' = T(\mathcal{L})$. Assume that $P_0 \in \mathcal{L}$ be an arbitrary point. We have $\|A - P_0)\| = \|B - P_0\|$. Since $T$ preserves distance, it follows that

$$\|T(A) - T(P_0)\| = \|A - P_0\| = \|B - P_0\| = \|T(B) - T(P_0)\|.$$

This implies that $T(P_0) \in \mathcal{L}'$.

Conversely, suppose that $Q_0 \in \mathcal{L}'$ be an arbitrary point. We have $\|T(A) - Q_0\| = \|T(B) - Q_0\|$. Since $T$ is onto, it follows that there exists a point $P_0$ such that $T(P_0) = Q_0$. Consequently, we can write $\|T(A) - T(P_0)\| = \|T(B) - T(P_0)\|$. So, we obtain

$$\|A - P_0)\| = \|T(A) - T(P_0)\| = \|T(B) - T(P_0)\| = \|B - P_0\|.$$

This yields that $P_0 \in \mathcal{L}$. Thus, we conclude that $Q_0 = T(P_0) \in T(\mathcal{L})$. This completes the proof.      ∎

**Corollary 7.23** *An isometry of $\mathbb{R}^2$ maps a triangle to a congruent triangle.*

**Proof** Since an isometry preserves distance, the length of the edges in the triangle does not change. Hence, an isometry maps a triangle to a congruent triangle.      ∎

**Corollary 7.24** *An isometry of* $\mathbb{R}^2$ *maps a polygon to a congruent polygon.*

***Proof*** Since a polygon can be decomposed into a union of triangles, it follows that an isometry maps a polygon to a congruent polygon.  ∎

**Theorem 7.25** *Isometries are collineations.*

***Proof*** The proof is left as an exercise for the reader.  ∎

An isometry is determined by how it maps three non-collinear points.

**Theorem 7.26** *Let* $T$ *and* $T'$ *be two isometries of* $\mathbb{R}^2$. *If* $T(A) = T'(A)$, $T(B) = T'(B)$, *and* $T(C) = T'(C)$, *for some non-collinear points* $A$, $B$, *and* $C$, *then* $T = T'$.

***Proof*** The isometry $T$ maps the triangle $\triangle ABC$ to a congruent triangle $\triangle T(A)T(B)$ $T(C)$. This yields that the points $T(A)$, $T(B)$, and $T(C)$ are also non-collinear. Assume that $T \neq T'$. Then, there exists a point $P$ such that $T(P) \neq T'(P)$. Let $\mathcal{L}$ be the line passing through $T(P)$ and $T'(P)$. We have

$$\|T(P) - T(A)\| = \|P - A\| = \|T'(P) - T'(A)\| = \|T'(P) - T(A)\|.$$

This shows that $T(A) \in \mathcal{L}$. In a similar way, we obtain $T(B) \in \mathcal{L}$ and $T(C) \in \mathcal{L}$. Therefore, we conclude that the points $T(A)$, $T(B)$, and $T(C)$ are collinear, and this is a contradiction.  ∎

An isometry of the plane can be recovered from its restriction to any triple of non-collinear points.

**Theorem 7.27** *Any isometry of the plane is a composition of at most three reflections.*

***Proof*** Let $T \colon \mathbb{R}^2 \to \mathbb{R}^2$ be an isometry, and $A$, $B$, and $C$ be three non-collinear points (see Fig. 7.1).

We consider the following steps;

(1) We find a reflection $R_1$ that maps $A$ to $T(A)$. The axis of such reflection is a perpendicular bisector of the segment $AT(A)$ (see Fig. 7.2). It is uniquely defined

**Fig. 7.1** Isometry $T$ of $\mathbb{R}^2$ and three non-collinear points

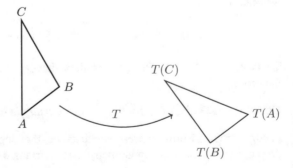

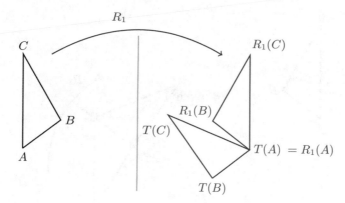

**Fig. 7.2** Reflection $R_1$

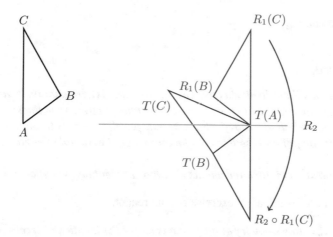

**Fig. 7.3** Reflection $R_2$

unless $T(A) = A$. If $T(A) = A$, then we can take either a reflection about any line passing through $A$ or take, instead of reflection an identity function for $R_1$.

(2) We determine a reflection $R_2$ which maps $R_1(B)$ to $T(B)$. The axis of such reflection is the perpendicular bisector of the segment $R_1(B)T(B)$ (see Fig. 7.3). The segment connecting $T(A)$ with $T(B)$ and $R_1(B)$ are congruent to the segment $AB$. So, we conclude that $T(A)$ belongs to the axis of the reflection $R_2$ and is not moved by $R_2$.

(3) Now, either $R_2\big(R_1(C)\big) = C$, or the triangles $R_2 \circ R_1(\triangle ABC)$ and $T(\triangle ABC)$ are symmetric in their common side $T(AB)$. In the former case $T = R_2 \circ R_1$, in the latter case denote by $R_3$ the reflection in $T(AB)$ (see Fig. 7.4), and observe that $T = R_3 \circ R_2 \circ R_1$. ∎

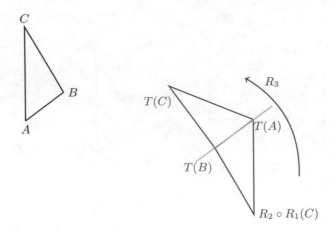

**Fig. 7.4** Reflection $R_3$

**Theorem 7.28**

*(1) The composition of two reflections in parallel lines is a translation in the direction perpendicular to the lines, by twice the distance between the lines;*

*(2) The composition of two reflections in non-parallel lines is a rotation about the intersection point of the lines by the angle equal to doubled angle between the lines;*

*(3) A composition of three reflections is either a reflection or a glide reflection.*

**Proof** The proof is left as an exercise for the reader.                              ∎

**Theorem 7.29** *Any isometry of the plane is either the identity, or a reflection about a line, or rotation, or translation, or gliding reflection.*

**Proof** By Theorem 7.27, an isometry is a composition of at most three reflections. By Theorem 7.28, it is either the identity, or a reflection about a line, or rotation, or translation, or gliding reflection.                              ∎

### Exercises

1. Check that translations, rotations, and reflections are indeed isometries of E2.
2. Which of the following transformations are collineation?

   (a) $T\left(\begin{bmatrix} x \\ y \end{bmatrix}\right) = \begin{bmatrix} -x \\ -y \end{bmatrix}$;

   (b) $T\left(\begin{bmatrix} x \\ y \end{bmatrix}\right) = \begin{bmatrix} \cos x \\ \sin y \end{bmatrix}$;

   (c) $T\left(\begin{bmatrix} x \\ y \end{bmatrix}\right) = \begin{bmatrix} 2y - x \\ x - 2 \end{bmatrix}$;

(d) $T\left(\begin{bmatrix} x \\ y \end{bmatrix}\right) = \begin{bmatrix} x^3 - x \\ y \end{bmatrix}$.

3. Find an example of a bijective transformation that is not a collineation.
4. Prove that

   (a) Each reflection is an isometry;
   (b) Each translation is an isometry;
   (c) Each rotation is an isometry;
   (d) Each glide reflection is an isometry.

5. Prove that any isometry of plane maps

   (a) an angle to a congruent angle;
   (b) a circles to a congruent circle.

6. The equation of line $\mathcal{L}$ is $x - 2y + 3 = 0$. Find the coordinates of the images of $\begin{bmatrix} 0 \\ 0 \end{bmatrix}$, $\begin{bmatrix} 3 \\ -2 \end{bmatrix}$, $\begin{bmatrix} -2 \\ 6 \end{bmatrix}$, and $\begin{bmatrix} 2 \\ 5 \end{bmatrix}$ under reflection in line $\mathcal{L}$.

7. Prove that $T : \mathbb{R}^2 \to \mathbb{R}^2$ defined by

$$T\left(\begin{bmatrix} x \\ y \end{bmatrix}\right) = \begin{bmatrix} 3 - \sqrt{3} - \dfrac{1}{2}x - \dfrac{\sqrt{3}}{2}y \\ -3 - \sqrt{3} + \dfrac{\sqrt{3}}{2}x - \dfrac{1}{2}y \end{bmatrix}$$

   is a rotation. Find its center and the angle through which points are rotated.

8. Prove that $T : \mathbb{R}^2 \to \mathbb{R}^2$ defined by

$$T\left(\begin{bmatrix} x \\ y \end{bmatrix}\right) = \begin{bmatrix} 6 - \dfrac{5}{13}x - \dfrac{12}{13}y \\ 4 - \dfrac{12}{13}x + \dfrac{5}{13}y \end{bmatrix}$$

   is a reflection and find its mirror.

9. Prove that the set of translations and rotations is closed under composition.
10. A *half-turn* $H_P$ for a point $P$ in plane is a rotation by the angle $\theta$ about the point $P$. Prove that

   (a) The composition of two half-turns is a translation;
   (b) Every translation is a composition of two half-turns;
   (c) If $M$ is the midpoint points $A$ and $B$, then $H_M \circ H_A = H_B \circ H_M$;
   (d) A composition of three half-turns is a half-turn;
   (e) $H_A \circ H_B \circ H_C = H_C \circ H_B \circ H_A$, for any three points $A$, $B$ and $C$.

11. Show that an isometry with exactly one fixed point is a rotation.
12. Show that the inverse of every glide reflection is also a glide reflection.

13. If $T$ is a glide reflection, show that $T^2$ is a translation.
14. Let $T$ be an isometry of $\mathbb{R}^2$. A line $\mathcal{L}$ is said to be *invariant under $T$* if $T(\mathcal{L}) = \mathcal{L}$. Show that

    (a) $T$ has no invariant lines if and only if it is a rotation by $a\pi$ for some $a \notin \mathbb{Z}$;
    (b) $T$ has a single invariant line if and only if it is a glide reflection with a nonzero shift;
    (c) If $T$ has multiple invariant lines, it has infinitely many ones.

15. Classify all the isometries $T$ of $\mathbb{R}^2$ such that $T^2$ is the identity function.
16. Classify all the isometries $T$ of $\mathbb{R}^2$ such that $T^6$ is the identity function.
17. Prove that there exists an isometry that cannot be expressed as a composition of one or two reflections.


## 7.4   Finite Rotation Groups

Let $F$ be a set of points in $\mathbb{R}^n$. The symmetry group of $F$ in $\mathbb{R}^n$ is the set of all isometries of $\mathbb{R}^n$ that carry $F$ onto itself. The group operation is function composition. We start with groups of rotations and with the following result.

**Theorem 7.30** *[Finite Symmetry Groups in the Plane] Every finite symmetry group $G$ in the plane is isomorphic to $\mathbb{Z}_n$ or to a dihedral group $D_n$, for some positive integer $n$.*

**Proof** Suppose that $G$ is a finite symmetry group of some figure. We first observe that $G$ cannot contain a translation or a glide reflection, because in otherwise $G$ would be infinite. Assume that $T_1, T_2, \ldots, T_m$ are distinct elements of $G$. Let

$$P_i = (x_i, y_i) = T_i(0, 0)$$

and

$$M = (\overline{x_i}, \overline{y_i}) = \left( \frac{1}{m} \sum_{i=1}^{m} x_i, \frac{1}{m} \sum_{i=1}^{m} y_i \right)$$

be the centroid of the points $P_1, P_2, \ldots, P_n$. Since $G$ is finite, it follows that $T_j \circ T_i = T_k$, for some $1 \le k \le m$. Hence, we obtain

$$T_j(P_i) = T_j(T_i(0, 0)) = T_k(0, 0) = P_k.$$

This yields that the elements of $G$ permute the points $P_1, P_2, \ldots, P_m$. Since every member of $G$ preserves the distance, it follows that every element of $G$ fixes the centroid $M$.

Looking at the possible kinds of isometries, only reflections and rotations fix a point. For any reflection in $G$, the axis of reflection must contain $(\overline{x}, \overline{y})$, and for any rotation in $G$, the center of rotation is $(\overline{x}, \overline{y})$. Let $H$ be the set of orientation preserving elements of $G$. These are the rotations. Since a rotation only fixes one point, it follows that the elements of $H$ are rotations about the centroid $(\overline{x}, \overline{y})$. Since the composition of two rotations about the same point is a rotation, it follows that $H$ is a subgroup of $G$. Now, among all rotations, let $\theta$ be the smallest positive angle of rotation. Note that such an angle exists, because $G$ is finite and $R_{2\pi}$ belongs to $G$. It is not difficult to see that every element of $H$ represents a rotation through a multiple of $\theta$. In other words, if $R_\theta$ represents rotation about $M$ through the angle $\theta$, then $H = \langle R_\theta \rangle$. In order to see this, suppose that $R_\alpha$ is in $H$. We may assume that $0 < \alpha \le 2\pi$. Then, $\theta \le \alpha$ and there is some integer $t$ such that $t\theta \le \alpha < (t+1)\theta$. But, then $R_{\alpha-t\theta} = R_\alpha \circ R_\theta^{-t}$ is in $H$ and $0 \le \alpha - t\theta < \theta$. Since $\theta$ represents the smallest positive angle of rotation among the elements of $H$, we must have $\alpha - t\theta = 0$. Therefore, we conclude that $R_\alpha = R_\theta^t$. Consequently, $H$ is a cyclic subgroup of $G$. Note that the product of two orientation reversing isometries is orientation preserving. For convenience, suppose that $o(R_\theta) = n$. Now, if $G$ has no reflections, then we have proved that $G = H = \langle R_\theta \rangle \cong \mathbb{Z}_n$. Otherwise, $G$ contains at least one reflection, say $S$ about a line through $M$. In this case,

$$S, \ R_\theta S, \ R_\theta^2 S, \ \ldots, \ R_\theta^{n-1} S$$

are also reflections. Moreover, this is the entire set of reflections of $G$. Because, if $S'$ is any reflection in $G$, then $S'S$ is a rotation, and so $S'S = R_\theta^k$, for some positive integer $k$. Thus, we have $S' = R_\theta^k S^{-1} = R_\theta^k S$. Hence, we obtain

$$G = \{R_0, \ R_\theta, \ R_\theta^2, \ \ldots, \ R_\theta^{n-1}, \ S, \ R_\theta S, \ R_\theta^2 S, \ \ldots, \ R_\theta^{n-1} S\},$$

and $G$ is generated by the pair of reflections $S$ and $R_\theta S$. Therefore, $G$ is isomorphic to the dihedral group $D_n$. ∎

*Example 7.31*  Fig. 7.5 gives an illustration of an organism whose plane symmetry group consists of four rotations and is isomorphic to $\mathbb{Z}_4$.

*Example 7.32*  Fig. 7.6 shows a sample of Iranian art whose plane symmetry group consists of eight rotations and is isomorphic to $\mathbb{Z}_8$.

### Exercises

1. Termeh is the name given to a specialty cloth that originated in different parts of Iran. One of the best samples of it belongs to Yazd. Traditionally, the cloth was hand-woven using natural silk. Yazd is the center of the design and producing of Termeh. Figure 7.7 shows some samples of Termeh design. Find any symmetry groups in each of them.

**Fig. 7.5** Aurelia insulinda.
An organism whose plane
symmetry group is $\mathbb{Z}_4$

**Fig. 7.6** Iranian art. A
sample of Iranian art whose
plane symmetry group is $\mathbb{Z}_8$

## 7.5 Worked-Out Problems

**Problem 7.33 (Cauchy–Schwarz Inequality).** If $X, Y \in \mathbb{R}^n$, prove that

$$|\langle X, Y \rangle| \leq \|X\| \, \|Y\|. \tag{7.4}$$

This inequality is an equality if and only if one of $X, Y$ is a scalar multiple of the other.

*Solution* Suppose that $X, Y \in \mathbb{R}^n$. If $Y = 0$, then both sides of (7.4) equal 0 and the desired inequality holds. So, we can assume that $Y$ is nonzero. Consider the orthogonal decomposition

$$X = \frac{\langle X, Y \rangle}{\|Y\|^2} Y + \left( X - \frac{\langle X, Y \rangle}{\|Y\|^2} Y \right) = \frac{\langle X, Y \rangle}{\|Y\|^2} Y + Z,$$

where $Z$ is orthogonal to $Y$. Then, we obtain

**Fig. 7.7** Some samples of Termeh

$$\|X\|^2 = \left\|\frac{\langle X, Y \rangle}{\|Y\|^2} Y \right\|^2 + \|Z\|^2 = \frac{|\langle X, Y \rangle|^2}{\|Y\|^2} + \|Z\|^2 \geq \frac{|\langle X, Y \rangle|^2}{\|Y\|^2}.$$

This gives the Cauchy–Schwarz inequality (7.4). Looking at the proof of the Cauchy–Schwarz inequality, note that (7.4) is an equality if and only if the last inequality above is an equality. Obviously, this happens if and only if $Z = 0$. But $Z = 0$ if and only if $X$ is a multiple of $Y$. Thus, the Cauchy–Schwarz inequality is an equality if and only if $X$ is a scalar multiple of $X$ or $Y$ is a scalar multiple of $X$ (or both; the phrasing has been chosen to cover cases in which either $X$ or $Y$ equals 0). ∎

**Problem 7.34** Given an isometry $T : \mathbb{R}^2 \to \mathbb{R}^2$, an *invariant line* $\mathcal{L} \subset \mathbb{R}^2$ is a line that gets mapped by $T$ onto itself, i.e., $T(\mathcal{L}) = \mathcal{L}$. Note that this does not mean that the points $P \in \mathcal{L}$ are fixed.

(a) Show that an isometry $T$ has exactly one of the following:

    (1) A line of fixed points;
    (2) A single fixed point;
    (3) No fixed points, and a parallel family of invariant lines;
    (4) No fixed points, and a single invariant line.

(b) In each of the four cases in part (a), describe the isometry $T$ as a product of one, two, or three reflections along lines.

*Solution*

(a) It is clear that $T$ cannot have more than one of these properties, so we only need to show that it has one of them. We know that $T$ is either a rotation, a translation, or a glide reflection. If we dispense with the possibility that $T$ is the identity (so $T$ has property (1)), then these three types of isometries are disjoint (because glide reflections reverse orientation while the others don't, and rotations preserve one point while translations don't). If $T$ is a rotation, say about the point $P$, then it fixes only the point $P$, so $T$ has property (2). If $T$ is a translation, then it has no fixed points. In this case, say $T = T_{a,b}$. Then, for any $c \in \mathbb{R}$, the line $\mathcal{L}_c$ defined by $ay - bx = c$ is invariant under $T$. The family $\{\mathcal{L}_c \mid c \in \mathbb{R}\}$ is then an infinite family of parallel invariant lines, so $T$ has property (3). Finally, suppose that $T$ is a glide reflection. If $T$ is a pure reflection, then it fixes the line in which it reflects, so $T$ has property (1). Otherwise, $T = T_P \circ S_{\mathcal{L}}$, for some $P \neq O$ such that the segment $PO$ is parallel to the line $\mathcal{L}$. First, consider the reflection $S_{\mathcal{L}}$. All points on a given side of $S_{\mathcal{L}}$ have been flipped to the other side. Performing the translation $T_P$ keeps them on this new side, because they only move parallel to $\mathcal{L}$. Therefore, $T$ does not fix any points not in $\mathcal{L}$. Nor does fix any points in $\mathcal{L}$, because such points are fixed by $S_{\mathcal{L}}$ and then moved downstream by $T_P \neq id$. However, this observation shows that $\mathcal{L}$ is an invariant line under $T$. Finally, we show that $T$ has no other invariant lines. Suppose that $\mathcal{L}'$ is a line distinct from $\mathcal{L}$. If $\mathcal{L}'$ crosses $\mathcal{L}$, then $S_{\mathcal{L}}$ changes the slope of $\mathcal{L}'$, and the subsequent translation leaves that new slope unchanged, so $\mathcal{L}'$ cannot be invariant under $T$. If $\mathcal{L}'$ is parallel to $\mathcal{L}$, then it is totally contained on one side of $\mathcal{L}$, and $T$ moves $\mathcal{L}'$ to the other side, so it again is not invariant. We conclude that $T$ has property (4).

(b) In case (1), $T$ is a single reflection or the identity (which is the product of two instances of the reflection through any single line).

In case (2), $T$ is a rotation $R_{\theta,P}$ for some $\theta \neq 0$. Then, $T = S_{\mathcal{L}'} \circ S_{\mathcal{L}}$, where $\mathcal{L}'$ and $\mathcal{L}$ are any two lines which meet at $P$ and have angle $\theta/2$ from $\mathcal{L}$ to $\mathcal{L}'$.

In case (3), $T$ is a translation $T_P$ for $P \neq O$. Then $T = S_{\mathcal{L}'} \circ S_{\mathcal{L}}$, where $\mathcal{L}'$ and $\mathcal{L}$ are any two parallel lines perpendicular to the segment $PO$ and separated by a distance $\|P\|/2$ from $\mathcal{L}$ to $\mathcal{L}'$.

In case (4), $T$ is a glide reflection $T_P \circ S_{\mathcal{L}''}$, where $P \neq O$ and the line $\mathcal{L}$ is parallel to the segment $PO$. Then, $T_P = S_{\mathcal{L}'} \circ S_{\mathcal{L}}$ can be deconstructed as above, so we have $T = S_{\mathcal{L}'} \circ S_{\mathcal{L}} \circ S_{\mathcal{L}''}$. Here $\mathcal{L}''$ is known from the original decomposition, and $\mathcal{L}'$ and $\mathcal{L}$ are any lines perpendicular to $\mathcal{L}$ such that $\mathcal{L}'$ is a distance $\|P\|/2$ from $\mathcal{L}$.

∎

**Problem 7.35** Let $S^2$ denote the sphere of radius 1 around the origin $O \in \mathbb{R}^3$. Given two pairs of vectors $(X_1, X_2)$ and $(Y_1, Y_2)$ in $\mathbb{R}^3$ we say they are *equidistant pairs* if $\|X_1 - X_2\| = \|Y_1 - Y_2\|$. Prove that the action of $SO_3(\mathbb{R})$ on $S^2$ is transitive, and moreover is transitive on equidistant pairs (whose points lie in $S^2$).

*Solution* In order to see that $SO_3(\mathbb{R})$ acts transitively on $S^2$, consider points $X, Y \in S^2$. We may assume that $X$ and $Y$ are distinct, in which case the define a plane in $\mathbb{R}^3$. Let $Z$ be a vector in $S^2$ perpendicular to this plane. Then it is clear that there is a rotation about the line through $Z$ which sends $X$ to $Y$ as required. For the second part, let $(X_1, X_2)$ and $(Y_1, Y_2)$ in $\mathbb{R}^3$ be an equidistant pair. First pick $T \in SO_3(\mathbb{R})$ such that $T(X_1) = Y_1$. Then we have

$$\|Y_1 - T(X_2)\| = \|T(X_1) - T(X_2)\| = \|Y_1 - Y_2\|,$$

so that $(Y_1, T(X_2))$ and $(Y_1, Y_2)$ are an equidistant pair. Thus, we need to find $T' \in \text{Stab}_{SO_3(\mathbb{R})}(Y_1)$ such that $T'(T(X_2)) = Y_2$. The stabilizer is exactly the subgroup of rotations about the axis given by $Y_1$. Then the equidistant criterion is exactly what is required. ∎

**Problem 7.36** Prove that the group $SO_3(\mathbb{R})$ is simple.

*Solution* We use Problem 7.35. The strategy is to show that a normal subgroup must contain certain kinds of elements, and then show that these elements in fact generate $SO_3(\mathbb{R})$. Suppose that $\{I_3\} \neq K \trianglelefteq SO_3(\mathbb{R})$ is a normal subgroup of $SO_3(\mathbb{R})$. Then pick $T$ as a non-identity element in $K$, so that there is a vector $U \in S^2$ such that $T$ is the rotation about $U$ by the angle $\theta$, $(0 \leq \theta \leq \pi)$. Then, if we pick $X \in S^2$ is the plane perpendicular to $U$, we have $\|X - T(X)\| = 2\sin\theta$. Now, suppose that $k$ is a positive integer such that $0 < \pi/k < \theta$. Then setting $b = \sin(\pi/k)/\sin\theta$ it follows that $Y = bX + \sqrt{1 + b^2}U$ has

$$\|Y - T(Y)\| = b\|X - T(X)\| = 2\sin(\pi/k).$$

Now, let $X_1 \in S^2$ be the vector perpendicular to $U$ between $X$ and $T(X)$ such that $\|X_1 - X\| = 2\sin(\pi/k)$. Then using the second part of Problem 7.35, we can find a $T' \in SO_3(\mathbb{R})$ such that $T'(Y) = X$ and $T'(T(Y)) = X_1$. Now since $K$ is normal, we have $T_1 = T' \circ T \circ T'^{-1} \in K$, and $T_1(X) = X_1$. Repeating this argument with $X_1$ instead of $X$, we find $T_2 \in K$ and $X_2 = T_2(X_1)$ such that $X_2 \in S^2$ is perpendicular to $U$ with $\|X_1 - X_2\| = 2\sin(\pi/k)$, and continuing in this way, we get $T, T_1, \ldots, T_k \in K$ and $X_1, X_2, \ldots, X_k \in \mathbb{R}^3$ all perpendicular to $U$ with $T_i(X_{i-1}) = X_i$ and $\|X_i - X_{i-1}\| = 2\sin(\pi/k)$. But if $\rho = T_k \circ T_{k-1} \circ \ldots \circ T_1 \circ T$, then $\rho(X) = -X$. But then $\rho \in K$ is a rotation by $\pi$, i.e., a rotation of order 2, and so since $K$ is normal, it contains all rotations of order 2. But these generate $SO_3(\mathbb{R})$, so that $K = SO_3(\mathbb{R})$ as required. ∎

## 7.6 Supplementary Exercises

1. **(Triangle Inequality).** If $X, Y \in \mathbb{R}^n$, prove that $\|X + Y\| \leq \|X\| + \|Y\|$. This inequality is an equality if and only if one of $X, Y$ is a non-negative multiple of the other.

2. Consider the lines

$$\mathcal{L}_1 = \{(x, y) \mid x + y = 2\} \text{ and } \mathcal{L}_2 = \{(x, y) \mid x - y = 2\}.$$

Find the effects on the point $P(1, 0)$ of the reflections $S_{\mathcal{L}_1}$ and $S_{\mathcal{L}_2}$.

3. Consider the lines

$$\mathcal{L}_1 = \{(x, y) \mid 2x + y = 0\} \text{ and } \mathcal{L}_2 = \{(x, y) \mid 2x + y = 2\}.$$

Express each of the isometries (reflections combination) $S_{\mathcal{L}_1} \circ S_{\mathcal{L}_2}$ and $S_{\mathcal{L}_2} \circ S_{\mathcal{L}_1}$ as translations.

4. Show that a triangle that has two axes of symmetry has three axes of symmetry.

5. Give an example of a bijection $\phi : \mathbb{R}^2 \to \mathbb{R}^2$ that preserves angles but not distances. Describe in general terms the effect of m on lines, circles, and triangles.

6. Verify that the translation subgroup is closed under conjugation in $E_n$.

7. Let $R$, $S$, and $T$ be a rotation, a reflection, and a translation, respectively. Find necessary and sufficient conditions on the parameters for the following to be a reflection.

   (a) $R \circ S$;
   (b) $R \circ T$.

8. Prove that $T : \mathbb{R}^2 \to \mathbb{R}^2$ defined by

$$T\left(\begin{bmatrix} x \\ y \end{bmatrix}\right) = \begin{bmatrix} \dfrac{3}{5}x - \dfrac{4}{5}y - 14 \\ \dfrac{4}{5}x - \dfrac{3}{5}y + 3 \end{bmatrix}$$

represents a glide reflection. Find the axis of the glide and the amount by which points are translated along this axis.

9. Let $G$ be a subgroup of some dihedral group. For each $x$ in $G$, define

$$f(x) = \begin{cases} 1 & \text{if } x \text{ is a rotation} \\ -1 & \text{if } x \text{ is a reflection} \end{cases}$$

Prove that $f$ is a homomorphism from $G$ to the multiplicative group $\{1, -1\}$. What is the kernel?

10. What is the probability that a randomly selected element from $D_4$ commutes with the vertical reflection $V$?

11. If $\psi$ is a glide reflection with axis $\mathcal{L}$, and if $T$ is an isometry, show that $T\psi T^{-1}$ is a glide along the line $T(\mathcal{L})$.

12. Prove that, in the decomposition of a glide reflection as the product $S \circ T$ of a reflection $S$ and a translation $T$ along the axis of $S$, the factors are unique.

13. Show that any glide reflection can be written as the product of reflections in the sides of an equilateral triangle.

14. Let $T$ denote translation by $v$ and $S$ be reflection in the line $\mathcal{L}$. Prove that $S \circ T$ is a reflection if $v$ is perpendicular to $\mathcal{L}$ and a glide reflection otherwise.

15. Let $S$ and $S'$ be distinct reflections in $D_{21}$. What are the possibilities for $o(S \circ S')$?

16. Suppose that $H$ is a proper subgroup of $D_{35}$ and $H$ contains at least two reflections. What are the possible orders of $H$? Explain your reasoning.

17. For $0 \leq m \leq n$, show the orthogonal matrix

$$
\begin{bmatrix}
-1 & 0 & \cdots & 0 & 0 \\
0 & -1 & \cdots & 0 & 0 \\
\vdots & \vdots & \ddots & \vdots & \vdots \\
0 & 0 & \cdots & 1 & 0 \\
0 & 0 & \cdots & 0 & 1
\end{bmatrix}
$$

with $m$ "1's" and $n - m$ "$-1$'s" on the diagonal is a composition of $m$ reflections in $O_n(\mathbb{R})$ and not less than $m$ refections in $O_n(\mathbb{R})$.

# Chapter 8
# Platonic Solids

We recall that a polygon is a geometric object consisting of a number of points, vertices, and an equal number of line segments and sides, namely a cyclically ordered set of points in a plane, with no three successive points collinear, together with the line segments joining consecutive pairs of the points. In this chapter we classify all possible finite subgroups of rotations in $\mathbb{R}^3$.

## 8.1 Platonic Solids

A regular polygon is a $p$-sided polygon in which the sides are all the same length and are symmetrically placed about a common center. A polygon is *convex* if the line connecting any two vertices remains inside or on the boundary of the polygon.

**Definition 8.1** A *polyhedron* is a three-dimensional solid which consists of a collection of polygons joined at their edges. A polyhedron is said to be *convex* if the planes that bound the solid do not enter its interior.

**Definition 8.2** A *regular polyhedron* is convex, with all of its faces congruent regular polygons and with the same number of faces at each vertex.

It is the case that there are only five solids in total that satisfy this definition. They are the five Platonic solids: tetrahedron, hexahedron, octahedron, dodecahedron, and icosahedron. These can be seen, respectively, in Fig. 8.1. The Platonic solids have a rich history. The original discovery of the Platonic solids is unknown. Indeed, the Platonic solids have been known since antiquity. Ornamented models of them can be found among the carved stone balls created by the late Neolithic people of Scotland at least 1000 years before Plato [9].

The five regular polyhedra all appear in nature whether in crystals or in living beings. They also appear all throughout history in children's toys, dice, art, and many

B. Davvaz, *Groups and Symmetry*, https://doi.org/10.1007/978-981-16-6108-2_8

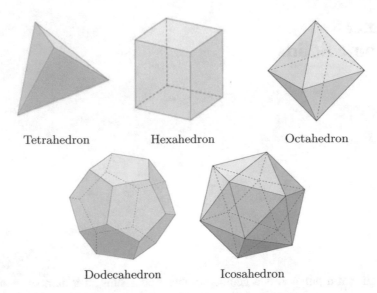

Tetrahedron            Hexahedron            Octahedron

Dodecahedron          Icosahedron

**Fig. 8.1**  Regular solid (or Platonic solid)

other areas. Around 360 BC the regular polyhedra are discussed in the dialogues of Plato, their namesake. The varying solids are likened to each of the elements, earth, fire, water, and air. He compared the tetrahedron to fire because the sharp stabbing points and edges reminded Plato of the stabbing heat from flames. The octahedron was associated with air because its many small smooth parts make it seem as though it is barely there. Plato saw the icosahedron as water because it flows, like water, out of one's hands. And the cube was associated with earth because it causes dirt to crumble and fall apart, and at the same time it resembles the solidity of earth. The fifth solid, the dodecahedron, was considered the shape that encompasses the whole universe and was used for arranging the constellations in the heavens [12, 19]; see Fig. 8.2.

In Euclid's famous text, Elements (published circa 300 BC), he begins with a discussion of constructing the equilateral triangle. Elements concludes with a detailed construction of the five regular solids. Euclid claims that these are in fact the only five with the necessary properties to be considered regular. Some people believe that Euclid created this text with the ultimate goal of establishing the five Platonic solids [35].

Similar to Plato, the German astronomer Johannes Kepler also searched for connections between the regular polyhedra and the natural world. In 1596 he published his astronomy book, Mysterium Cosmographicum, translated as The Cosmographic Mystery, where he proposed his model for our solar system. At the time they knew of five planets other than earth. Kepler predicted connections between these five planets and the five Platonic solids. His model had each planet's orbit associated with a sphere, and the distance between the spheres was determined by a Platonic

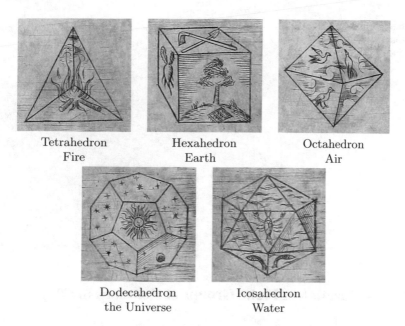

|  |  |  |
|---|---|---|
| Tetrahedron | Hexahedron | Octahedron |
| Fire | Earth | Air |

| | |
|---|---|
| Dodecahedron | Icosahedron |
| the Universe | Water |

**Fig. 8.2** Five regular solids as depicted by Juhannese Kepler in Harmonices Mundi, Book II (1619)

solid, as seen in Fig. 8.3. The spheres of orbits circumscribed and inscribed each Platonic solid. The outmost sphere represented the orbit of Saturn. The remaining planets, moving in toward the sun, were Jupiter, Mars, Earth, Venus, and Mercury. The Platonic solids, again moving inward to the sun, were the cube, tetrahedron, dodecahedron, icosahedron, and octahedron. Kepler ultimately determined this model to be incorrect; however his astronomical research was quite fruitful. Notably, he developed Kepler's laws of planetary motion. The first of these laws asserts that a planet's orbit is in fact an ellipse.

**Exercises**

1. Give an example of convex regular polygon.
2. If a sphere is circumscribed around a cube with side length 2, what is the surface area of the sphere?
3. What is the volume of a sphere inscribed in a cube with surface area 64?
4. If an ant starts at a vertex of an octahedron and only walks on the edges, is it possible for the ant to walk along each edge exactly once without repeating any? Which of Platonic solids is this possible on?
5. Let $a$ be the side length of a regular polygon or edge length of a Platonic solid. Find the surface area of each Platonic solid.

**Fig. 8.3** Kepler's model of
the solar system using the
Platonic solids

## 8.2   Classification of Finite Groups of Rotations in $\mathbb{R}^3$

Now we turn to the classification of finite three-dimensional rotation groups.

**Theorem 8.3** *If $G$ is a finite subgroup of $SO_3(\mathbb{R})$, then $G$ is isomorphic to precisely one of the following groups: $\mathbb{Z}_n$, $D_n$, $A_4$, $S_4$, and $A_5$.*

**Proof** Let $G$ be a finite subgroup of $SO_3(\mathbb{R})$. Each non-identity element of $G$ represents a rotation in $\mathbb{R}^3$ around an axis passing through the origin. Take the unit sphere centered at the origin $(0, 0, 0)$. Then each rotation gives two poles on the unit sphere which are the intersection of the axis of rotation with the unit sphere. Suppose that $\Omega$ is the set of all poles of all the non-identity elements in $G$. Since $G$ is finite, it follows that $\Omega$ is finite. We show that $G$ acts on the set *Omega*. Indeed, if $X \in Omega$ is a pole for $A \in G$ (i.e., $A(X) = X$) and $B \in G$, then $(BAB^{-1})(B(X)) = BA(X) = B(X)$. Further, we have $\|B(X)\| = \|X\|$. This yields that $B(X)$ is a pole for $BAB^{-1}$, and so $B(X) \in \Omega$. The identity element fixes every pole, and each non-identity element of $G$ fixes exactly two poles. Let $X_1, \ldots, X_n$ be a representative from each distinct orbit. Hence, by Burnside's lemma, we have

$$k = \frac{1}{|G|}\left(2(|G| - 1) + \sum_{i=1}^{k} |\mathrm{Orb}_G(X_i)|\right).$$

By using Orbit-Stabilizer theorem, we conclude that

$$2\left(1 - \frac{1}{|G|}\right) = k - \sum_{i=1}^{k} \frac{|\text{Orb}_G(X_i)|}{|G|} = k - \sum_{i=1}^{k} \frac{1}{|\text{Stab}_G(X_i)|}$$
$$= \sum_{i=1}^{k} \left(1 - \frac{1}{|\text{Stab}_G(X_i)|}\right). \tag{8.1}$$

Suppose that $G$ is non-trivial, i.e., $|G| > 1$. Then, we obtain

$$1 \le 2\left(1 - \frac{1}{|G|}\right) < 2. \tag{8.2}$$

Since $\text{Stab}_G(X_i)$ contains at least the identity element and one rotation, it follows that $|\text{Stab}_G(X_i)| \ge 2$, for all $1 \le i \le k$. Hence, we obtain

$$\frac{1}{2} \le 1 - \frac{1}{|\text{Stab}_G(X_i)|} < 1, \tag{8.3}$$

for all $1 \le i \le k$. Therefore, by (8.1), (8.2), and (8.3), we deduce that $2 \le k \le 4$, which implies that $k = 2$ or $3$.

If $k = 2$, then we obtain $|\text{Stab}_G(X_1)| + |\text{Stab}_G(X_2)| = 2$. This implies that $|\text{Stab}_G(X_1)| = |\text{Stab}_G(X_2)| = 1$, and so each orbit contains one pole. In this case, we have only two poles. Hence, each rotation has the same axis. The plane passing through the origin and perpendicular to this axis is preserved by $G$. Consequently, $G$ is isomorphic to a subgroup of $SO_2(\mathbb{R})$. Now, by Theorem 7.30, we conclude that $G \cong \mathbb{Z}_n$, for some integer $n$.

If $k = 3$, then

$$1 + \frac{2}{|G|} = \frac{1}{|\text{Stab}_G(X_1)|} + \frac{1}{|\text{Stab}_G(X_2)|} + \frac{1}{|\text{Stab}_G(X_3)|} > 1. \tag{8.4}$$

Note that if $|\text{Stab}_G(X_i)| \ge 3$, for all $1 \le i \le 3$, then $1 + 2/|G| \le 1/3 + 1/3 + 1/3 = 1$, and this is a contradiction. Hence, we conclude that $|\text{Stab}_G(X_i)| = 2$, for some $1 \le i \le 3$. Without loss of generality, assume that $|\text{Stab}_G(X_1)| = 2$. Therefore, we have

$$\frac{1}{|\text{Stab}_G(X_2)|} + \frac{1}{|\text{Stab}_G(X_3)|} = \frac{1}{2} + \frac{2}{|G|}.$$

Both $|\text{Stab}_G(X_2)|$ and $|\text{Stab}_G(X_3)|$ cannot be greater than 3, and so assume that $|\text{Stab}_G(X_2)| = 2$ or $3$.

If $|\text{Stab}_G(X_1)| = 2$ and $|\text{Stab}_G(X_3)| = 3$, then

$$\frac{1}{|\text{Stab}_G(X_3)|} = \frac{1}{6} + \frac{2}{|G|}. \tag{8.5}$$

We observe that the solutions of (8.5) are as follows:

$$|\text{Stab}_G(X_3)| = 3 \text{ and } |G| = 12,$$
$$|\text{Stab}_G(X_3)| = 4 \text{ and } |G| = 24,$$
$$|\text{Stab}_G(X_3)| = 5 \text{ and } |G| = 60.$$

Hence, we can consider the following cases:

(1) $|\text{Stab}_G(X_1)| = |\text{Stab}_G(X_2)| = 2$ and $|\text{Stab}_G(X_3)| = n$, for some $n \geq 2$;
(2) $|\text{Stab}_G(X_1)| = 2$ and $|\text{Stab}_G(X_2)| = |\text{Stab}_G(X_3)| = 3$;
(3) $|\text{Stab}_G(X_1)| = 2$, $|\text{Stab}_G(X_2)| = 3$, and $|\text{Stab}_G(X_3)| = 4$;
(4) $|\text{Stab}_G(X_1)| = 2$, $|\text{Stab}_G(X_2)| = 3$, and $|\text{Stab}_G(X_3)| = 5$.

*Case 1*: If $|\text{Stab}_G(X_3)| = 2$, then by (8.4), we get $|G| = 4$. We know that, up to isomorphism, there exist only two groups of order 4, namely $\mathbb{Z}_4$ and $D_2 \cong \mathbb{Z}_2 \times \mathbb{Z}_2$.

If $|\text{Stab}_G(X_3)| = n \geq 3$, then by (8.4), we get $|G| = 2n$. Since $[G : \text{Stab}_G(X_3)] = 2$, it follows that $\text{Stab}_G(X_3) \trianglelefteq G$ and

$$|\text{Orb}_G(X_3)| = |G|/|\text{Stab}_G(X_3)| = 2.$$

This means that $\text{Orb}_G(X_3)$ contains exactly two poles. The axis through $X_3$ is fixed by $A \in \text{Stab}_G(X_3)$, and so $\text{Stab}_G(X_3)$ is cyclic of order $n$. Take $A$ to be the minimal rotation in $\text{Stab}_G(X_3)$, the generator of $\text{Stab}_G(X_3)$. We claim that $X_1, A(X_1), \ldots, A^{n-1}(X)$ are all distinct. To see this suppose that $A^i(X_1) = A^j(X_1)$, for some $i > j$. Then, we have $A^{i-j}(X_1) = X_1$. But $X_3$ and $-X_3$ are the only two poles fixed by $\text{Stab}_G(X_3)$ and $X_1$ cannot be equal to $-X_3$. Hence, $A^i(X_1)$ cannot be equal to $A^j(X_1)$. Finally, we use the fact that $A$ preserves distance, and we write

$$\|X_1 - A(X_1)\| = \|A(X_1) - A^2(X_1)\| = \cdots = \|A^{n-1}(X_1) - X_1\|.$$

This yields that the points $X_1, A(X_1), \ldots, A^{n-1}(X)$ lie in the same plane and they must be the vertices of a regular $n$-gon. Other elements of $G$ reflect the regular $n$-gon about an axis in the its plane. Therefore, $G$ is the rotational symmetry group of the $n$-gon, $D_n$.

*Case 2*: In this case by the Orbit-Stabilizer theorem, we obtain $\text{Orb}_G(X_3)| = 4$. Let $A$ be the generator of $\text{Stab}_G(X_3)$, i.e., $\text{Stab}_G(X_3) = \langle A \rangle$, and let $Y \in \text{Stab}_G(X_3)$. Clearly, $Y \neq X_3$ and $Y \neq -X_3$. Further, we observe that $Y$, $A(Y)$, and $A^2(Y)$ are distinct and equidistant from $X_3$, and lie at the corner of an equilateral triangle, as Fig. 8.4. Now, the orbit $\text{Orb}_G(X_3) = \{X_3, Y, A(Y), A^2(Y)\}$ is preserved under the action of $G$. For $B \in \text{Stab}_G(Y)$, we have $B(Y) = Y$ and $B$ permutes $X_3$, $A(Y)$ and $A^2(Y)$. Since $B$ preserves distances, it follows that the distance from $Y$ to $X_3$, $A(Y)$, and $A^2(Y)$ is equal. Consequently, $\{X_3, Y, A(Y), A^2(Y)\}$ form a regular tetrahedron $\mathcal{T}$. Now, we can consider a homomorphism $\varphi : G \to H$, where $H$ is the rotational symmetry group of $\mathcal{T}$. Since no non-identity rotation fixes $\mathcal{T}$, it follows that the kernel of $\varphi$ is singleton, and hence $\varphi$ is one to one. Since $|G| = |H| = 12$, we conclude that $\varphi$ is an isomorphism and $G \cong H \cong A_4$.

**Fig. 8.4** Rotational
symmetries of a regular
tetrahedron

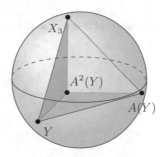

*Case 3*: In this case, by the Orbit-Stabilizer theorem, the orbit of $X_3$ has 6 elements. As $\text{Stab}_G(X_3)$ is a cyclic group of order 4, let $\text{Stab}_G(X_3) = \langle A \rangle$, and choose $Y \in \text{Orb}_G(X_3)$ such that $Y$ is neither $X_3$ nor $-X_3$. As before, we observe that $Y$, $A(Y)$, $A^2(Y)$, and $A^3(Y)$ are distinct and equidistant from $X_3$, and lie at the corner of a square. As $-X_3 \notin \text{Orb}_G(X_1) \cup \text{Orb}_G(X_2)$ (otherwise $|\text{Orb}_G(-X_3)| = |\text{Orb}_G(X_1)|$ or $|\text{Orb}_G(X_2)|$), we have

$$\text{Orb}_G(X_3) = \{X_3, -X_3, Y, A(Y), A^2(Y), A^3(Y)\}$$

Now, $-Y \in \text{Orb}_G(X_3)$ (as $|\text{Orb}_G(-Y)| = |\text{Orb}_G(Y)| = |\text{Orb}_G(X_3)|$) and $-Y \neq X_3$ and $-Y \neq -X_3$. Since $Y$, $A(Y)$, $A^2(Y)$, and $A^3(Y)$ form a square, it follows that $\|A(Y) - Y\| = \|A^3(Y) - Y\| < 2$. Hence, we conclude that $-Y = A^2(Y)$. We can look at the orbit starting at any one of the points $X_3$, $-X_3$, $Y$, $A(Y)$, $A^2(Y)$, and $A^3(Y)$, and we always see four equidistant points forming a square. Hence, we see that the points are the vertices of a regular octahedron (see Fig. 8.5). Let $H$ be the rotational symmetry group of this regular octahedron. Then we can define a one-to-one homomorphism $\varphi : G \to H$. Since $|G| = |H| = 24$, it follows that $\varphi$ is an isomorphism, and so $G \cong H \cong S_4$. This shows that $G$ is also the rotational symmetry group of a cube.

*Case 4*: In this case, by the Orbit-Stabilizer theorem, the orbit of $X_3$ has 12 elements. As $\text{Stab}_G(X_3)$ is a cyclic group of order 5, suppose that $A$ is the minimal rotation which generates $\text{Stab}_G(X_3)$. Let $Y$ and $Z$ be in the orbit of $X_3$ such that $0 < \|X_3 - Y\| < \|X_3 - Z\| < 2$. This keeps everything in the unit sphere. Moreover, $Y$, $A(Y)$, $A^2(Y)$, $A^3(Y)$, and $A^4(Y)$ are distinct and equidistant from $X_3$. Similarly, $Z$, $A(Z)$, $A^2(Z)$, $A^3(Z)$, and $A^4(Z)$. These two sets of points lie at the vertices of two regular pentagons, and since the two sets of points are not the same distance away from $X_3$ we have two separate regular pentagons. Moreover, we can check that $\{X_3, -X_3, Y, A(Y), A^2(Y), A^3(Y), A^4(Y), Z, A(Z), A^2(Z), A^3(Z), A^4(Z)\}$ form a regular icosahedron; see Fig. 8.6, where each element (corner) is labeled by a number. Now, if $H$ is the rotational symmetry group of this regular icosahedron, then we obtain a one-to-one homomorphism $\varphi : G \to H$. Since $|G| = |H| = 60$, it follows that $\varphi$ is an isomorphism, and we conclude that $G \cong H \cong A_5$.                                  ∎

**Fig. 8.5** Rotational
symmetries of an octahedron

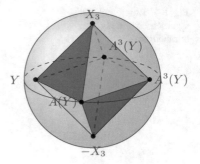

**Fig. 8.6** Rotational
symmetries of an
icosahedron

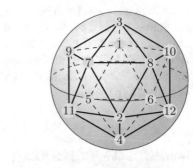

**Fig. 8.7** A 3-prism

## Exercises

1. Glue two dodecahedral faces together along a pentagonal face, and find the rotational symmetry group of this new solid. What is its full symmetry group?
2. Realize each of $\mathbb{Z}_{2n}$, $D_{2n}$, and $A_4 \times \mathbb{Z}_2$ as the (full) symmetry group of an appropriate solid.
3. Show that the group of rotations in $\mathbb{R}^3$ of a 3-prism (i.e., a prism with equilateral ends, as in Fig. 8.7) is isomorphic to $D_3$.
4. What is the order of the (entire) symmetry group in $\mathbb{R}^3$ of a 3-prism?
5. What is the order of the symmetry group in $\mathbb{R}^3$ of a 4-prism (a box with square ends that is not a cube)?
6. What is the order of the symmetry group in $\mathbb{R}^3$ of an $n$-prism?

## 8.3 Worked-Out Problems

**Problem 8.4** Determine the group $G$ of rotations of the solid in Fig. 8.8, which is composed of six congruent squares and eight congruent equilateral triangles.

*Solution* We start by separating out any one of the squares. It is clear that there exist four rotations that map this square to itself, and the selected square can be rotated to the position of any of the other five. Hence, by the Orbit-Stabilizer theorem, the rotation group has order 24. By Theorem 8.3, $G$ is one of $\mathbb{Z}_{24}$, $D_{12}$, and $S_4$. But each of the first two groups has exactly two elements of order 4, whereas $G$ has more than two. Therefore, we conclude that $G$ is isomorphic to $S_4$. ∎

**Problem 8.5** Consider a square, centered at the origin in the $xy$-plane (Fig. 8.9) in $\mathbb{R}^3$ with equation $z = 0$. Find its group of rotations.

*Solution* The square has four rotational symmetries, with axis of rotation the $z$-axis, represented by the matrices of orders 1, 4, 2, and 4, respectively,

**Fig. 8.8** A solid which is composed of six congruent squares and eight congruent equilateral triangles

**Fig. 8.9** A square centered at the origin in the $xy$-plane

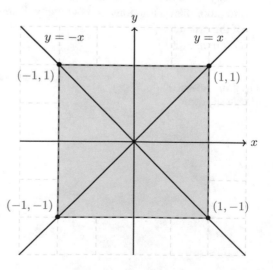

$$\begin{bmatrix} 1 & 0 & 0 \\ 0 & 1 & 0 \\ 0 & 0 & 1 \end{bmatrix}, \begin{bmatrix} 0 & -1 & 0 \\ 1 & 0 & 0 \\ 0 & 0 & 1 \end{bmatrix}, \begin{bmatrix} -1 & 0 & 0 \\ 0 & -1 & 0 \\ 0 & 0 & 1 \end{bmatrix}, \begin{bmatrix} 0 & 1 & 0 \\ -1 & 0 & 0 \\ 0 & 0 & 1 \end{bmatrix}.$$

Also, the square has four rotational symmetries of order 2 given by

$$\begin{bmatrix} 1 & 0 & 0 \\ 0 & -1 & 0 \\ 0 & 0 & -1 \end{bmatrix}, \begin{bmatrix} 0 & 1 & 0 \\ 1 & 0 & 0 \\ 0 & 0 & -1 \end{bmatrix}, \begin{bmatrix} -1 & 0 & 0 \\ 0 & 1 & 0 \\ 0 & 0 & -1 \end{bmatrix}, \begin{bmatrix} 0 & -1 & 0 \\ -1 & 0 & 0 \\ 0 & 0 & -1 \end{bmatrix},$$

with axes of rotation the $x$-axis; the line $y = x$ with $z = 0$; the $y$-axis; and the line $y = -x$ with $z = 0$, respectively. Assume that

$$R = \begin{bmatrix} 0 & -1 & 0 \\ 1 & 0 & 0 \\ 0 & 0 & 1 \end{bmatrix} \text{ and } S = \begin{bmatrix} 1 & 0 & 0 \\ 0 & -1 & 0 \\ 0 & 0 & -1 \end{bmatrix}.$$

It is easy to check that

$$R^4 = I_3, \ S^2 = I_3, \text{ and } SRS = R^3.$$

The above eight rotations do in fact a group isomorphic to $D_4$. ∎

## 8.4  Supplementary Exercises

1. Consider a regular hexagonal prism (see Fig. 8.10). It has twelve vertices. We want to color each vertex black or white. We consider two colorings as the same if they can be obtained from each other via a rigid motion in $\mathbb{R}^3$. Note that there are more than 6 rotations. In how many different ways can we color the vertices of the prism?
   *Solution*: 382.

**Fig. 8.10**  A regular hexagonal prism

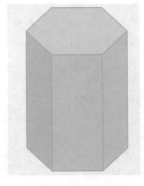

**Fig. 8.11** A stellar octangular

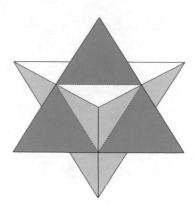

2. A stellar octangular is a solid, which is made up of two regular tetrahedra arranged so that their edges bisect one another at right angles (see Fig. 8.11). Find its rotational symmetry group.

3. Then $G$ acts on the sphere $S$ of radius one centered at the origin. We define a point of $S$ to be a *pole* of $G$ if it is fixed by at least one $g \in G$ with $g \neq e$. Let $P$ be the set of poles. Consider a regular tetrahedron inscribed in the unit sphere $S$. Show that the set of poles $P$ of the symmetry group $T$ of the tetrahedron consists of the vertices, midpoints of edges, and midpoints of faces extended to $S$. Show that the $T$ action on $P$ has three orbits, where one of them has a stabilizer of order 2 and the remaining two have stabilizers of order 3.

4. Determine the poles of the symmetry group of the cube, and determine the orbits and stabilizers as in Exercise 3.

5. Find solids whose rotational symmetry groups are isomorphic to $SO_2(\mathbb{R})$, $SO_2(\mathbb{R}) \times \mathbb{Z}_2$ and $O_2(\mathbb{R})$, respectively.

6. Prove that $SO_3(\mathbb{R})$ does not contain a subgroup which is isomorphic to $SO_2(\mathbb{R}) \times SO_2(\mathbb{R})$.

7. Let $G$ be a finite subgroup of $SO_3(\mathbb{R})$ and $P$ be the set of poles. Show that if $x \in P$, and $g \in G$, then $gx \in P$.

# Chapter 9
# Frieze and Wallpaper Symmetry Groups

Frieze and wallpaper groups are discovered and studied more than one hundred years ago. In this chapter we study classification of two-dimensional repeated patterns in terms of their respective symmetry groups, the well-known seven frieze groups, and seven wallpaper groups.

## 9.1 Point Groups and Rosette Groups

Elements or operations of symmetry, either singly or in combination with one another, may intersect at a point and constitute a point group. Thus, a point group is a group of symmetry operations that leaves a point invariant.

**Definition 9.1** A *point group* is a group of geometric symmetries (isometries) that keep at least one point fixed. Point groups can exist in a Euclidean space with any dimension, and every point group in dimension $n$ is a subgroup of the orthogonal group $O_n(\mathbb{R})$.

*Example 9.2* In nature the flowers follow the rule of symmetry. In Fig. 9.1, you can see one point fixed.

Point groups can be realized as sets of orthogonal matrices $M$ that transform point $X$ into point $Y$ by $Y = MX$, where the origin is the fixed point. Point-group elements can be either rotations (determinant of $M = 1$) or else reflections, or improper rotations (determinant of $M = -1$).

**Definition 9.3** A *rosette* is a plane figure $F$ with the following properties:

(1) There exists a non-identity rotational symmetry of $F$ with minimal positive rotation angle $\theta$.
(2) All non-identity rotational symmetries of $F$ have the same center.

**Fig. 9.1** Symmetry in flowers

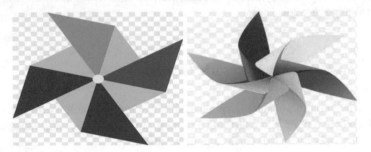

**Fig. 9.2** Rosettes as flowers

The symmetry group of a rosette is called a *rosette group*.

*Example 9.4* Typically we can think of a rosette as a pinwheel (see Fig. 9.2) or a flower with $n$-petals (see Fig. 9.3).

**Theorem 9.5** *The rosette groups are either dihedral $D_n$ or finite cyclic $C_n$ with $n \geq 2$.*

**Proof** It is an immediate consequence of Theorem 7.30.                          ∎

**Definition 9.6** Let $F_1$ and $F_2$ be rosettes with respective symmetry groups $G_1$ and $G_2$. Rosettes $F_1$ and $F_2$ have the same symmetry type if there is an inner automorphism of the group of all plane isometries that restricts to an isomorphism $f: G_1 \to G_2$.

**Fig. 9.3** A typical rosette
with 32 petals

**Corollary 9.7** *Two rosettes have the same symmetry type if and only if their respective symmetry groups are isomorphic.*

**Proof** It is straightforward. ∎

**Exercises**

1. Let $H$ be the translation subgroup of $E_2$ and $G$ be a subgroup of $E_2$ such that $G \cap H$ contains exactly the identity. Prove that there is a point in the plane fixed by every element of $G$.
2. Prove that a point in or on a plane triangle of maximal distance from one of the vertices can only be one (or both) of the other two.

## 9.2 Frieze Groups

Frieze patterns are typically the familiar decorative borders often seen on walls or facades extended infinitely far in either direction (see Fig. 9.4).

**Fig. 9.4** Two typical frieze patterns

We recall that if $v$ is a vector in $\mathbb{R}^2$, then a function $T_v : \mathbb{R}^2 \to \mathbb{R}^2$ defined by $T_v(r) = r + v$, for all $r \in \mathbb{R}^2$, is a translation by the vector $v$. Let $T_v$ be a translation. The length of $T_v$, denoted by $\|T_v\|$, is the length of the vector of $v$.

**Definition 9.8** A *frieze pattern* is a plane figure $F$ with the following properties:

(1) There is a translational symmetry $T_v$ of $F$ with minimal length; i.e., if $T_w$ is any non-identity translational symmetry of $F$, then $0 < \|T_v\| \leq \|T_w\|$.
(2) All non-identity translational symmetries of $F$ fix the same lines.

**Definition 9.9** The symmetry group of a frieze pattern is called a *frieze group*.

Consider a row of equally spaced letter $F$'s extending infinitely far in either direction (see Fig. 9.5). The frieze pattern of type 1 has only translational symmetry. There are two translational symmetries of minimal length (the distance between centroids of consecutive $F$'s), one shifting left and the other shifting right. If $T_v$ is a translational symmetry of shortest length, then $T_v^n \neq id$, for all $n \neq 0$, and the frieze group of $F_1$ is the infinite cyclic group $F_1 = \langle T_v \rangle$.

The frieze pattern of type 2 has a glide reflection symmetry (see Fig. 9.6). The frieze pattern of type 2 like that of frieze pattern of type 1 is infinite cyclic group. Let $T$ be a glide reflection such that $T^2$ is a translational symmetry of shortest length. Then, $T^n$ is non-identity for all nonzero integer $n$ and $T^2$ generates just the translation subgroup. We may write the frieze group $F_2$ as $F_2 = \{T^n \mid n \in \mathbb{Z}\}$.

The frieze group $F_3$ of pattern of type 3 (see Fig. 9.7) is generated by a translation $T_v$ and a reflection $S_\ell$ about the dotted vertical line $\ell$. (There are infinitely many axes of reflection symmetry including those midway between constructive $F$'s facing the same direction. And one will do.) In this case, the frieze group is $F_3 = \langle T_v, S_\ell \rangle = \{T_v^n S_\ell^m \mid n \in \mathbb{Z} \text{ and } m = 0, 1\}$, which is the infinite dihedral group $D_\infty$. Note that the pair of elements $T_v \circ S_\ell$ and $S_\ell$ have order 2, they generate $F_3$, and their product $(T_v \circ S_\ell) \circ S_\ell = T_v$ has infinite order. A geometrical fact about frieze pattern $F_3$

FFFFFFFFF

**Fig. 9.5** Frieze pattern of type 1

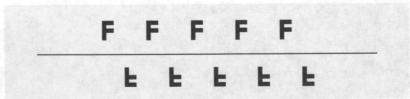

**Fig. 9.6** Frieze pattern of type 2

**Fig. 9.7** Frieze pattern of type 3

**Fig. 9.8** Frieze pattern of type 4

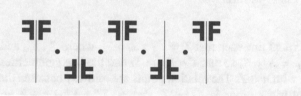

**Fig. 9.9** Frieze pattern of type 5

worth mentioning is that the distance between consecutive pairs of vertical reflection axis is half the length of the $T_v$.

In the frieze pattern of type 4 (see Fig. 9.8) the frieze group is generated by a translation and a $\pi$ rotation $R_P$ about a point $P$ midway between consecutive $F$'s (such a rotation is called a *half-turn*). Choose a point $Q$ such that $T_v = R_Q \circ R_P$ is a translational symmetry of shortest length. Then, $Q$ is also a point of symmetry because $R_Q = T_v \circ R_P$. In general, the half-turn $T_v^n \circ R_P$ is a symmetry for all $n \in \mathbb{Z}$; these half-turns determine all points of symmetry. As in frieze pattern of type 3, the distance between consecutive points of rotational symmetry is half of the length of the smallest translation vector. The frieze group is $F_4 = \langle T_v, R_P \rangle = \{T_v^n \circ R_P^m \mid n \in \mathbb{Z} \text{ and } m = 0, 1\}$. This group, like $F_3$, is also infinite dihedral.

The frieze pattern of type 5 can be identified by its half-turn symmetry and glide reflection symmetry (see Fig. 9.9). Let $P$ be a point of symmetry and suppose that $T$ is a glide reflection such that $T^2$ is a translational symmetry of shortest length. Choose a point $Q$ such that $T^2 = R_Q \circ R_P$, where $R_Q$ and $R_P$ are $\pi$ rotations about $Q$ and $P$, respectively. Since $R_Q = T^2 \circ R_P$, it follows that $Q$ is also a point of symmetry. Moreover, the half-turn $T^{2n} \circ R_P$ is a symmetry for all integer $n$; these half-turns determine all points of symmetry. Now, the line symmetries can be obtained from $T$ and $R_P$. Let $c$ be the horizontal axis of $T$, $a$ be the vertical line through $P$, and $\ell$

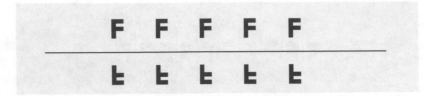

**Fig. 9.10**  Frieze pattern of type 6

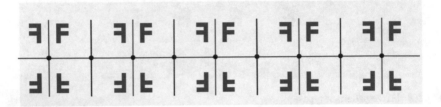

**Fig. 9.11**  Frieze pattern of type 7

be the vertical line such that $T = S_\ell \circ S_a \circ S_c$, where $S_\ell$, $S_a$, and $S_c$ are reflections. Then, $R_P = S_c \circ S_a$ so that $T \circ R_P = S_\ell$ and the line symmetries are the reflections $T^{2n} \circ S_\ell$ with $n \in \mathbb{Z}$. The rotation points are midway between the vertical reflection axes. The frieze group is $F_5 = \langle T, R_P \rangle = \{T^n \circ R_P^m \mid n \in \mathbb{Z}$ and $m = 0, 1\}$. Note that the frieze groups $F_3$, $F_4$, and $F_5$ are isomorphic groups.

The frieze pattern of type 6 is generated by a translation $T_v$ and a horizontal reflection (see Fig. 9.10). It has a unique horizontal line of symmetry $c$. Thus, the frieze group is $F_6 = \langle T_v, S_c \rangle = \{T_v^n \circ S_c^m \mid n \in \mathbb{Z}$ and $m = 0, 1\}$. Note that $T_v$ and $S_c$ commute, and it follows that $F_6$ is not infinite dihedral group. Indeed, $F_6$ is isomorphic to $\mathbb{Z} \times \mathbb{Z}_2$. Frieze pattern VI is left invariant under a glide reflection as well, but in this case the glide reflection is considered trivial since it is the product of $T_v$ and $S_c$. Conversely, a glide reflection is non-trivial if its translation component and reflection component are not elements of the symmetry group.

The frieze group of frieze pattern of type 7 has vertical line symmetry and a unique horizontal line of symmetry $c$ (see Fig. 9.11). Let $\ell$ be a vertical line of symmetry, and let $T_v$ be a translational symmetry of shortest length. Then, the vertical line symmetries are the reflections $T_v^n \circ S_\ell$ with $n\mathbb{Z}$ and the point $P = c \cap \ell$ is a point of symmetry since $R_P = S_c \circ S_\ell$. Hence, the half-turn symmetries are the half-turns $T_v^n \circ R_P$ with $n \in \mathbb{Z}$. The frieze group is $F_7 = \langle T_v, S_c, S_\ell \rangle = \{T_v^n \circ S_c^m \circ S_\ell^k \mid n \in \mathbb{Z}$ and $m, k = 0, 1\}$. It is isomorphic to the direct product of infinite dihedral group and $\mathbb{Z}_2$. The combination of $S_c$ and $S_\ell$ is a $\pi$ rotation.

We collect the observations above as a theorem.

**Theorem 9.10** *Every frieze group is one of the following:*

- $F_1 = \langle T_v \rangle$;
- $F_2 = \langle T \rangle$;

- $F_3 = \langle T_v, S_\ell \rangle$;
- $F_4 = \langle T_v, R_P \rangle$;
- $F_5 = \langle T, R_P \rangle$;
- $F_6 = \langle T_v, S_c \rangle$;
- $F_7 = \langle T_v, S_c, S_\ell \rangle$;

where $T_v$ is a translation of shortest length, $T$ is a glide reflection such that $T^2$ is a translational symmetry, $\ell$ is a vertical line of symmetry, $P$ is a point of symmetry, and $c$ is the unique horizontal line of symmetry.

**Proof**  We left to the readers to show that this list exhausts all possibilities (Fig. 9.12).

∎

Table 9.1 provides an identification algorithm for the frieze patterns:

## Exercises

1. Show that the frieze group $F_6$ is isomorphic to $\mathbb{Z} \times \mathbb{Z}_2$.
2. Prove that in $F_7$ the cyclic subgroup generated by $T_v$ is a normal subgroup.
3. Find at least two friezes in your campus architecture, and identify their frieze groups.
4. Describe the center of each frieze group $F_i$, $i = 1, \ldots, 7$.
5. For each frieze group $F_i$ (for $i = 1, \ldots, 7$), throw enough commutators into its presentation to be able to describe $F_i / F_i'$, and then identify $F_i$ (for $i = 1, \ldots, 7$).
6. Say which of the frieze groups contain glide reflections.
7. Identify the frieze groups for each of patterns in Fig. 9.13.

## 9.3  Wallpaper Groups

This section introduces the symmetry groups of wallpaper patterns and provides the vocabulary and techniques necessary to identify them.

**Definition 9.11**  A *wallpaper pattern* is a plane figure $W$ whose group of translational symmetries $T_W$ has two independent generators $u$ and $v$. Hence, we can write

$$T_W = \{T_{mu+nv} \mid m, n \in \mathbb{Z}\} = \langle T_u, T_v \rangle,$$

with neither of $u$ and $v$ being a multiple of the other. This means that the angle between $u$ and $v$ satisfies the following inequalities

$$0 < \cos^{-1}\left(\frac{\langle u, v \rangle}{\|u\|\,\|v\|}\right) < \pi,$$

or equivalently

$$-1 < \frac{\langle u, v \rangle}{\|u\|\,\|v\|} < 1.$$

**Fig. 9.12** The seven frieze groups (by Andrew Glassner)

**Table 9.1** An identification algorithm for the frieze patterns

| Type | $\pi$ rotation | Horizontal reflection | Vertical reflection | Non-trivial glide reflection |
|---|---|---|---|---|
| 1 | No | No | No | No |
| 2 | No | No | No | Yes |
| 3 | No | No | Yes | No |
| 4 | Yes | No | No | No |
| 5 | Yes | No | Yes | Yes |
| 6 | No | Yes | No | No |
| 7 | Yes | Yes | Yes | No |

**Fig. 9.13** Frieze patterns

The set of points $Span_{\mathbb{Z}}(u, v) = \{mu + nv \mid m, n \in \mathbb{Z}\}$ is called the *lattice spanned* the vectors $u$ and $v$, and it forms an abelian group under addition. So, we can write $T_W = \{T_w \mid w \in Span_{\mathbb{Z}}(u, v)\}$, and we call $Span_{\mathbb{Z}}(u, v)$ the *translation lattice* of $W$. Hence, we can say a wallpaper pattern is a plane figure $W$ with independent translational symmetries $T_u$ and $T_v$ satisfying the following property: Given any translational symmetry $T$, there exist integers $m$ and $n$ such that $T = mT_u + nT_v$. Translations $T_u$ and $T_v$ are called *basic translations*.

**Definition 9.12** The symmetry group of a wallpaper pattern is called a *wallpaper group*.

Hence, $\langle T_u, T_v \rangle$ is the subgroup of translational symmetries in the wallpaper group of $W$.

**Definition 9.13** Let $W$ be a wallpaper pattern with basic translations $T_u$ and $T_v$. Given any point $A$, suppose that $B = T_u(A)$, $C = T_v(B)$, and $D = T_v(A)$. The *unit cell* of $W$ with respect to $A$, $T_u$, and $T_v$ is the plane region bounded by parallelogram $ABCD$.

The translation lattice of $W$ determined by $A$ is the set of points $\{(T_v^n \circ T_u^m)(A) \mid m, n \in \mathbb{Z}\}$. This lattice is square, rectangular, or rhombic if and only if the unit cell of $W$ with respect to $A$, $T_u$, and $T_v$ is square, rectangular, or rhombic. The lattice of wallpaper groups can be divided into five classes. We notice that

- Since $u \pm v$ are skew to $u$, it follows that $\|v\| \leq \|u + v\|$ and $\|v\| \leq \|u - v\|$.
- We can arrange $\|u - v\| \leq \|u + v\|$ by replacing $v$ by $-v$, if necessary. Then, we have

$$\|u\| \leq \|v\| \leq \|u - v\| \leq \|u + v\|. \tag{9.1}$$

So, we obtain eight cases by considering "=" or "<" for each of the "$\leq$" in (9.1), where they are indicated in Table 9.2.

**Definition 9.14** Let $W$ be a wallpaper pattern. A point $P$ is an *n-center* of $W$ if and only if the group of rotational symmetries of $W$ centered at $P$ is cyclic group $C_n$ with $n > 1$.

**Table 9.2** Eight cases by considering "=" or "<" for each of the "$\leq$" in (9.1)

| Case | Lattice |
|---|---|
| $\|u\| = \|v\| = \|u - v\| = \|u + v\|$ | Impossible |
| $\|u\| = \|v\| = \|u - v\| < \|u + v\|$ | Hexagonal |
| $\|u\| = \|v\| < \|u - v\| = \|u + v\|$ | Square |
| $\|u\| = \|v\| < \|u - v\| < \|u + v\|$ | Centered rectangular |
| $\|u\| < \|v\| = \|u - v\| = \|u + v\|$ | Impossible |
| $\|u\| < \|v\| = \|u - v\| < \|u + v\|$ | Centered rectangular |
| $\|u\| < \|v\| < \|u - v\| = \|u + v\|$ | Rectangular |
| $\|u\| < \|v\| < \|u - v\| < \|u + v\|$ | Oblique |

**Theorem 9.15** *The symmetries of a wallpaper pattern fix the set of n-centers. In other words, if P is an n-center of W and T is a symmetry of W, then T(P) is an n-center of W.*

**Proof** Suppose that $W$ is a wallpaper pattern with symmetry group $W$ and $P$ is an $n$-center of $W$. Since $C_n$ is the subgroup of rotational symmetries with center $P$, it follows that there is smallest positive real number $\theta$ such that $R^n_{P,\theta} = id$. Now, if $T \in W$ and $Q = T(P)$, then we conclude that $T \circ R_{P,\theta} \circ T^{-1} = R_{Q,\pm\theta} \in W$ and

$$R^n_{Q,\pm\theta} = \left(T \circ R_{P,\theta} \circ T^{-1}\right)^n = T \circ R^n_{P,\theta} \circ T^{-1} = id.$$

It is clear that $R_{Q,\theta} \in W$ if and only if $R_{Q,\;\theta} \in W$. This yields that $R^n_{Q,\theta} = id$. Consequently, $Q$ is an $m$-center, for some $m \leq n$. Similarly, we have $(T^{-1}) \circ R_{Q,\theta} \circ (T^{-1})^{-1} = R_{P,\pm\theta} \in W$. This gives that $R^m_{P,\theta} = id$, and so $P$ is an $n$-center with $n \leq m$. Therefore, we obtain $n = m$ and $Q$ is an $n$-center, as desired. ∎

**Theorem 9.16 (Crystallographic Restriction).** *Let G be a plane symmetry group. Each rotation of G necessary has order 1, 2, 3, 4, or 6. In other words, there are only 5 possible rotational symmetries: 60°, 90°, 120°, 180°, and no rotational symmetry.*

**Proof** In order to determine why these are the only rotations we first determine the regular polygons with internal angles which divide 360°, as they will be the possible rotationally symmetrical angles. There are only two capabilities of rotations not represented by these polygons, and they are fairly intuitive ones. The first is one in which there is no rotational symmetry, as we cannot have a regular polygon with internal angles of 0° or 180°. The other way we can have rotational symmetry not representable with the internal angles is one of 180° as this would result in no angle the edge would just be continuing in a linear direction. Hence, we start looking at the internal angles of regular polygons which we know can be represented with the following formula:

$$\theta = \frac{n-2}{n}(180°).$$

Since the angle has to divide 360° and our $n$ has to be a minimum of 3, then we can very quickly find out what possible angles there are. First we will note $n = 3$ is our first possible n and it results in an angle of 60°, which divides 360°. Consequently, a rotational symmetry of 60° is possible. Now, we notice the only other angles that divide 360° that are greater than 60° are: 72°, 90°, and 120°. Next, we look at a table of the first five $n$ values and their resulting interior angles (Fig. 9.14):

| $n$ | 3 | 4 | 5 | 6 | 7 |
|---|---|---|---|---|---|
| Angle | 60° | 90° | 108° | 120° | $128 \cdot \frac{4}{7}°$ |

Using these values, we know the angle with keep growing and approach 180°. Therefore, we observe the only $n$ values which result in a number that divides 360° are 3,

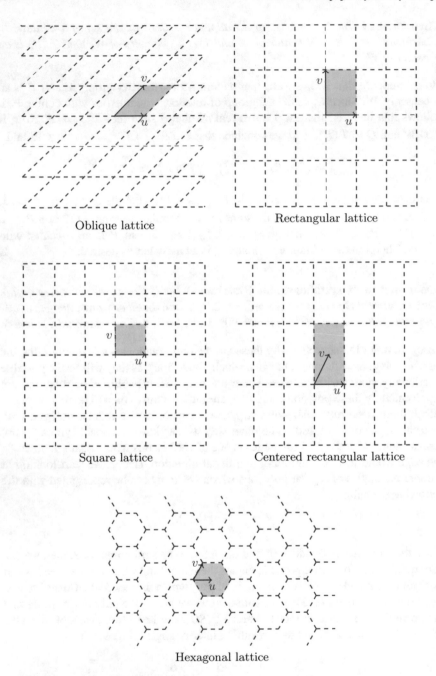

**Fig. 9.14** Lattice of wallpaper groups

4, and 6. This yields that only 5 rotational symmetries exist: 60°, 90°, 120°, 180°, and no rotational symmetry. Now that we have set the classifications of rotational symmetries, we proceed to look at the reflections and glide reflections within each of these classes as they are the only other possible motions.                                    ■

**Corollary 9.17**  *A wallpaper pattern with a 4-center has no 3- or 6-centers.*

**Proof**  If $P$ is a 3-center and $Q$ is a 4-center of a wallpaper pattern $W$, the corresponding wallpaper group $W$ contains the rotations $R_{P,2\pi/3}$ and $R_{Q,-\pi/2}$. This yields that $W$ also contains the $\pi/6$ rotation $R_{P,2\pi/3} \circ R_{Q,-\pi/2}$, which generates $C_{12}$. Therefore, there exists an $n$-center of $W$ with $n \geq 12$. But this contradicts Theorem 9.16. Similarly, if $Q$ is a 4-center and $Q'$ is a 6-center of $W$, then there exists an $n$-center of $W$ with $n \geq 12$ because $R_{Q',-\pi/3} \circ R_{Q,\pi/2}$ is a $\pi/6$ rotation.                                    ■

In addition to translational symmetry, wallpaper patterns can have line of symmetry, glide reflection symmetry, and $\pi$, $2\pi/3$, $\pi/2$, or $\pi/3$ rotational symmetry. Since the only rotational symmetries in a frieze group are half-turns, it is not surprising to find more wallpaper groups than frieze groups. We shall identify seventeen distinct wallpaper groups, but we omit the proof that every wallpaper group is one of these seventeen. Throughout this discussion, $W$ denotes a wallpaper pattern. We use the international standard notation to denote the various wallpaper groups. Each symbol is a string of letters and integers selected from $p$, $c$, $m$, $g$ and 1, 2, 3, 4, 6. The letter $p$ stands for *primitive translation lattice*. The points in a primitive translation lattice are the vertices of parallelograms with no interior points of symmetry. When a point of symmetry lies at the center of some unit cell, we use the letter $c$. The letter $m$ stands for mirror and indicates lines of symmetry; the letter $g$ indicates glide reflection symmetry. Integers indicate the maximum order of the rotational symmetries of $W$. There are four symmetry types of wallpaper patterns with no $n$-centers. These are analyzed as follows: If $W$ has no line of symmetry or glide reflection symmetry, the corresponding wallpaper group consists only of translations and is denoted by $p1$. If $W$ has glide reflection symmetry but no lines of symmetry, the corresponding wallpaper group is denoted by $pg$. There are two ways that both line of symmetry and glide reflection symmetry can appear in $W$: (1) The axis of some glide reflection symmetry is not a line of symmetry, and (2) the axis of every glide reflection symmetry is a line of symmetry. The corresponding wallpaper groups are denoted by $cm$ and $pm$, respectively. There are five symmetry types whose $n$-centers are all 2-centers. If $W$ has neither lines of symmetry nor glide reflection symmetries, the corresponding wallpaper group is denoted by $p2$. If $W$ has no line of symmetry but has glide reflection symmetry, the corresponding group is denoted by $pgg$. If $W$ has parallel lines of symmetry, the corresponding group is denoted by $pmg$. If $W$ has lines of symmetry in two directions, there are two ways to configure them relative to the 2-centers in $W$: (1) All 2-centers lie on a line of symmetry, and (2) not all 2-centers lie on a line of symmetry. The corresponding wallpaper groups are denoted by *pmm* and *cmm*, respectively. Three wallpaper patterns have *n*-centers whose smallest rotation angle is $\pi/2$. Those with no lines of symmetry have wallpaper group *p4*. Those with lines of symmetry in four directions have wallpaper group *p4m*; other patterns with lines of symmetry have wallpaper group *p4g*. Three symmetry types have *n*-centers whose smallest rotation angle is $2\pi/3$. Those with no lines of symmetry have wallpaper group *p3*. Those whose 3-centers lie on lines of symmetry have wallpaper group *p3m1*; those with some 3-centers off lines of symmetry have wallpaper group *p31m*. Finally, two symmetry types have n-centers whose smallest rotation angle is $\pi/3$. Those with line of symmetry have wallpaper group *p6m*; those with no line of symmetry have wallpaper group *p6* (Figs. 9.15, 9.16, 9.17, 9.18, 9.19, 9.20, 9.21, 9.22, 9.23, 9.24, 9.25, 9.26, 9.27, 9.28, 9.29, 9.30, 9.31, 9.32, 9.33, 9.34, 9.35, 9.36, 9.37, 9.38, 9.39, 9.40, 9.41, 9.42, 9.43, 9.44, 9.45, 9.46, 9.47, and 9.48).

**Theorem 9.18** *Every wallpaper group is one of the following:*

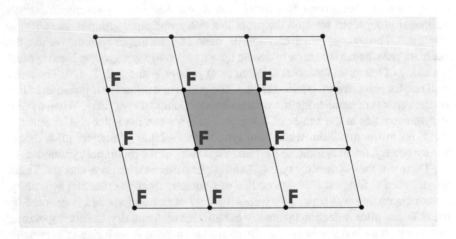

**Fig. 9.15**  *p*1 generated by two translations

**Fig. 9.16**  Wallpaper group *p*1

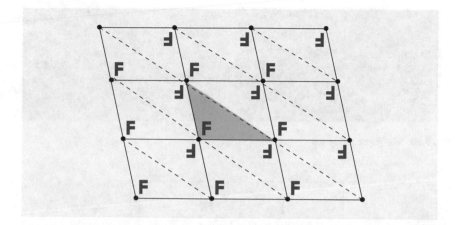

**Fig. 9.17** $p2$ generated by three half-turns

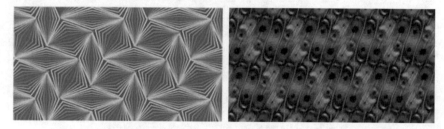

**Fig. 9.18** Wallpaper group $p2$

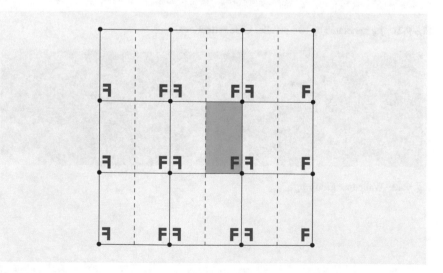

**Fig. 9.19** $pm$ generated by two reflections and a translation

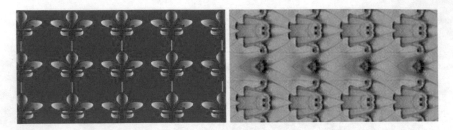

**Fig. 9.20**  Wallpaper group *pm*

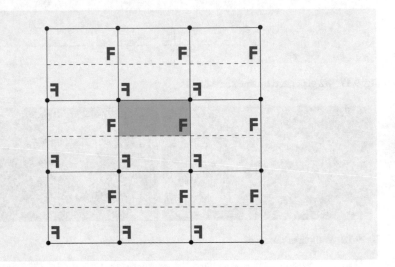

**Fig. 9.21**  *pg* generated by two parallel glide reflections

**Fig. 9.22**  Wallpaper group *pg*

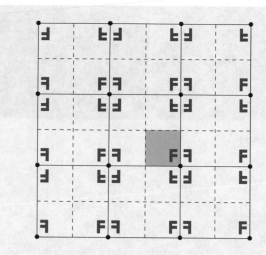

**Fig. 9.23**  *pmm* generated by four reflections

**Fig. 9.24**  Wallpaper group *pmm*

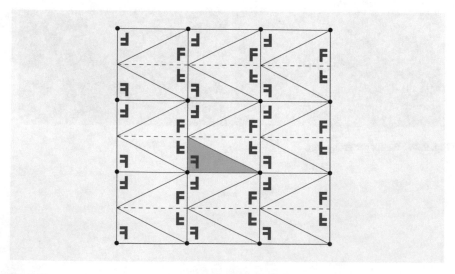

**Fig. 9.25**  *pmg* generated by two perpendicular glide reflections

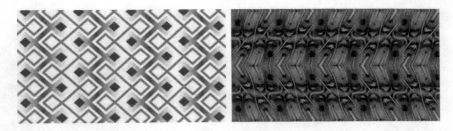

**Fig. 9.26**  Wallpaper group *pmg*

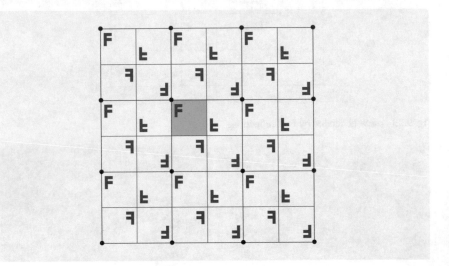

**Fig. 9.27**  *pgg* generated by two perpendicular glide reflections

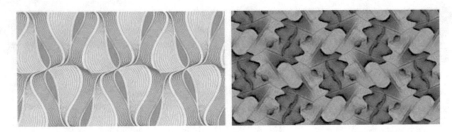

**Fig. 9.28**  Wallpaper group *pgg*

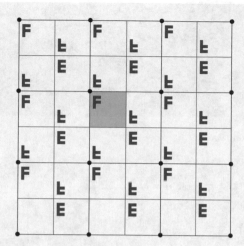

**Fig. 9.29** *cm* generated by a reflection and a parallel glide reflection

**Fig. 9.30** Wallpaper group *cm*

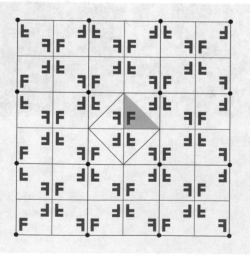

**Fig. 9.31** *cmm* generated by two perpendicular reflections and a half-turn

**Fig. 9.32**  Wallpaper group *cmm*

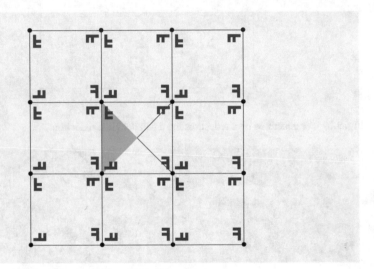

**Fig. 9.33**  *p*4 generated by a half-turn and a quarter-turn

**Fig. 9.34**  Wallpaper group *p*4

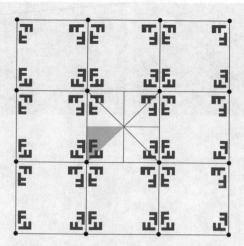

**Fig. 9.35** $p4m$ generated by reflections in the sides of a $(\pi/4,\ \pi/4,\ \pi/2)$ triangle

**Fig. 9.36** Wallpaper group $p4m$

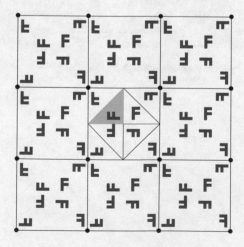

**Fig. 9.37** $p4g$ generated by a reflection and a quarter-turn

**Fig. 9.38**   Wallpaper group $p4g$

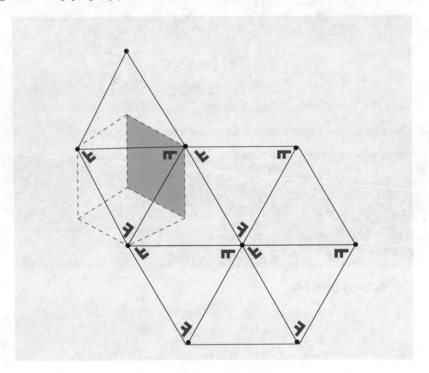

**Fig. 9.39**   $p3$ generated by three rotations through $2\pi/3$

**Fig. 9.40** Wallpaper group $p3$

**Table 9.3** Identification table for plane patterns

| Type | Lattice | Highest order of rotation | Reflections | Non-trivial glide reflections | Generating region |
|---|---|---|---|---|---|
| $p1$ | Parallelogram | 1 | No | No | 1 unit |
| $p2$ | Parallelogram | 2 | No | No | 1/2 units |
| $pm$ | Rectangular | 1 | Yes | No | 1/2 units |
| $pg$ | Rectangular | 1 | No | Yes | 1/2 units |
| $pmm$ | Rectangular | 2 | Yes | No | 1/4 units |
| $pmg$ | Rectangular | 2 | Yes | Yes | 1/4 units |
| $pgg$ | Rectangular | 2 | No | Yes | 1/4 units |
| $cm$ | Rhombic | 1 | Yes | Yes | 1/2 units |
| $cmm$ | Rhombic | 2 | Yes | Yes | 1/4 units |
| $p4$ | Square | 4 | No | No | 1/4 units |
| $p4m$ | Square | 4 | Yes | Yes | 1/8 units |
| $p4g$ | Square | 4 | Yes | Yes | 1/8 units |
| $p3$ | Hexagonal | 3 | No | No | 1/3 units |
| $p3m1$ | Hexagonal | 3 | Yes | Yes | 1/6 units |
| $p31m$ | Hexagonal | 3 | Yes | Yes | 1/6 units |
| $p6$ | Hexagonal | 6 | No | No | 1/6 units |
| $p6m$ | Hexagonal | 6 | Yes | Yes | 1/12 units |

Table 9.3 can be used to identify the wallpaper group associated with a particular wallpaper pattern.

**Exercises**

1. Determine which of the 17 wallpaper groups is the symmetry group of each pattern in Fig. 9.49.
2. Determine the plane symmetry group for each of the patterns in Fig. 9.50.

## 9.4    Worked-Out Problems

**Problem 9.19**  Prove that if a symmetry group of a frieze pattern contains a reflection
in a vertical mirror and a rotation by $\pi$, then it must also contain a glide reflection.

*Solution*  If we denote the reflection in a vertical mirror by $V$ and the rotation by $\pi$
by $R$, then the symmetry group of the frieze pattern must also contain $R \circ V$. This
composition is an opposite isometry, and it is easy to see that it is either reflection in
a horizontal mirror or a glide reflection along a horizontal axis. (Remember that the
center of rotation for $R$ does not have to lie on the mirror for $V$.) Since the symmetry
group of a frieze pattern contains a horizontal translation by definition, it follows
that in either case, the symmetry group of the frieze pattern must contain a glide
reflection along a horizontal axis.

So how do you know that the composition $R \circ V$ is a reflection in a horizontal
mirror or a glide reflection along a horizontal axis? One way to see this is to consider
what happens to the picture of a left footprint walking right, above the center of
rotation of $R$. After applying the reflection $V$, this becomes a right footprint walking
left, above the center of rotation of $R$. And then after applying $R$, this becomes a
right footprint walking right below the center of rotation of $R$. The only way to start

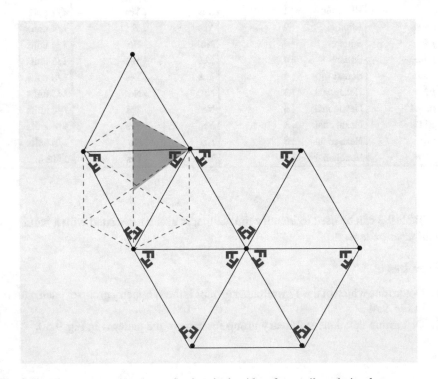

**Fig. 9.41**  $p3m1$ generated by three reflections in the sides of an equilateral triangle

**Fig. 9.42** Wallpaper group $p3m1$

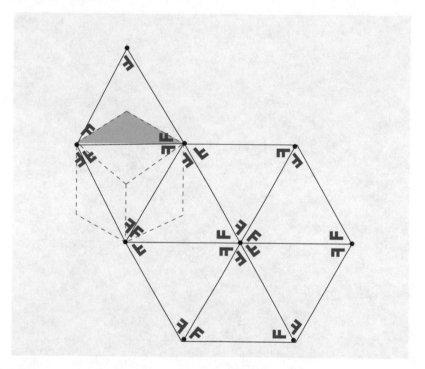

**Fig. 9.43** $p31m$ generated by a reflection and a rotation through $2\pi/3$

with a left footprint walking right and end up with a right footprint walking right is via a reflection in a horizontal mirror or a glide reflection along a horizontal axis. ∎

**Problem 9.20** A point $P$ in the plane has polar coordinates $(r, \theta)$, where $r$ is the distance of the point from the origin and $\theta$ is the angle that the ray $OP$ makes with the positive $x$-axis (see Fig. 9.51). We also allow $r$ to take negative values. This is done by defining $P(r, \theta)$, where $r < 0$, to be the reflection of the point $P'(|r|, \theta)$ in the origin $O$.

**Fig. 9.44**  Wallpaper group *p*31*m*

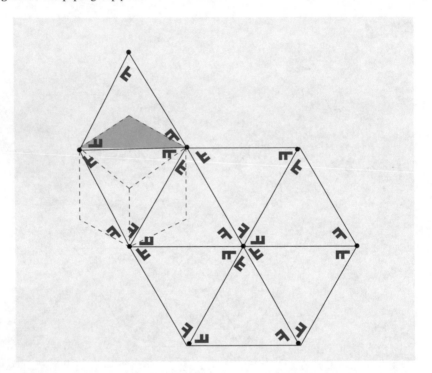

**Fig. 9.45**  *p*6 generated by a half-turn and a rotation through $2\pi/3$

**Fig. 9.46**  Wallpaper group *p*6

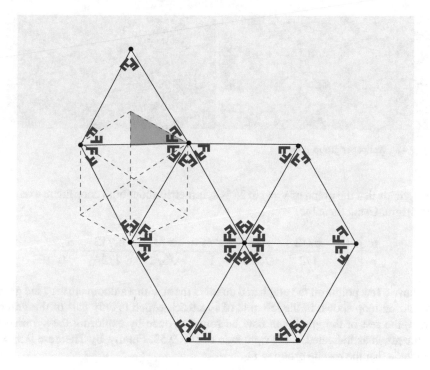

**Fig. 9.47** *p6m* generated by reflections in the sides of a $(\pi/6, \pi/3, \pi/2)$ triangle

In graphing a polar equation $f(r, \theta) = 0$, always be on the lookout for possible symmetries of its graph. Let $P = (r, \theta)$ be any point other than the pole (origin) itself; let $P'$, $Q$, and $P''$ be the image of $P$ under reflection in the $x$-axis, the origin, and the $y$-axis, respectively. Then, it follows from the polar representations of the points $P'$, $Q$, and $P''$ given in Fig. 9.52. The graph of the equation $f(r, \theta) = 0$ is

(1) Symmetric about the $x$-axis (the polar axis) if the set of solutions of $f(r, \theta) = 0$ is the same as that of $f(r, -\theta) = 0$ or $f(-r, \pi - \theta) = 0$;
(2) Symmetric about the origin $O$ (the pole) if the set of solutions of $f(r, \theta) = 0$ is the same as that of $f(-r, \theta) = 0$ or $f(r, \pi + \theta) = 0$;
(3) Symmetric about the $y$-axis if the set of solutions of $f(r, \theta) = 0$ is the same as that of $f(-r, -\theta) = 0$ or $f(r, \pi - \theta) = 0$.

Show that the graph of $r = \sin 2\theta$ is a rosette with symmetry group $D_4$.

*Solution* First, we look for symmetries. We write $f(r, \theta) = r - \sin 2\theta = 0$. It is easy to check that

$$f(-r, \pi - \theta) = -f(r, \theta),$$
$$f(r, \pi + \theta) = f(r, \theta),$$
$$f(-r, -\theta) = -f(r, \theta).$$

**Fig. 9.48** Wallpaper group *p6m*

This yields that the graph of $y = \sin 2\theta$ is symmetric about both coordinate axes and the origin. Using the table

| $\theta$ | 0 | $\pi/12$ | $\pi/6$ | $\pi/4$ | $\pi/3$ | $5\pi/12$ | $\pi/2$ |
|---|---|---|---|---|---|---|---|
| $r$ | 0 | $1/2$ | $\sqrt{3}/2$ | 1 | $\sqrt{3}/2$ | $1/2$ | 0 |

We have a few points on the graph and connect them with a smooth curve. This gives the closed loop shown in the left side of Fig. 9.53, which is only part of the graph. Then, the rest of the graph can now be found at once by exploring the symmetry of the graph as indicated in the right side of Fig. 9.53. Finally, by Theorem 9.5, we conclude that the rosette group is $D_4$. ■

**Problem 9.21** Let $S$ and $S'$ be two reflections across distinct lines $\ell$ and $\ell'$, respectively.

(1) If $\ell$ and $\ell'$ are not parallel, prove that $S \circ S'$ is a rotation about the point of intersection of $\ell$ and $\ell'$ through an angle twice that from $\ell$ to $\ell'$.
(2) If $\ell$ and $\ell'$ are parallel, prove that $SS'$ is a translation in the direction normal to $\ell$ and $\ell'$ through distance twice that from $\ell$ to $\ell'$.

*Solution* Suppose that $S$ and $S'$ are reflections across distinct lines $\ell$ and $\ell'$ such that $\ell \neq \ell'$.

(1) Let $\ell$ and $\ell'$ not be parallel. Thus, $\ell \cap \ell' = \{O\}$, where $O$ is the point of intersection of lines $\ell$ and $\ell'$ (see Fig. 9.54). Taking $O$ as the origin and $\ell$ as the axis, suppose that $P$ is a point in $\mathbb{R}^2$. We can write $P$ in terms of polar coordinates $P = (\rho, \theta)$. Hence, $S : (\rho, \theta) \mapsto (\rho, -\theta)$. Suppose that $S'(P) = P' = (\rho, \phi)$. Since $\ell'$ bisects the angle $P'OP$, it follows that $\alpha = (\theta + \phi)/2$, and so $\phi = 2\alpha - \theta$. This yields that $S' \circ S : (\rho, \theta) \mapsto (\rho, 2\alpha + \theta)$. Consequently, $S' \circ S$ is a rotation about $O$ through twice the angle from $\ell$ to $\ell'$.

(2) Let $\ell$ and $\ell'$ be parallel. Assume that $P \in \mathbb{R}^2$. Taking $\ell$ as the $x$-axis, we can write $P$ in terms of Cartesian coordinates $P = (x, y)$. Hence, $S : (x, y) \mapsto (x, -y)$ and $S' : (x, y) \mapsto (x, z)$. In Fig. 9.54, we have $(y + z)/2 = a$, and so $z = 2a - y$. This implies that $S' \circ S : (x, y) \mapsto (x, 2a + y)$. Therefore, $S' \circ S$ is a translation through twice the distance between lines $\ell$ and $\ell'$. ■

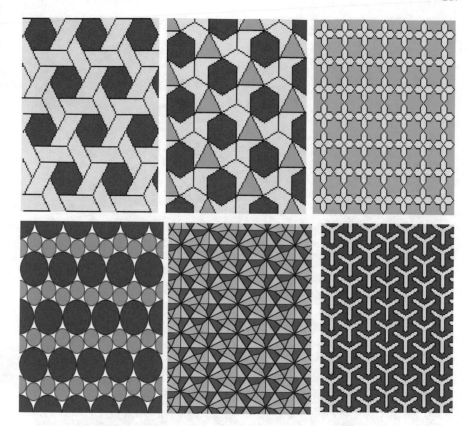

**Fig. 9.49** Some wallpaper patterns

**Problem 9.22** If $G$ is a wallpaper group corresponding to a pattern with an oblique lattice, prove that the point group of $G$ is $\{I_2, -I_2\}$ or $\{I_2\}$.

*Solution* Let $G$ be a wallpaper group corresponding to a pattern with an oblique lattice. Figure 9.55 illustrates an oblique lattice unit. We see that the only possible non-trivial rotations in $G$ are of order two. That is, if we rotate it $\pi$, we have the same figure. However, there are no possible reflections. Hence, the point group of $G$ must be a subgroup of $\{I_2, -I_2\} \cong \mathbb{Z}_2$. If we eliminate $\pi$ rotation, then $\{I_2\}$ is the point group of $G$. Since the only subgroup of $\mathbb{Z}_2$ is $\{I_2\}$, we have established all possible point groups of $G$. $\blacksquare$

**Problem 9.23** If $G$ is a wallpaper group corresponding to a pattern with a rectangular lattice, then the possible point groups of $G$ are $\{I_2\}$, $\{I_2, -I_2\}$, $\{I_2, A_0\}$, $\{I_2, A_\pi\}$ or $\{I_2, -I-2, A_0, A_\pi\}$, where

$$A_\theta = \begin{bmatrix} \cos\theta & \sin\theta \\ \sin\theta & -\cos\theta \end{bmatrix}.$$

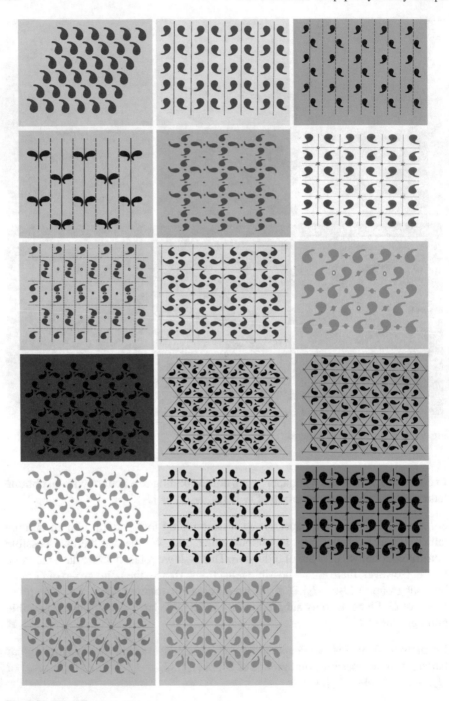

**Fig. 9.50** 17 wallpaper patterns

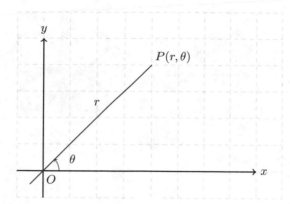

**Fig. 9.51** Polar coordinates

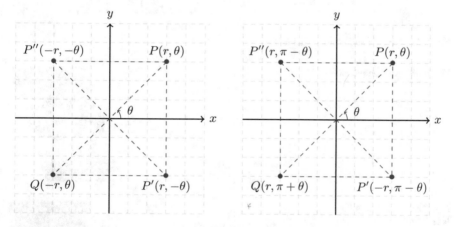

**Fig. 9.52** Symmetry tests for polar coordinates

*Solution* Let $G$ be a wallpaper group corresponding to a pattern with a rectangular lattice and having point group $H$. Figure 9.56 illustrates a lattice unit.

From Fig. 9.56, we observe that $H$ can have $\pi$ rotation, vertical reflections, and horizontal reflections. Thus, $H$ must be a subgroup of $\{I_2, -I_2, A_0, A_\pi\} \cong K_4$. The subgroups of $K_4$ are $K_4$, $\mathbb{Z}_2$ and $\{I_2\}$. If $H$ is isomorphic $K_4$, then $H = \{I_2, -I_2, A_0, A_\pi\}$. If $H$ is isomorphic to $\mathbb{Z}_2$, then we have three possibilities: $\{I_2, A_0\}$, $\{I_2, A_\pi\}$, and $\{I_2, -I_2\}$. Since $A_0 A_0 = I_2$, $A_\pi A_\pi = I_2$, and $(-I_2)(-I_2) = I_2$, it follows that these sets are closed and are groups. The only other possible point group left is $\{I_2\}$. This completes the proof. ∎

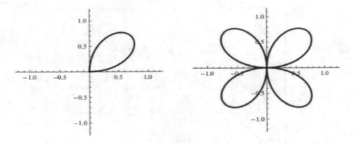

**Fig. 9.53**   Graph of $r = \sin 2\theta$

**Fig. 9.54**   Parallel and non-parallel lines

**Fig. 9.55**   An oblique lattice unit

**Fig. 9.56**   A rectangular lattice unit

## 9.5   Supplementary Exercises

1. Show that

   (a) If $n \geq 2$ is even, then the graph of $r = \cos n\theta$ is a rosette with symmetry group $D_{2n}$.

   (b) If $n \geq 2$ is odd, then the graph of $r = \cos n\theta$ is a rosette with symmetry group $D_n$.

**Fig. 9.57** Frieze patterns

2. Explain why the graph of the equation $r = \cos\theta$ in polar coordinates is not a rosette.
3. Determine which of the seven frieze groups in the symmetry group of each of patterns in Fig. 9.57.
4. Show that the seven frieze groups fall into exactly four isomorphism classes.
5. If $G$ is a wallpaper group corresponding to a pattern with a centered rectangular lattice, find the possible point groups of $G$.
6. If $G$ is a wallpaper group corresponding to a pattern with a square lattice, find the possible point groups of $G$.
7. If $G$ is a wallpaper group corresponding to a patter with a hexagonal lattice, find the possible point groups of $G$.
8. Describe all wallpaper groups in terms of generators and relations.
9. Two plane groups $G$ and $H$ with lattices $L_G$ and $L_H$ are *equivalent* if there is an isomorphism $G \to H$ which maps the subgroup $L_G$ onto $L_H$. Prove that

   (a) There are 5 equivalence classes of plane group $G$ whose point group contains no reflection.
   (b) There are 3 equivalence classes of plane group $G$ whose point group contains a single reflection.
   (c) There are 9 equivalence classes of plane group $G$ whose point group contains more than one reflection.

# Chapter 10
# Representations and Characters of Groups

In this chapter, we will introduce the notions of representations of groups. The idea is then to establish a viable connection between representations and modules, so that we can transfer results back and forth. We will then go on to discuss what is meant by reducibility. This gives us the opportunity to break up representations and modules into indivisible components.

## 10.1 Modules

Modules over a ring are generalization of abelian groups (which are modules over $\mathbb{Z}$).

**Definition 10.1** Let $R$ be a ring. A non-empty subset $M$ is said to be an *R-module* (or, a *a module over R*) if $M$ is an abelian group under an operation $+$ such that for every $r \in R$ and $m \in M$ there exists an element $rm$ in $M$ subject to

(1) $r(a + b) = ra + rb$;
(2) $r(sa) = (rs)a$;
(3) $(r + s)a = ra + sa$;

for all $a, b \in M$ and $r, s \in R$. If $R$ has a unit element 1, and if $1m = m$ for every element $m$ in $M$, then $M$ is called a *unitary R-module*.

**Remark 10.2** Note that if the ring $R$ is a field, a unitary $R$-module is nothing more than a vector space over $R$.

**Example 10.3** Every abelian group $G$ is a module over the ring of integers. Addition is carried out according to the group structure of $G$; the key point is that we can multiply $x \in G$ by the integer $n$. If $n > 0$, then $nx = x + x + \cdots + x$ ($n$ times); if $n < 0$, then $nx = -x - x - \cdots - x$ ($|n|$ times).

© The Author(s), under exclusive license to Springer Nature Singapore Pte Ltd. 2021 213
B. Davvaz, *Groups and Symmetry*, https://doi.org/10.1007/978-981-16-6108-2_10

**Example 10.4**  Any ring $R$ is an $R$-module over itself.

**Example 10.5**  Let $R$ be a ring and $M = Mat_{m \times n}(R)$ be the set of all $m \times n$ matrices with entries in $R$. Then, $M$ is an $R$-module, where addition is ordinary matrix addition, and multiplication of the scalar $c$ by the matrix $A$ means multiplication of each entry of $A$ by $c$.

$M = \{0\}$ is called *zero module* and is often written simply as $0$.

Let $M$ be an $R$-module. The following results hold for any $x \in M$ and $r \in R$. We distinguish the vector $0_M$ from the zero scalar $0_R$.

(1) $r0_M = 0_M$;
(2) $0_R x = 0_M$;
(3) $(-r)x = r(-x) = -(rx)$;
(4) If $R$ is a field, or more generally a division ring, then $rx = 0_M$ implies that either $r = 0_R$ or $x = 0_M$.

**Definition 10.6**  An additive subgroup $A$ of the $R$-module $M$ is called a *submodule* (or *R-submodule*) of $M$ if whenever $r \in R$ and $a \in A$, then $ra \in A$.

If $\{N_i \mid i \in I\}$ is a family of submodules of an $R$-module $M$, then $\bigcap_{i \in I} N_i$ is a submodule of $M$. If $X$ is a subset of an $R$-module $M$, then the intersection of all submodules of $M$ containing $X$ is called the *submodule generated by* $X$. If $X$ is finite, and $X$ generates the module $N$, then $N$ is said to be *finitely generated*. If $X$ consists of a single element, $X = \{a\}$, then the submodule generated by $X$ is called the *cyclic submodule* generated by $a$. An $R$-module $M$ is *cyclic* if it is generated by a single element $a$.

**Theorem 10.7**  *Let $R$ be a ring, $M$ be an $R$-module, $X$ be a subset of $M$, $\{N_i \mid i \in I\}$ is a family of submodules $M$, and $a \in M$.*

*(1) Let $Ra = \{ra \mid r \in R\}$. Then, $Ra$ is a submodule of $M$ and the map $R \to Ra$ given by $r \mapsto ra$ is an $R$-module epimorphism;*

*(2) The cyclic submodule $A$ generated by $a$ is $\{ra + na \mid r \in R, n \in \mathbb{Z}\}$. If $R$ has an identity and $A$ is unitary, then $A = Ra$;*

*(3) The submodule $N$ generated by $X$ is*

$$\left\{ \sum_{i=1}^{s} r_i a_i + \sum_{j=1}^{t} n_j b_j \mid s, t \in \mathbb{N}, a_i, b_j \in X, r_i \in R, n_j \in \mathbb{Z} \right\}.$$

*If $R$ has an identity and $M$ is unitary, then*

$$N = RX = \left\{ \sum_{i=1}^{s} r_i a_i \mid s \in \mathbb{N}, a_i \in X, r_i \in R \right\};$$

***Proof*** It is straightforward. Note that if $R$ has a unit $1_R$ and $M$ is unitary, then $n1_R \in R$, for all $n \in \mathbb{Z}$ and $na = (n1_R)a$, for all $a \in M$. ∎

Given an $R$-module $M$ and a submodule $A$, we could construct the factor module $M/A$ in a manner similar to the way we constructed factor group. One could also talk about homomorphisms of one $R$-module into another and prove the appropriate homomorphism theorems.

**Theorem 10.8** *Let $A$ be a submodule of an $R$-module $M$. Then, the factor group $M/A$ is an $R$-module with the following external operation $r(m + A) = rm + A$, for all $r \in R$ and $m \in M$.*

***Proof*** Since $M$ is an additive abelian group, it follows that $A$ is a normal subgroup, and $M/A$ is well-defined abelian group. If $m + A = m' + A$, then $m - m' \in A$. Since $A$ is a submodule, it follows that $rm - rm' = r(m - m') \in A$, for all $r \in R$. Thus, $rm + A = rm' + A$ and the external operation of $R$ on $M/A$ is well defined. The remainder of the proof is now easy. ∎

**Definition 10.9** Let $M$ and $M'$ be two $R$-modules. A function $f: M \to M'$ is an *$R$-module homomorphism* provided that $f(a + b) = f(a) + f(b)$ and $f(ra) = rf(a)$, for all $a, b \in M$ and $r \in R$.

An $R$-module homomorphism $f: M \to M'$ is called an *epimorphism* in case it is onto (surjective). It is called a *monomorphism* in case it is one to one (injective). Observe that an $R$-module homomorphism $f: M \to M'$ is necessary a homomorphism of additive abelian groups. Consequently, the same terminology is used. $f$ is an $R$-module isomorphism if it is one to one and onto too. The *kernel* of $f$ is its kernel as a homomorphism of abelian groups. Every homomorphism of abelian groups is a $\mathbb{Z}$-module homomorphism.

A homomorphism $f: M \to N$ that is the composite of homomorphisms $f = g \circ h$ is said to *factor through* $g$ and $h$. The following result essentially says that a homomorphism $f$ factors uniquely through every epimorphism whole kernel is contained in that of $f$ and through every monomorphism whose image contains the image of $f$.

**Theorem 10.10** *Let $M$, $M'$, $N$, and $N'$ be $R$-modules, and let $f: M \to N$ be an $R$-module homomorphism.*

*(1) If $g: M \to M'$ is an epimorphism with $Ker\, g \subseteq Ker\, f$, then there exists a unique homomorphism $h: M' \to N$ such that $f = h \circ g$. Moreover, $Ker\, h = g(Ker\, f)$ and $Im\, h = Im\, f$, so that $h$ is monomorphism if and only if $Ker\, g = Ker\, f$ and $h$ is epimorphism if and only if $f$ is epimorphism;*

*(2) If $g: N' \to N$ is a monomorphism with $Im\, f \subseteq g$, then there exists a unique homomorphism $h: M \to N'$ such that $f = g \circ h$. Moreover, $Ker\, h = Ker\, f$ and $Im\, h = g^{-1}(Im\, f)$, so that $h$ is monomorphism if and only if $f$ is monomorphism and $h$ is epimorphism if and only if $Im\, g = Im\, f$.*

**Proof** (1) Since $g: M \to M'$ is epimorphism, for each $m' \in M'$ there is at least one $m \in M$ with $g(m) = m'$. Also, if $l \in M$ with $g(l) = m'$, then clearly $m - l \in Ker g$. But since $Ker g \subseteq Ker f$, we have that $f(m) = f(l)$. Thus, there is a well-defined function $h: M' \to N$ such that $f = hg$. To see that $h$ is actually an $R$-module homomorphism, let $x', y' \in M'$ and $x, y \in M$ with $g(x) = x'$, $g(y) = y'$. Then, for each $r \in R$, $g(rx + y) = rx' + y'$, so that $h(rx' + y') = f(rx + y) = rf(x) + f(y) = rh(x') + h(y')$. The uniqueness of $h$ with these properties is assured, since $g$ is epimorphism. The final assertion is trivial.

(2) For each $m \in M$, $f(m) \in Im f \subseteq Im g$. So since $g$ is monomorphism, there is a unique $n' \in N'$ such that $g(n') = f(m)$. Therefore, there is a function $h: M \to N'$ with $m \mapsto n'$ such that $f = gh$. The rest of the proof is also easy.  ∎

In view of the preceding results it is not surprising that the various isomorphism theorems for groups are valid for modules. One need only check at each stage of the proof to see that every subgroup or homomorphism is in fact a submodule or module homomorphism. For convenience we list these results here.

- Let $R$ be a ring, $M$ and $M'$ be two $R$-modules. If $f: M \to M'$ is an $R$-module homomorphism and $N$ is a submodule of $Ker f$, then there is a unique $R$-module homomorphism $\phi : M/N \to M'$ such that $\phi(a + N) = f(a)$ for all $a \in M$; $Im \phi = Im f$ and $Ker \phi = Ker f/N$. Moreover, $\phi$ is an $R$-module isomorphism if and only if $f$ is an onto $R$-module homomorphism and $N = Ker f$. In particular, $M/Ker f \cong Im f$.
- Let $R$ be a ring, $M$ and $M'$ be two $R$-modules. If $N$ is a submodule of $M$, $N'$ is a submodule of $M'$ and $f: M \to M'$ is an $R$-module homomorphism such that $f(N) \subseteq N'$, then $f$ induces an $R$-module isomorphism $\phi : M/N \to M'/N'$ given by $a + N \mapsto f(a) + N'$. Moreover, $\phi$ is an $R$-module isomorphism if and only if $Im f + N' = M'$ and $f^{-1}(N') \subseteq N$. In particular, if $f$ is an onto $R$-module homomorphism such that $f(N) = N'$ and $Ker f \subseteq N$, then $\phi$ is an $R$-module isomorphism.
- Let $A$ and $B$ be two submodules of an $R$-module $M$.

(1) There is an $R$-module isomorphism $A/(A \cap B) \cong (A + B)/B$;
(2) If $B \subseteq A$, then $A/B$ is a submodule of $M/B$, and there is an $R$-module isomorphism $(M/B)/(A/B) \cong M/A$.

Already, we studied direct products of groups, and the basic idea seems to carry over to modules. Suppose that we have an $R$-module $M_i$ for each $i$ in some index set $I$. The elements of the direct product of the $M_i$, denoted by $\prod_{i \in I} M_i$ are $(a_i)_{i \in I}$, where $a_i \in M_i$. Addition is described by $(a_i)_{i \in I} + (b_i)_{i \in I} = (a_i + b_i)_{i \in I}$ and scalar multiplication by $r(a_i)_{i \in I} = (ra_i)_{i \in I}$.

**Definition 10.11** The *external direct sum of the modules* $M_i$, for $i \in I$, denoted by $\oplus \sum_{i \in I} M_i$, consists of all families $(a_i)_{i \in I}$ with $a_i \in M_i$ such that $a_i = 0$ for all but finitely many $i$. Addition and scalar multiplication are defined exactly as for the direct product, so that the external direct sum coincides with the direct product when the index set $I$ is finite. The $R$-module $M$ is the *internal direct sum of the submodules*

$M_i$ if each $x \in M$ can be expressed uniquely as $x_{i_1} + \cdots + x_{i_n}$ where $0 \neq x_{i_k} \in M_{i_k}$, for all $k = 1, \ldots, n$. Just as with groups, the internal and external direct sums are isomorphic.

**Theorem 10.12** *The module $M$ is the direct sum of submodules $M_i$ if and only if both of the following conditions are satisfied:*

*(1)* $M = \sum_{i \in I} M_i$, *i.e, each $x \in M$ is a finite sum of the form $x_{i_1} + \cdots + x_{i_n}$, where $x_{i_k} \in M_{i_k}$;*
*(2)* $M_i \cap \sum_{j \neq i} M_j = \{0\}$, *for all $i$.*

**Proof** The proof is similar to groups, and we left it to reader. ∎

**Definition 10.13** Let $R$ be a ring and $M$ be a non-trivial $R$-module.

(1) $M$ is said to be *irreducible* if the only submodules are $\{0\}$ and $M$.
(2) $M$ is said to be *reducible* if there is a proper non-trivial submodule $N$ of $M$.
(3) $M$ is *completely reducible* (or *semisimple*) if $M$ is a direct sum of irreducible $R$-modules.

**Definition 10.14** A ring $R$ is *semisiple* if $R$ is itself a complete reducible $R$-module.

If $M$ and $N$ are $R$-modules, then the set $Hom_R(M, N)$ of all $R$-module homomorphisms $M \to N$ is an abelian group with $f + g$ given by $(f + g)(a) = f(a) + g(a)$, for all $a \in M$. The identity element is the zero function. Moreover, $Hom_R(M, N)$ is a ring with identity, where multiplication is composition of functions.

$Hom_R(M, M)$ is called the *endomorphism ring* and is denoted by $End_R(M)$.

**Lemma 10.15 (Schur's Lemma).** *If $R$ is a ring and $M, N$ are irreducible $R$-modules, then*

$$Hom_R(M, N) = \begin{cases} \{0\} & \text{if } M \ncong N \\ \text{a division ring} & \text{if } M \cong N. \end{cases}$$

**Proof** Suppose that $f$ is a nonzero element of $Hom_R(M, N)$. Since $Kerf$ is a submodule of $M$, it follows that $Kerf = \{0\}$ or $Kerf = M$. As $f$ is nonzero and $M$ is irreducible, we conclude that $Kerf = \{0\}$, and so $f$ is one to one. Since $Imf$ is a submodule of $N$, it follows that $Imf = \{0\}$ or $Imf = N$. As $f$ is nonzero and $N$ is irreducible, we obtain $Imf = W$. This shows that $f$ is onto. Therefore, $f$ is an isomorphism and must have an inverse, proving that $Hom_R(M, M) = End_R(M)$ is a division ring. ∎

**Lemma 10.16** *Let $R$ be a ring, and $M = M_1 \oplus \cdots \oplus M_n$ and $N = N_1 \oplus \cdots \oplus N_m$ be $R$-modules. Let $\iota_j : M_j \to M$ be the natural injection $\iota_j : a \mapsto a$ and let $\pi_i : N \to N_i$ be the natural projection.*

*(1) If for each $i$ and $j$, $\varphi_{ij} \in Hom_R(M_j, N_i)$, then we can define $\varphi \in Hom_R(M, N)$ by*

$$\varphi(a_1 + \cdots + a_n) = \begin{bmatrix} \varphi_{11} & \cdots & \varphi_{1n} \\ \vdots & \vdots & \vdots \\ \varphi_{m1} & \cdots & \varphi_{mn} \end{bmatrix} \begin{bmatrix} a_1 \\ \vdots \\ a_n \end{bmatrix}$$

$$= \underbrace{\varphi_{11}(a_1) + \cdots + \varphi_{1n}(a_n)}_{\text{lies in } W_1} + \cdots + \underbrace{\varphi_{m1}(a_1) + \cdots + \varphi_{mn}(a_n)}_{\text{lies in } W_m},$$

for all $a_i \in M_i$ $(i = 1, \ldots, n)$;

(2)  If $\varphi \in Hom_R(M, N)$, then $\varphi_{ij} = \pi_i \circ \varphi \circ \iota_j \in Hom_R(M_j, N_i)$ and

$$\varphi(a_1 + \cdots + a_n) = \begin{bmatrix} \varphi_{11} & \cdots & \varphi_{1n} \\ \vdots & \vdots & \vdots \\ \varphi_{m1} & \cdots & \varphi_{mn} \end{bmatrix} \begin{bmatrix} a_1 \\ \vdots \\ a_n \end{bmatrix};$$

(3)  We have

$$Hom_R(M, N) = \begin{bmatrix} Hom_R(M_1, N_1) & \cdots & Hom_R(M_n, N_1) \\ \vdots & \vdots & \vdots \\ Hom_R(M_1, N_m) & \cdots & Hom_R(M_n, N_m) \end{bmatrix},$$

as additive groups;

(4)  In particular, if $M^{(n)} = \underbrace{M \oplus \cdots \oplus M}_{n \text{ copies}}$, then $End_R\big(M^{(n)}\big) \cong Mat_{n \times n}\big(End_R(V)\big)$

as rings.

**Proof**  It is clear, and we left it to reader.  ∎

**Theorem 10.17**  *Let $R$ be a ring and $M$ be an $R$-module. Then, the following statements are equivalent:*

(1)  *$M$ is a sum of irreducible submodules.*
(2)  *$M$ is completely reducible.*
(3)  *If $N$ is any submodule of $M$, then $M = N \oplus N'$, for some submodule $N'$.*

**Proof**  Note that Zorn's lemma is used to obtain several maximal objects in the following proof.

$(1 \Rightarrow 2)$: Suppose that $M = \sum_{i \in I} M_i$, where $M_i$'s are irreducible and let $J$ be a maximal subset of $I$ such that $M' = \sum_{j \in J} M_j$ is a direct sum. If $i \notin J$, then $M_i \cap M'$ is a submodule of $M_i$. Since $M_i$ is irreducible, it follows that $M_i \cap M' = \{0\}$ or $M_i \cap M' = M_i$. If $M_i \cap M' = \{0\}$, then $\sum_{j \in J \cup \{i\}} M_j$ is a direct sum, contradicting the maximality of $J$. So, $M_i \cap M' = M_i$. This yields that all $M_i \subseteq M'$, and so $M' = M$.

$(2 \Rightarrow 3)$: Suppose that $M = \oplus \sum_{i \in I} M_i$ such that $M_i$'s are irreducible and $N$ is a submodule of $M$. We consider a maximal subset $J$ of $I$ such that $N + \big( \oplus \sum_{j \in J} M_j \big)$ is a direct sum. For any $i \in I - J$, $M_i \cap \big( N + \big( \oplus \sum_{j \in J} M_j \big) \big)$ is a submodule of $M_i$, and so is $\{0\}$ or $M_i$. Similar to the previous proof, we observe that $M_i \cap \big( N + \big( \oplus$

$\sum_{j \in J} M_j)) \neq \{0\}$. Consequently, we obtain $M_i \subseteq N \oplus (\oplus \sum_{j \in J} M_j)$. Therefore, we can take $N' = \oplus \sum_{j \in J} M_j$.

$(3 \Rightarrow 1)$: First we consider the following two claims:

*Claim 1:* If $M$ satisfies (3), so does any submodule $M_0$ of $M$.
To prove the claim, assume that $N$ is any submodule of $M_0$. Then, by (3), we have $M = N \oplus N'$, for some submodule $N'$ of $M$. Now, let $a_0 \in M_0$ be an arbitrary element. Then, $a_0 = x + y$, for some $x \in N$ and $y \in N'$. Hence, we can write $y = a_0 - x \in M_0$, and so $y \in N' \cap M_0$. This shows that $M_0 = N \oplus (N' \cap M_0)$.

*Claim 2:* Any nonzero submodule $M_0$ of $M$ contains an irreducible submodule.
To prove the claim, suppose that $0 \neq a_0 \in M_0$ and $N_0$ is a maximal submodule of $M_0$ such that $a_0 \notin N_0$. By Claim 1, we have $M_0 = N_0 \oplus N_1$, for some submodule $N_1$. If $N_1$ is not irreducible, then by Claim 1, $N_1 = N_2 \oplus N_3$. Hence, we have $M_0 = N_0 \oplus N_2 \oplus N_3$, which implies that $N_0 = N_0 + \{0\} = N_0 + (N_2 \cap N_3) = (N_0 + N_2) \cap (N_0 + N_3)$. Since $a_0 \notin N_0$, it follows that $a_0 \notin N_0 + N_2$ or $a_0 \notin N_0 + N_3$, a contradiction to the maximality of $N_0$. Therefore, we conclude that $N_1$ is irreducible.

Now, suppose that (3) holds for $M$. Let $M'$ be the sum of all irreducible submodules of $M$. By (3), we can write $M = M' + N$, for some submodule $N$. If $N \neq \{0\}$, then by Claim 2, there exists an irreducible submodule $N_0$ of $N$. But $N_0 \subseteq M'$, a contradiction. Consequently, we have $N = \{0\}$, and so $M = M'$, i.e., $M$ is a sum of irreducible submodules. ∎

**Lemma 10.18** *Let $R$ be a ring and $M$ be a completely reducible $R$-module. Suppose that $A = End_R(M)$, where $M$ is an $A$-module under $\varphi \cdot a = \varphi(a)$, for all $\varphi \in A$ and $a \in M$. Then, for any $a \in M$ and any $f \in End_A(M)$, there is $x \in R$ such that $f(a) = xa$.*

**Proof** For each $a \in M$, $Ra = \{ra \mid r \in R\}$ is a submodule of $M$. Since $M$ is completely reducible, it follows that $M = Ra \oplus N$, for some submodule $N$ of $M$. Suppose that $\pi : M \to Ra$ is the projection. Then, $\pi \in End_R(M) = A$. Since $M$ is an $A$-module, it follows that $f(a) = f(\pi(a)) = \pi f(a) \in \pi(M) = Ra$. This implies that $f(a) = xa$, for some $x \in R$. ∎

**Theorem 10.19 (Jacobson Density Theorem).** *Let $M$ be an irreducible $R$-module, $A = End_R(M)$, $f \in End_A(M)$ and $a_1, \ldots, a_n \in M$. Then, there is $x \in R$ such that $f(a_i) = xa_i$, for all $1 \leq i \leq n$.*

**Proof** Denote $M^{(n)} = \underbrace{M \oplus \cdots \oplus M}_{n \text{ copies}}$ and define $f^{(n)} : M^{(n)} \to M^{(n)}$ by $f^{(n)}(a_1 + \cdots + a_n) = f(a_1) + \cdots + f(a_n)$, for all $a_i, f(a_i)$ belong to the $i$th summand $M$. Let $B = End_R(M^{(n)})$ and $\varphi \in B$ be arbitrary. By Lemma 10.16, we can write

$$\varphi(a_1 + \cdots + a_n) = \begin{bmatrix} \varphi_{11} & \cdots & \varphi_{1n} \\ \vdots & \vdots & \vdots \\ \varphi_{n1} & \cdots & \varphi_{nn} \end{bmatrix} \begin{bmatrix} a_1 \\ \vdots \\ a_n \end{bmatrix},$$

where all $\varphi_{ij} \in End_R(M) = A$. Thus, we observe that

$$
\begin{aligned}
f^{(n)}&\big(\varphi(a_1 + \cdots + a_n)\big)\\
&= f^{(n)}\big(\varphi_{11}(a_1) + \cdots + \varphi_{1n}(a_n) + \cdots + \varphi_{n1}(a_1) + \cdots + \varphi_{nn}(a_n)\big)\\
&= f^{(n)}\big(\big(\varphi_{11}(a_1) + \cdots + \varphi_{n1}(a_1)\big) + \cdots + \big(\varphi_{1n}(a_n) + \cdots + \varphi_{nn}(a_n)\big)\big)\\
&= f\big(\varphi_{11}(a_1) + \cdots + \varphi_{n1}(a_1)\big) + \cdots + f\big(\varphi_{1n}(a_n) + \cdots + \varphi_{nn}(a_n)\big)\\
&= f\big(\varphi_{11}(a_1)\big) + \cdots + f\big(\varphi_{n1}(a_1)\big) + \cdots + f\big(\varphi_{1n}(a_n)\big) + \cdots + f\big(\varphi_{nn}(a_n)\big)\\
&= \varphi_{11}\big(f(a_1)\big) + \cdots + \varphi_{1n}\big(f(a_n)\big) + \cdots + \varphi_{n1}\big(f(a_1)\big) + \cdots + \varphi_{nn}\big(f(a_n)\big)\\
&= \varphi\big(f^{(n)}(a_1 + \cdots + a_n)\big).
\end{aligned}
$$

Hence, $f^{(n)} \in End_B\big(M^{(n)}\big)$. Now, by Lemma 10.18, there is $x \in R$ such that $f^{(n)}(a_1 + \cdots + a_n) = x(a_1 + \cdots + a_n)$. This implies that $f(a_1) + \cdots + f(a_n) = xa_1 + \cdots + xa_n$, all $a_i, f(a_i)$ belong to $i$th summand. Therefore, we obtain $f(a_i) = xa_i$, for all $1 \le i \le n$.                                                             ∎

**Exercises**

1. Prove that a finitely generated $R$-module need not be finitely generated as an abelian group.
2. If $f: M \to M$ is an $R$-module homomorphism such that $f \circ f = f$, show that $M = Ker f \oplus Im f$.
3. Prove that any unitary, irreducible $R$-module is cyclic.
4. Let $M, M_1, \ldots, M_n$ be $R$-modules. Prove that $M \cong M_1 \oplus \ldots \oplus M_n$ if and only if for each $i = 1, \ldots, n$ there is an $R$-module homomorphism $\varphi_i : M \to M$ such that $Im \varphi_i \cong M_i$, $\varphi_i \circ \varphi_j = 0$ for $i \ne j$, and $\varphi_1 + \ldots + \varphi_n = id_M$.
5. Let $R$ be a ring and $M$ be a completely reducible $R$-module. Show that if $End_R(M)$ is a division ring, then $M$ is irreducible.
6. Show that if $R$ is a commutative ring, then $Hom_R(R, M) \cong M$ as $R$-modules.
7. If a ring $R$ is semisimple, prove that every $R$-module is completely reducible.
8. Show that a semisimple commutative ring is a direct sum of fields.
9. Let $M$ be an $R$-module. Show that $M$ is completely irreducible if and only if the intersection of all the maximal submodules of $M$ is trivial.

## 10.2  Group Representations

First of all we need the notion of a representation. That is, we want to create a map that takes us from some finite group $G$ into a subgroup of $GL_n(\mathbb{F})$, where $\mathbb{F}$ is the underlying field.

**Definition 10.20**  A *representation* of $G$ over a field $\mathbb{F}$ is a homomorphism $T: G \to GL(V, \mathbb{F})$, where $V$ is a vector space of dimension $n$ over $\mathbb{F}$, for some positive integer $n$. The *degree of representation* $T$ is the integer $n$.

Since $GL(V, \mathbb{F}) \cong GL_n(\mathbb{F})$, we can also consider $T$ as a homomorphism from $G$ to $GL_n(\mathbb{F})$.

**Example 10.21** For any group $G$ and any field $\mathbb{F}$, the function $T: G \to GL_1(\mathbb{F})$ with $T(a) = I_1$, for all $a \in G$, is a a representation. This is called the *trivial representation* of $G$ (over $\mathbb{F}$).

**Example 10.22** The dihedral group $D_n = \langle x, y \mid x^2 = y^n = e, \, xyx = y^{-1} \rangle$ naturally acts on the regular $n$-gon, and this induces an action on $\mathbb{R}^2$ or $\mathbb{C}^2$ by

$$
x \mapsto \begin{bmatrix} 0 & 1 \\ 1 & 0 \end{bmatrix} \text{ and } y \mapsto \begin{bmatrix} \cos\left(\dfrac{2\pi}{n}\right) & \sin\left(\dfrac{2\pi}{n}\right) \\ -\sin\left(\dfrac{2\pi}{n}\right) & \cos\left(\dfrac{2\pi}{n}\right) \end{bmatrix},
$$

which defines a representation of $D_n$ of degree 2.

**Example 10.23** The symmetric group $S_3$, the smallest non-abelian group of order 6, is generated by elements $x, y$ with relations $x^2 = y^3 = e$, and $yx = xy^2$. Let $\mathbb{F} = \mathbb{C}$ and $\omega$ be a cube root of 1. Then

$$
\begin{bmatrix} 0 & 1 \\ 1 & 0 \end{bmatrix}^2 = I_2 \text{ and } \begin{bmatrix} \omega & 0 \\ 0 & \omega^2 \end{bmatrix}^3 = I_2,
$$

and

$$
\begin{bmatrix} \omega & 0 \\ 0 & \omega^2 \end{bmatrix}\begin{bmatrix} 0 & 1 \\ 1 & 0 \end{bmatrix} = \begin{bmatrix} 0 & 1 \\ 1 & 0 \end{bmatrix}\begin{bmatrix} \omega & 0 \\ 0 & \omega^2 \end{bmatrix}^2.
$$

So, we may construct a representation of $S_3$ by

$$
x \mapsto \begin{bmatrix} 0 & 1 \\ 1 & 0 \end{bmatrix} \text{ and } y \mapsto \begin{bmatrix} \omega & 0 \\ 0 & \omega^2 \end{bmatrix}.
$$

**Example 10.24** Let $p$ be a prime and $C_p = \langle a \rangle$ be the cyclic group of order $p$. It is easy to check that $T: C_p \to GL_2(\mathbb{F}_p)$ with the assignment

$$
a^k \mapsto \begin{bmatrix} 1 & 0 \\ k & 1 \end{bmatrix}
$$

is a representation. In this case the fact that we have a representation is very much dependent on the choice of the field as the field $\mathbb{F}_p$; in any other characteristic it would not work, because the matrix shown would no longer have order $p$.

**Definition 10.25** Let $T: G \to GL_m(\mathbb{F})$ and $T' : G \to GL_n(\mathbb{F})$ be representations of a group $G$ over $\mathbb{F}$. We say that $T$ is equivalent to $T'$ if $n = m$ and there exists $A \in GL_n(\mathbb{F})$ such that $A^{-1}T(a)A = T'(a)$, for all $a \in G$.

**Lemma 10.26** *The above relation is an equivalence relation.*

***Proof*** It is straightforward.                                          ∎

**Example 10.27** Let $D_4 = \langle x, y \mid x^2 = y^4 = e, \ xyx = y^3 \rangle$ and let $T: D_4 \to GL_2(\mathbb{C})$ be a representation of $D_4$ defined by $T(x) = \begin{bmatrix} 1 & 0 \\ 0 & -1 \end{bmatrix}$ and $T(y) = \begin{bmatrix} 0 & 1 \\ -1 & 0 \end{bmatrix}$. If $A = 1/\sqrt{2} \begin{bmatrix} 1 & 1 \\ i & -i \end{bmatrix}$, then $A^{-1} = 1/\sqrt{2} \begin{bmatrix} 1 & -i \\ 1 & i \end{bmatrix}$, and so we obtain

$$A^{-1}T(x)A = \begin{bmatrix} 0 & 1 \\ 1 & 0 \end{bmatrix} \text{ and } A^{-1}T(y)A = \begin{bmatrix} i & 0 \\ 0 & -i \end{bmatrix}.$$

Consequently, we obtain a representation $T': D_4 \to GL_2(\mathbb{C})$ with

$$T'(x) = \begin{bmatrix} 0 & 1 \\ 1 & 0 \end{bmatrix} \text{ and } T'(y) = \begin{bmatrix} i & 0 \\ 0 & -i \end{bmatrix}.$$

The representations $T$ and $T'$ are equivalent.

**Definition 10.28** Let $G$ be a group and $R$ a ring. Firstly, we define the set $RG$ to be one of the following:

- the set of all formal $R$-linear combinations of elements of $G$;
- the set of all functions $f: G \to R$ with $f(g) = 0$ for all but finitely many $g$ in $G$.

No matter which definition is used, we can write the elements of $RG$ in the form $\sum_{g \in G} a_g g$, with all but finitely many of the $a_g$ being $0$, and the addition on $RG$ is the addition of formal linear combinations or addition of functions, respectively. The multiplication of elements of $RG$ is defined by setting

$$\left( \sum_{g \in G} a_g g \right) \left( \sum_{h \in G} b_h h \right) = \sum_{g, h \in G} (a_g b_h) gh.$$

If $R$ has an identity element, this is the unique bilinear multiplication for which $(1g)(1h) = (1gh)$. In this case, $G$ is commonly identified with the set of elements $1g$ of $RG$. The identity element of $G$ then serves as the $1$ in $RG$.

It is not difficult to verify that $RG$ is a ring. This ring is called the *group ring* of $G$ over $R$. If $R$ and $G$ are both commutative (i.e., $R$ is commutative and $G$ is an abelian group), then $RG$ is commutative.

Let $G$ be a group and $\mathbb{F}$ be a field. According to Definition 10.28, we can construct the ring $\mathbb{F}G$, where we called it the group ring. For any $\sum_{g \in G} a_g g \in \mathbb{F}G$ and $c \in \mathbb{F}$, we define

$$c\left( \sum_{g \in G} a_g g \right) = \sum_{g \in G} (c a_g) g.$$

It is easy to check that $\mathbb{F}G$ is a vector space over $\mathbb{F}$. Indeed, $\mathbb{F}G$ is an algebra over $\mathbb{F}G$. When $G = \{g_1, g_2, \ldots, g_n\}$, then the dimension of $\mathbb{F}G$ over $\mathbb{F}$ is $|G|$, and $\{g_1, g_2, \ldots, g_n\}$ is a basis for it. Note that $\mathbb{F}G$ is itself an $\mathbb{F}G$-module.

The center of $\mathbb{F}G$ is $Z(\mathbb{F}G) = \{z \in \mathbb{F}G \mid zr = rz, \text{ for all } r \in \mathbb{F}G\}$. It is clear that $Z(\mathbb{F}G)$ is a subspace.

**Definition 10.29** Let $C_1, \ldots, C_h$ be the distinct conjugacy class of $G$. For each $1 \leq i \leq h$, we define

$$\overline{C_i} = \sum_{g \in C_i} g.$$

The elements $\overline{C_1}, \ldots, \overline{C_h}$ are called *class sums*.

If $\overline{C_i}$ consists of $k$ distinct conjugates $x_1^{-1}gx_1, \ldots, x_k^{-1}gx_k$ of an element $g \in G$, then $\overline{C_i} = \sum_{i=1}^k x_i^{-1}gx_i$. Hence, for each $y \in G$, we have

$$y^{-1}\overline{C_i}y = \sum_{i=1}^k y^{-1}x_i^{-1}gx_iy.$$

Note that $y^{-1}x_r^{-1}gx_ry = y^{-1}x_s^{-1}gx_sy$ if and only if $x_r^{-1}gx_r = x_s^{-1}gx_s$. This yields that $y^{-1}\overline{C_i}y = \overline{C_i}$. Hence, we conclude that $\overline{C_i}$ commutes with all $y \in G$, and so for all elements of $\mathbb{F}G$. This shows that $\overline{C_i} \in Z(\mathbb{F}G)$, for all $1 \leq i \leq h$.

**Theorem 10.30** *The class sums $\overline{C_1}, \ldots, \overline{C_h}$ form a basis of $Z(\mathbb{F}G)$.*

**Proof** First, we show that $\overline{C_1}, \ldots, \overline{C_h}$ are linearly independent. To do this, assume that $a_1\overline{C_1} + \cdots + a_h\overline{C_h} = 0$, for some $a_1, \ldots, a_h \in \mathbb{F}$. Since $G$ is a union of conjugacy classes and distinct conjugacy classes are disjoint, it follows that $a_i = 0$. Next, we show that $\overline{C_1}, \ldots, \overline{C_h}$ span $Z(\mathbb{F}G)$. Suppose that $b = \sum_{g \in G} a_g g$ is an arbitrary element of $\mathbb{F}G$. For each $x \in G$, we have $xb = bx$, and so $x^{-1}bx = b$. This means that

$$\sum_{g \in G} a_g x^{-1}gx = \sum_{g \in G} a_g g.$$

Thus, the function $g \mapsto a_g$ is constant on conjugacy classes of $G$. This gives that $b = \sum_{i=1}^h \overline{C_i}$, where $a_i$ is the coefficient $a_{g_i}$, for some $g_i \in C_i$.  ∎

**Definition 10.31** Let $G = \{e = g_1, g_2, \ldots, g_n\}$. We define $T: G \to GL_n(\mathbb{F})$ by $T(g) = (a_{ij})_{n \times n}$, for all $g \in G$, such that

$$a_{ij} = \begin{cases} 1 & \text{if } gg_i = g_j \\ 0 & \text{otherwise} \end{cases}$$

This is a *permutation matrix* with one 1 in each row and each column. This representation is called the *regular representation* of $G$ and can be constructed for any finite group.

**Example 10.32** Let $G = C_3 = \langle a \mid a^3 = e \rangle$. The elements of $\mathbb{F}G$ have the form $c_1 e + c_2 a + c_3 a^2$, where $c_i, c_2, c_3 \in \mathbb{F}$. We have

$$(c_1 e + c_2 a + c_3 a^2)e = c_1 e + c_2 a + c_3 a^2,$$
$$(c_1 e + c_2 a + c_3 a^2)a = c_3 e + c_1 a + c_2 a^2,$$
$$(c_1 e + c_2 a + c_3 a^2)a^2 = c_2 e + c_3 a + c_1 a^2.$$

By taking matrices relative to the basis $\{e, a, a^2\}$ of $\mathbb{F}G$, we obtain the regular representation of $G$ as follows:

$$e \to I_3, \; a \to \begin{bmatrix} 0 & 1 & 0 \\ 0 & 0 & 1 \\ 1 & 0 & 0 \end{bmatrix} \text{ and } a^2 \to \begin{bmatrix} 0 & 0 & 1 \\ 1 & 0 & 0 \\ 0 & 1 & 0 \end{bmatrix}$$

**Definition 10.33** A representation $T \colon G \to GL_n(\mathbb{F})$ is said to be *faithful* if $KerT = \{e\}$.

**Example 10.34** Regular representation is faithful.

**Theorem 10.35** *Studding representations of the group $G$ over $\mathbb{F}$ is equivalent to studding $\mathbb{F}G$-modules.*

**Proof** Suppose that $T \colon G \to GL(V, \mathbb{F})$ is a representation of $G$. If we define $gv := T(g)v$, for all $g \in G$ and $v \in V$, then $V$ becomes an $\mathbb{F}G$-module.

Conversely, let $V$ be an $\mathbb{F}G$-module and $g \in G$. We define $T(g) \colon V \to V$ by $T(g)v := gv$, for all $v \in V$. Then, we have $T(g) \in GL(V, \mathbb{F})$. Since $V$ is an $\mathbb{F}G$-module, it follows that

$$T(g_1 g_2) = g_1 g_2 v = g_1(g_2 v) = g_1\big(T(g_2)v\big) = T(g_1)T(g_2)v,$$

for all $g_1, g_2 \in G$ and $v \in V$. This yields that $T(g_1 g_2) = T(g_1)T(g_2)$, and so $T$ is a representation. This representation is called the *representation afforded by* $V$. ∎

**Exercises**

1. Let $G$ be a group and $\mathbb{F}$ be a field. Prove that if $V$ is an $\mathbb{F}G$-module, then the vector space $V^* = Hom_{\mathbb{F}}(V, \mathbb{F})$ has an $\mathbb{F}G$-module structure such that

$$\left( \left( \sum_{g \in G} a_g g \right)(f) \right)(v) = f\left( \left( \sum_{g \in G} a_g g^{-1} \right)(v) \right),$$

for any $f \in V^*$ and $v \in V$.

2. Let $G = \langle a \mid a^m = e \rangle$ be the cyclic group of order $m$. Suppose that $A \in GL_n(\mathbb{C})$ and define $T \colon G \to GL_n(\mathbb{C})$ by $T(a^k) = A^k$, for all $0 \le k \le m - 1$. Show that $T$ is a representation of $G$ over $\mathbb{C}$ if and only if $A^m = I_n$.

3. Suppose that $T: G \to GL_n(\mathbb{C})$ is a representation of a finite group $G$ over $\mathbb{C}$. Show that for each $g \in G$ the matrix $T(g)$ is diagonalizable.
4. Let $V$ and $W$ be $\mathbb{C}G$-modules. Prove that $V \times W$ equipped with the map $G \times (V \times W) \to (V \times W)$ such that $(g, (v, w)) \mapsto (gv, gw)$ is also a $\mathbb{C}G$-module.
5. Let $V$ be a $\mathbb{C}G$-module and $W$ be a $\mathbb{C}H$-module. Prove that $V \times W$ equipped with the map $(G \times H) \times (V \times W) : ((g, h), (v, w)) \mapsto (gv, hw)$ is a $\mathbb{C}(G \times H)$-module.
6. Let $G = D_6 = \langle a, b \mid a^6 = b^2 = e, \ b^{-1}ab = a^{-1} \rangle$. Define the matrices $A$, $B$, $C$, and $D$ over $\mathbb{C}$ by

$$A = \begin{bmatrix} e^{i\pi/3} & 0 \\ 0 & e^{i\pi/3} \end{bmatrix}, \quad B - \begin{bmatrix} 0 & 1 \\ 1 & 0 \end{bmatrix},$$

$$C = \begin{bmatrix} \dfrac{1}{2} & \dfrac{\sqrt{3}}{2} \\ -\dfrac{\sqrt{3}}{2} & \dfrac{1}{2} \end{bmatrix}, \quad D = \begin{bmatrix} 1 & 0 \\ 0 & -1 \end{bmatrix}.$$

Prove that each of the functions $T_k : G \to GL_2(\mathbb{C})$ $(k = 1, 2, 3, 4)$, given by

$$T_1 : a^r b^s \mapsto A^r B^s,$$
$$T_2 : a^r b^s \mapsto A^{3r}(-B)^s,$$
$$T_3 : a^r b^s \mapsto (-A)^r B^s,$$
$$T_4 : a^r b^s \mapsto C^r D^s \quad (0 \le r \le 5, \ 0 \le s \le 1)$$

is a representation of $G$. Which of these representations are faithful? Which are equivalent?
7. Give an example of a faithful representation of $D_4$ of degree 3.
8. Let $G = S_n$ and let $V$ be a vector space over $\mathbb{F}$. Show that $V$ becomes an $\mathbb{F}G$-module if we define, for all $v \in V$,

$$gv := \begin{cases} v & \text{if } g \text{ is even permutation} \\ -v & \text{if } g \text{ is odd permutation.} \end{cases}$$

9. Let $G$ be a group, and $H$ be a subgroup of index $k$. Explain how we can use $H$ to construct a representation of $G$ of degree $k$. If $G = S_n$ and $H = A_n$, what representation do we get?

## 10.3 Reducible and Irreducible Representation

Let $G$ be a group and $V$ be a finite vector space on a field $\mathbb{F}$.

**Definition 10.36** A representation $T: G \to GL(V, \mathbb{F})$ is *reducible, irreducible,* or *completely reducible (semisimple)*, according to whether $V$ is reducible, irreducible, or completely reducible (semisimple) as $\mathbb{F}G$-module.

If $V$ is reducible with submodule $W$, then we can write $V = W \oplus W'$, where $W'$ is a subspace of $V$. Suppose that $\{v_1, \ldots, v_m\}$ is an ordered basis for $W$ and $v_{m+1}, \ldots, v_n\}$ is an ordered basis for $W'$. In the basis $\{v_1, \ldots, v_m, v_{m+1}, \ldots, v_n\}$ of $V$, we observe that matrices of elements of $G$ have the form

$$\begin{bmatrix} A & B \\ 0 & C \end{bmatrix},$$

where $A$ is $m \times m$ matrix, $B$ an $m \times (n - m)$ matrix, $C$ an $(n - m) \times (n - m)$ matrix, and $0$ the $(n - m) \times m$ zero matrix.

If $V$ is completely reducible, let $V = W_1 \oplus \cdots \oplus W_k$, where $W_i$'s are irreducible submodules of $V$. By choosing bases for $W_i$'s and combining them, we observe that the matrices of elements of $G$ have the form

$$\begin{bmatrix} A_1 & 0 & \cdots & 0 \\ 0 & A_2 & \cdots & 0 \\ \vdots & \vdots & \ddots & \vdots \\ 0 & 0 & \cdots & A_k \end{bmatrix}$$

We see that the presence of submodule enables us to greatly simplify matrix representation. Now, our attention is directed at two problems:

- Determine which representations are completely reducible.
- Determine all irreducible representations.

By far and away the most important condition of complete reducibility is Maschke's theorem.

**Theorem 10.37 (Maschke's Theorem).** *Let $G$ be a finite group, $\mathbb{F}$ be a field whose characteristic does not divide the order of $G$, and let $V$ be an $\mathbb{F}G$-module of finite dimension over $\mathbb{F}$. If $U$ is an $\mathbb{F}G$-submodule of $V$, then there is an $\mathbb{F}G$-submodule $W$ of $V$ such that $V = U \oplus W$.*

***Proof*** Note that since $V$ is a vector space, we can certainly write $V = U \oplus W_0$, where $W_0$ is a subspace (but perhaps not an $\mathbb{F}G$-submodule).

We are given $U$ as an $\mathbb{F}G$-submodule of an $\mathbb{F}G$-module $V$. Let $\{v_1, \ldots, v_m\}$ be a basis for $U$ and extend it to a basis $\{v_1, \ldots, v_n\}$ of $V$ and let $W_0$ be the subspace spanned by $\{v_{m+1}, \ldots, v_n\}$. Then, $V = U \oplus W$. For $v \in V$, we have $v = u + w$ for unique vectors $u \in U$ and $w \in W_0$. We define $\pi : V \to U$ by setting $\pi(v) = u$. Then, $\pi$ is a projection of $V$ with image $U$. In fact, $\pi$ is a linear transformation on $\mathbb{F}$, but not necessarily a homomorphism of $\mathbb{F}G$-modules. We aim to modify the projection $\pi$ to create a homomorphism of $R$-modules from $V$ to $V$ with image $U$. To this end, we define $\pi^* : V \to V$ by

$$\pi^*(v) = \frac{1}{|G|} \sum_{g \in G} g^{-1}\pi(gv).$$

By our hypothesis, $1/|G|$ exists in $\mathbb{F}$. It is clear that $\pi^*$ is an endomorphism of $V$ and $Im\pi^* \subseteq U$. First, we show that $\pi^*$ is a homomorphism of $\mathbb{F}G$-modules. Assume that $v \in V$ and $x \in G$. Then, we have

$$x^{-1}\pi^*(xv) = \frac{1}{|G|}\sum_{g \in G} x^{-1}g^{-1}\pi(gxv).$$

As $g$ runs over the elements of $G$, so does $y = gx$. Hence, we can write

$$x^{-1}\pi^*(xv) = \frac{1}{|G|}\sum_{y \in G} y^{-1}\pi(yv) = \pi^*(v),$$

which implies that

$$\pi^*(xv) = x\pi^*(v).$$

Consequently, $\pi^*$ is a homomorphism of $\mathbb{F}G$-modules. Now, we prove that $(\pi^*)^2 = \pi^*$. Note that for $u \in U$ and $g \in G$, we have $gu \in U$, and so $\pi(gu) = gu$. Using this, we have

$$\pi^*(u) = \frac{1}{|G|}\sum_{g \in G} g^{-1}\pi(gu) = \frac{1}{|G|}\sum_{g \in G} g^{-1}gu = \frac{1}{|G|} \cdot |G|u = u. \qquad (10.1)$$

Now, suppose that $v \in V$ is arbitrary. Then, we have $\pi^*(v) \in U$. So, by (10.1), we get $\pi^*\big(\pi^*(v)\big) = \pi^*(v)$. This implies that $(\pi^*)^2 = \pi^*$, as claimed. We established that $\pi^* : V \to V$ is a projection and a projection and a homomorphism of $\mathbb{F}G$-modules. Moreover, (10.1) shows that $Im\pi^* = U$. It is now easy to check that

$$V = \pi^*(V) \oplus (1 - \pi^*)(V) = U \oplus (1 - \pi^*)(V).$$

It remains to prove that $W = (1 - \pi^*)(V)$ is an $\mathbb{F}G$-submodule of $V$. In fact, any element $(1 - \pi^*)(v) \in (1 - \pi^*)(V)$ satisfies

$$x\big((1 - \pi^*)(v)\big) = xv - x\pi^*(v) = xv - \pi^*(xv)$$
$$= (1 - \pi^*)(xv) \in (1 - \pi^*)(V),$$

and we are done. ∎

**Corollary 10.38** *If $G$ is a finite group, then every nonzero $\mathbb{F}G$-module is completely irreducible.*

***Proof*** Suppose that $V$ is an $\mathbb{F}G$-module. We apply mathematical induction on $dimV$. If $dimeV = 1$, then the result is true, because $V$ is irreducible in this case. If $V$ is irreducible, then we are done. Otherwise, assume that $V$ is reducible. So, there is a nonzero proper submodule $U$ of $V$. Next, by Maschke's theorem, there exists an $\mathbb{F}G$-module$W$ such that $V = U \oplus W$. Since $dimU < dimV$ and $dimW < dimV$, by induction hypothesis we conclude that $U = U_1 \oplus \cdots \oplus U_m$ and $W = W_1 \oplus \cdots \oplus W_n$. Hence, $V$ is a direct sum of irreducible $\mathbb{F}G$-modules.                                      ∎

There is another criterion for complete reducibility which has the advantage of making no restriction on field or group.

**Theorem 10.39** (**Cliford's Theorem**). *Let $G$ be any group, $\mathbb{F}$ be any field, $M$ be an irreducible $\mathbb{F}G$-module with finite dimension on $\mathbb{F}$, and $N$ be a normal subgroup of $G$. Of course, for $g \in G$ and $W$ an $\mathbb{F}N$-submodule of $M$, we denote $gW = \{gw \mid w \in W\} \subseteq M$.*

*(1) If $0 \neq W$ is an $\mathbb{F}N$-submodule of $M$, then $M = \sum_{g \in G} gW$. If $W$ is irreducible, then so is every $gW$. Thus, $M$ is completely reducible $\mathbb{F}N$-module;*

*(2) Let $W_1, \ldots, W_m$ be representatives of the isomorphism types of irreducible $\mathbb{F}N$-submodules of $M$. Define $V_i$ to be the all of $\mathbb{F}N$-submodules of $M$ that are isomorphic to $W_i$. Then, $M = V_1 \oplus \cdots \oplus V_m$ and $V_i$ is a direct sum of $\mathbb{F}N$-modules isomorphic to $W_i$;*

*(3) If $g \in G$, then each $gV_i$ is some $V_j$. The group $G$ is a transitive permutation group on $\{V_1, \ldots, V_m\}$;*

*(4) If $H_i = \{g \in G \mid gV_i = V_i\}$, then $V_i$ is irreducible as $\mathbb{F}H_i$-module.*

***Proof*** (1) Obviously, $\sum_{g \in G} gW$ is an $\mathbb{F}G$-submodule. Since $M$ is irreducible, it follows that $M = \sum_{g \in G} gW$. If $x \in N$ and $gw \in gW$, then $xgw = gg^{-1}xgw = gx^g w \in gW$. So, each $gW$ is an $\mathbb{F}N$-module. If $gW$ had a proper $\mathbb{F}N$-submodule $W_0$, then $g^{-1}W_0$ would be a proper submodule of $W$, contradicting its irreducibility. Hence, all $gW$,s are irreducible and $M$ is completely reducible.

(2) First of all, we observe that there are only finitely many isomorphism types: Since $M$ has finite dimension on $\mathbb{F}$, we apply the Jordan–Hölder theorem. Since $M$ is completely reducible, we obtain $M = V_1 + \cdots + V_m$. Also, $V_i$ is a direct sum of $\mathbb{F}N$-modules isomorphic to $W_i$. If $V_i \cap \sum_{j \neq i} V_j \neq \{0\}$, then this intersection would contain an irreducible $\mathbb{F}N$-submodule which, by the Jordan–Hölder theorem, would be isomorphic to $W_i$ and also with some $W_j, j \neq i$, and this is impossible. Therefore, we conclude that $M = V_1 \oplus \cdots \oplus V_m$.

(3) It is clear that

$$U_i := \sum_{\substack{x \in G \\ xW_1 \cong W_i}} xW_1 \subseteq W_i$$

and by (1) we obtain $M = U_1 + \cdots + U_m$. Now, by (2), we conclude that $U_i = V_i$, i.e.,

$$V_i = \sum_{\substack{x \in G \\ xW_1 \cong W_i}} xW_1. \tag{10.2}$$

Next, we prove that

$$xW_i \cong yW_i \Rightarrow gxW_i \cong gyW_i. \tag{10.3}$$

Indeed, if $\varphi : xW_i \to yW_i$ is an isomorphism, then $g\varphi g^{-1} : gxW_i \to gyW_i$ is an $\mathbb{F}$-isomorphism. And for $a \in N$, we have

$$(g\varphi g^{-1})(a(gxw_i)) = g\varphi((g-1ag)xw_i) = g(g^{-1}ag)\varphi(xw_i)$$
$$= a(g\varphi g^{-1})(gxw_i).$$

By (10.2) and (10.3), we obtain $g_i \subseteq V_j$, for some $V_j$. Also, $V_i \subseteq g^{-1}V_j$. This yields that $gV_i = V_j$. Irreducibility of $M$ as $\mathbb{F}G$-module forces $G$ to be transitive on $\{V_1, \ldots, V_m\}$.

(4) By (10.3), we have $[G : H_i] = m$. Suppose that $\{x_1, \ldots, x_m\}$ is the set of representatives of disjoint left cosets of $H_i$ in $G$. Then, $V_i = x_i V_1$, for $i = 1, \ldots, m$ after that $x_i$ have been suitably labelled. Hence, we have $M = x_1 V_1 \oplus \cdots \oplus x_m V_1$. Suppose that $A_1$ is a nonzero proper $\mathbb{F}H_1$-submodule of $V_1$ and write $A = \sum_{i=1}^{m} x_i A_1$. Now, $gx_i = x_j h$, for some $h \in H_1$ and $j$. This yields that $g(x_i A_1) = x_j(hA_1) = x_j A_1$. Consequently, $A$ is an $\mathbb{F}G$-module and $M = A$. But $x_i A_1 \subseteq x_i V_1 = V_i$, so that $A_1 = V_1$, a contradiction. Therefore, $V_1$ is an irreducible $\mathbb{F}H_1$-module. This implies that $V_i = x_i V_1$ is an irreducible $\mathbb{F}H_i$-module.  ∎

### Exercises

1. If $f: G \to H$ is a homomorphism of groups and $T: H \to GL(V, \mathbb{F})$ is a representation of $H$, the *pullback* of $T$ by $f$ is the representation

$$f^*T := T \circ f : G \to GL(V, \mathbb{F})$$

   of $G$. Prove that the pullback representation $f^*T$ is irreducible if and only if $T$ is irreducible.

2. Find a group $G$, a $\mathbb{C}G$-module $M$ and a $\mathbb{C}G$-module homomorphism $f: M \to M$ such that $M \neq Kerf \oplus Imf$.

3. Show that if $M$ is an irreducible $\mathbb{C}G$-module, then there exists $\lambda \in \mathbb{C}$ such that

$$x\left(\sum_{g \in G} g\right) = \lambda x,$$

   for all $x \in M$.

4. Which of the following groups have a faithful irreducible representation?

   (a) $C_n$ ($n$ is a positive integer);
   (b) $D_4$, the dihedral group of order 8;
   (c) $C_2 \times D_4$;
   (d) $C_3 \times D_4$.

5. Let $\langle x \rangle$ be the cyclic group of order $p$ generated by $x$, where $p$ is a prime. Let $\mathbb{F}$ be a field of characteristic $p$. Show that

$$x \mapsto \begin{bmatrix} 1 & 1 \\ 0 & 1 \end{bmatrix}$$

defines a two-dimensional representation of $\langle x \rangle$ over $\mathbb{F}$ which is reducible but not completely reducible. (This shows that the hypothesis $p \nmid |G|$ is necessary in Maschke's Theorem.)

6. Let $G$ be a (possibly infinite) group and let $H$ be a subgroup of $G$ with finite index. Suppose that $\mathbb{F}$ is a field whose characteristic does not divide $[G : H]$ and that $M$ is an $\mathbb{F}G$-module which is completely reducible as an $\mathbb{F}H$-module. Prove that $M$ is completely reducible as an $\mathbb{F}G$-module.
   *Hint:* Imitate the proof of Maschke's theorem.

7. Suppose that $G$ is the infinite group

$$\left\{ \begin{bmatrix} 1 & 0 \\ n & 0 \end{bmatrix} \mid n \in \mathbb{Z} \right\}.$$

and let $V$ be the $\mathbb{C}G$-module $\mathbb{C}^2$, with the natural multiplication by elements of $G$ (so that for $v \in V$, $g \in G$, the vector $vg$ is just the product of the row vector $v$ with the matrix $g$). Show that $V$ is not completely reducible. (This shows that Maschke's theorem fails for infinite groups.)

8. Let $T\colon G \to GL(V, \mathbb{F})$ and $T'\colon G \to GL(W, \mathbb{F})$ be two equivalent representations of $G$. Prove that $V$ is irreducible if and only if $W$ is irreducible.

9. Let $T\colon G \to GL_n(\mathbb{C})$ be a representation of a group $G$. Prove that the following are equivalent:

   (a) $T$ is reducible;
   (b) There are positive integers $k, l$ with $k + l = n$, and a representation $T'$ of $G$, equivalent to $T$, such that the following holds for all $a \in G$. No nonzero entry of the matrix $T'(a)$ is both in one of the first $k$ columns and in one of the last $l$ rows.

## 10.4   Characters

Characters are an extremely important tool for handling the irreducible representations of a group. In this section, we will see them in the form that applies to representations over a field of characteristic zero, and these are called ordinary characters. Since representations of finite groups in characteristic zero are completely reducible, knowing about the irreducible representations in some sense tells us about all representations.

**Definition 10.40** Let $G$ be a group, $\mathbb{F}$ be a field, and $f\colon G \to \mathbb{F}$ be a function. If $f$ is constant on conjugacy classes, i.e., $f(x^{-1}yx) = f(y)$, for all $x, y \in G$, then $f$ is called a *class function*.

**Lemma 10.41** *If $G$ has $h$ conjugacy classes, then the class functions on $G$ form a vector space $Cf(G, \mathbb{F})$ of dimension $h$ over $\mathbb{F}$*

**Proof** It is easy to check that $Cf(G, \mathbb{F})$ is a vector space. Suppose that $\mathcal{C}_1, \ldots, \mathcal{C}_h$ are conjugacy classes. We define $f_i : G \to \mathbb{F}$ by

$$f_i(x) = \begin{cases} 1 & \text{if } x \in \mathcal{C}_i \\ 0 & \text{if } x \notin \mathcal{C}_i, \end{cases}$$

for all $x \in G$ and $1 \leq i \leq h$. Let $\lambda_1 f_1 + \cdots + \lambda_h f_h = 0$, where $\lambda_i \in \mathbb{F}$ and $1 \leq i \leq h$. Suppose that $x \in \mathcal{C}_i$ is arbitrary. Then, $\lambda_1 f_1(x) + \cdots + \lambda_h f_h(x) = 0$, which implies that $\lambda_i = 0$. This yields that $f_1, \ldots, f_h$ are linearly independent. Next, assume that $f$ is an arbitrary class function. Then, for any $x_i \in \mathcal{C}_i$ $(1 \leq i \leq h)$, we can write $\big(f(x_1)f_1 + \cdots + f(x_h)f_h\big)(x) = f(x_1)f_1(x) + \cdots f(x_h)f_h(x)$. If we choose $x \in \mathcal{C}_i$, then we obtain $\big(f(x_1)f_1 + \cdots + f(x_h)f_h\big)(x) = f(x_i) = f(x)$, and so $f = f(x_1)f_1 + \cdots + f(x_h)f_h$. Therefore, we conclude that $\{f_1, \ldots, f_h\}$ is a basis for $Cf(G, \mathbb{F})$. ∎

We know that similar matrices have the same trace, and the matrices of the same linear transformation in two different bases are similar. Thus, for any representation $T: G \to GL(V, \mathbb{F})$ and any $x \in G$, the trace $T(x)$ is well defined.

**Definition 10.42** Suppose that $G$ is a group, $\mathbb{F}$ is a field, and $V$ is a finite-dimensional vector space over $\mathbb{F}$. If $T: G \to GL(V, \mathbb{F})$ is a representation, then the function $\chi : G \to \mathbb{F}$ defined by

$$\chi(x) = tr\big(T(x)\big),$$

for all $x \in G$, is called a *character*; it is an *irreducible character* if the representation $T$ is irreducible.

Clearly a one-dimensional representation over any field coincides with its character. Such characters are called *linear character*.

**Lemma 10.43** *Characters are class functions. Moreover, isomorphic $\mathbb{F}G$-modules or equivalent representations afford the same character.*

**Proof** For any $x, y \in G$, we have

$$\chi(x^{-1}yx) = tr\big(T(x^{-1}yx)\big) = tr\big(T(x^{-1})\big)tr\big(T(y)\big)tr\big(T(x)\big)$$
$$= tr\big(T(y)\big) = \chi(y).$$

This shows that $\chi$ is a class function. Isomorphic modules afford equivalent representations, and equivalent representations have similar matrices. ∎

**Lemma 10.44** *Let $x, y \in G$. Then, $x$ is conjugate to $y$ if and only if $\chi(x) = \chi(y)$, for all character $\chi$ of $G$.*

**Proof** Suppose that $x$ is conjugate to $y$. By Lemma 10.43, since characters are class functions, it follows that $\chi(x) = \chi(y)$, for all characters $\chi$ of $G$.

Conversely, let $\chi(x) = \chi(y)$, for all characters $\chi$. Hence, we conclude that $f(x) = f(y)$, for all class functions $f$ on $G$. In particular, it is true for the class function $f_0$ which takes the value 1 on the conjugacy class of $x$ and takes the value 0 elsewhere. Then, $f_0(x) = f_0(y) = 1$, and so $x$ is conjugate to $y$.                                   ∎

**Theorem 10.45** *Every character is a sum of irreducible characters.*

**Proof** Suppose that $M$ is an $\mathbb{F}G$-module with finite dimension over $\mathbb{F}$ affording a representation $T$. Let $\{0\} = M_0 \subset M_1 \subset \cdots \subset M_k = M$ be a composition series of $M$, i.e., $M_i/M_{i-1}$ is irreducible, for each $i$. Suppose that $\chi$ is the character of $M$ and $\chi_i$ for the character of irreducible module $M_i/M_{i-1}$. Choose a basis for $M_1$, extend it to a basis of $M_2$, then to a basis of $M_3$, and so on. This results in a basis of $M$ with respect to which the associated matrix representation is given by

$$T(x) = \begin{bmatrix} T_1(x) & & & \\ & T_2(x) & & \mathbf{0} \\ & \bigstar & \ddots & \\ & & & T_k(x) \end{bmatrix},$$

where $T_i$ is the representation afforded by $M_i/M_{i-1}$. Consequently, we have $tr\big(T(x)\big) = tr\big(T_1(x)\big) + \cdots + tr\big(T_k(x)\big)$, for all $x \in G$. This shows that $\chi = \chi_1 + \cdots + \chi_k$.   ∎

**Theorem 10.46** *Let $G$ be a finite group, $\mathbb{F}$ be a field, and $T$, $T'$ be irreducible representations afforded, respectively, by the irreducible $\mathbb{F}G$-modules $V$ and $W$. For fixed bases, let $T$ and $T'$ afford the matrix representations $T(x) = \big(a_{ij}(x)\big)$ and $T'(x) = \big(b_{ij}(x)\big)$, respectively. Then,*

*(1) If $V \not\equiv W$, then for every $i, j, r$ and $s$,*

$$\sum_{x \in G} b_{ij}(x^{-1}) a_{rs}(x) = 0;$$

*(2) If $\mathbb{F}$ is algebraically closed and its characteristic does not divide $|G|$, then*

$$\sum_{x \in G} a_{ij}(x^{-1}) a_{rs}(x) = \delta_{is} \delta_{jr} \frac{|G|}{n},$$

*where $n$ is the degree of $T$.*

**Proof** For any $f \in Hom_{\mathbb{F}}(V, W)$, suppose that $f$ has matrix $(c_{ij})$ in the chosen of $V$ and $W$. We define $f^* : V \to W$ by

$$f^* = \sum_{x \in G} T'(x^{-1}) f T(x). \tag{10.4}$$

For any $g \in G$, we can write

$$f^*T(g) = \sum_{x \in G} T'(x^{-1})fT(x)T(g) = \sum_{x \in G} T'(x^{-1})fT(xg)$$
$$= T'(g) \sum_{x \in G} T'((xg)^{-1})fT(xg) = T'(g)f^*.$$

Hence, for any $v \in V$, we have

$$f^*(gv) = f^*T(g)(v) = T'(g)f^*(v) = gf^*(v).$$

This yields that $f^* \in Hom_{\mathbb{F}G}(V, W)$.

(1) If $V \not\cong W$, then by Schur's lemma $f^* = 0$, for any $f$. Now, choose $f$ to be the linear transformation which maps the $j$th basis element of $V$ to the $r$th basis of element of $W$, and all other basis elements of $V$ to 0. Then, $f$ is represented by a matrix $(c_{kl})$ whose $kl$th entry is $\delta_{jk}\delta_{rl}$. Taking the matrix form of (10.4) we obtain

$$0 = \sum_{x \in G} \sum_{k,l} b_{ik}(x^{-1})c_{kl}a_{ls}(x) = \sum_{x \in G} \sum_{k,l} b_{ik}(x^{-1})\delta_{jk}\delta_{rl}a_{ls}(x)$$
$$= \sum_{x \in G} b_{ij}(x^{-1})a_{rs}(x).$$

(2) In order to prove this part, we put $T = T'$. By Lemma 10.18, $End_{\mathbb{F}G}(V) = \{aI_n \mid a \in \mathbb{F}\}$. So, for any $f \in End_{\mathbb{F}}(V)$, we have $f^* = a_f I_n$, for some $a_f \in \mathbb{F}$. Moreover, we have

$$tr(f^*) = \sum_{x \in G} tr(T(x^{-1})fT(x)) = |G|tr(f).$$

On the other hand, if $f$ has matrix $(\delta_{jk}\delta_{rl})$, then $tr(f) = \delta_{jr}$. Also, we have $tr(f^*) = tr(a_f I_n) = na_f$. Therefore, we conclude that $na_f = |G|\delta_{jr}$. Since $|G|$, and hence, $n$ is not divisible by the characteristic of $\mathbb{F}$, it follows that $|G|/n \in \mathbb{F}$. Consequently, we have shown that

$$\sum_{x \in G} T(x^{-1})fT(x) = f^* = a_f I_n = \frac{|G|}{n}\delta_{jr}I_n.$$

Taking the entry in this equation, gives

$$\sum_{x \in G} \sum_{k,l} a_{ij}(x^{-1})\delta_{jk}\delta_{rl}a_{rs}(x) = \frac{|G|}{n}\delta_{is}\delta_{jr}.$$

This implies that

$$\sum_{x \in G} a_{ij}(x^{-1})a_{rs}(x) = \frac{|G|}{n}\delta_{is}\delta_{jr},$$

and we are done.                                                                                        ∎

From this, result may deduce the fundamental orthogonality relations which connect the irreducible characters.

**Theorem 10.47 (Orthogonality Relations).** *Let G be a finite group and $\mathbb{F}$ be a field. Suppose that $\chi$ and $\psi$ are distinct irreducible characters on $\mathbb{F}$. Then,*

*(1)* $\sum_{x \in G} \chi(x)\psi(x^{-1}) = 0$;
*(2) If $\mathbb{F}$ is algebraically closed and its characteristic does not divide $|G|$, then*

$$\sum_{x \in G} \chi(x)\chi(x^{-1}) = |G|;$$

*(3) If $\mathbb{F}$ has characteristic 0, then $1/|G| \sum_{x \in G} \chi(x)\chi(x^{-1})$ is always a positive integer.*

**Proof** Suppose that $T$ and $T'$ are irreducible representations with characters $\chi$ and $\psi$, respectively. Let $(a_{ij})$ and $(b_{ij})$ be the associated matrix functions with respect to fixed bases. Then, we have $\chi = \sum a_{ii}$ and $\psi = \sum b_{jj}$.

(1) By Theorem 10.46 (1), we can write

$$\sum_{x \in G} \chi(x)\psi(x^{-1}) = \sum_{x \in G} \sum_i a_{ii}(x) \sum_j b_{jj}(x^{-1})$$
$$= \sum_i \sum_j \left( \sum_{x \in G} a_{ii}(x) b_{jj}(x^{-1}) \right) = 0.$$

(2) Suppose that $\mathbb{F}$ is algebraically closed and $char\mathbb{F} \nmid |G|$. If $n$ is the degree of $T$, then by using Theorem 10.46 (2), we obtain

$$\sum_{x \in G} \chi(x)\chi(x^{-1}) = \sum_i \sum_j \left( \sum_{x \in G} a_{ii}(x) a_{jj}(x^{-1}) \right)$$
$$= \sum_i \sum_j \frac{|G|}{n} \delta_{ij}^2 = |G|.$$

(3) Suppose that $\chi_1, \ldots, \chi_h$ are irreducible characters of $G$ over $\mathbb{F}$. By Theorem 10.45, we can write $\chi = m_1 \chi_1 + \cdots + m_h \chi_h$, where $m_j$'s are nonnegative integers. Then, by applying the results of (1) and (2) we have

$$\sum_{x \in G} \chi(x)\chi(x^{-1}) = \sum_j \sum_k \left( \sum_{x \in G} \chi_j(x) \chi_k(x^{-1}) \right)$$
$$= \left( \sum_j m_j^2 \right) |G|,$$

which implies (3).  ∎

**Definition 10.48** Let $V$ be a vector space on a field $\mathbb{F}$. Recall that a bilinear form on $V$ is a function $\langle\ ,\ \rangle : V \times V \to \mathbb{F}$, that is linear in each argument separately. A bilinear form $\langle\ ,\ \rangle$ is symmetric if $\langle v, w \rangle = \langle w, v \rangle$, for all $v, w \in V$. If $\langle\ ,\ \rangle$ is symmetric, we denote $V^{\perp} = \{v \in V \mid \langle v, w \rangle = 0,\ \text{for all } w \in V\}$. We say $\langle\ ,\ \rangle$ is *non-singular* if $V^{\perp} = \{0\}$ A basis $\{v_1, \ldots, v_n\}$ of $V$ is called *orthogonal* if $i \neq j$ implies $\langle v_i, v_j \rangle = 0$; it is *orthonormal*, if it is orthogonal and $\langle v_i, v_i \rangle = 1$, for all $i$.

Let $\mathbb{F}$ be an algebraically closed field and its characteristic does not divide $|G|$, and suppose that $Cf(G, \mathbb{F})$ be the vector space of class functions. If we define $\langle\ ,\ \rangle :$ $Cf(G, \mathbb{F}) \times Cf(G, \mathbb{F}) \to Cf(G, \mathbb{F})$ by

$$\langle f_1, f_2 \rangle = \frac{1}{|G|} \sum_{x \in G} f_1(x) f_2(x^{-1}), \tag{10.5}$$

then $\langle\ ,\ \rangle$ is a symmetric bilinear form, called the *inner product*.

**Theorem 10.49** *Let $\mathbb{F}$ be an algebraically closed field and its characteristic does not divide $|G|$. The number of irreducible characters of $G$ is equal to the number of conjugacy classes of $G$.*

**Proof** Suppose that $\chi_1, \ldots, \chi_k$ are the irreducible characters of $G$ and $h$ is the number of conjugacy classes of $G$. Assume that $c_1\chi_1 + \cdots + c_k\chi_k = 0$, for some $c_1, \ldots, c_k \in \mathbb{F}$. Then, we can write $0 = \langle c_1\chi_1 + \cdots + c_k\chi_k, \chi_i \rangle = c_i$, for all $1 \leq i \leq k$. This shows that $\chi_1, \ldots, \chi_k$ are linear independent, and so $k \leq l$. Now, let $V_1, \ldots, V_k$ is a complete set of non-isomorphic irreducible $\mathbb{F}G$-modules. By Corollary 10.38, we know that $\mathbb{F}G = W_1 \oplus \cdots \oplus W_k$, where for each $i$, $W_i$ is isomorphic to a direct sum of copies of $V_i$. Since $\mathbb{F}G$ contains the identity element $e$, we can write $e = f_1 + \cdots + f_k$ with $f_i \in W_i$, for $1 \leq i \leq k$. Now, suppose that $z \in Z(\mathbb{F}G)$. Then, for all $r \in \mathbb{F}G$ and $v \in V_i$, we have $vrz = vzr$, and so the function $v \mapsto vz$ from $V$ to $V$ is a homomorphism of $\mathbb{F}G$-modules. This homomorphism is equal to $c_i id_V$, for some $c_i \in \mathbb{F}$. Hence, we have $vz = c_i v$. Thus, $wz = c_i w$, for all $w \in W_i$, and in particular we have $f_i z = c_i f_i$, for all $1 \leq i \leq k$. This yields that $z = ez = (f_1 + \cdots + f_k)z = f_1 z + \cdots + f_k z = c_1 f_1 + \cdots + c_k f_k$, which implies that $Z(\mathbb{F}G)$ is contained in the subspace of $\mathbb{F}G$ spanned by $f_1, \ldots, f_k$. By Theorem 10.30, $dim Z(\mathbb{F}G) = h$, so we conclude that $h \leq k$. Therefore, we get $h = k$. ∎

**Corollary 10.50** *If $\chi_1, \ldots, \chi_h$ are the irreducible characters of $G$, then $\chi_1, \ldots, \chi_h$ form an orthonormal basis for $Cf(G, \mathbb{F})$.*

**Proof** By Theorem 10.47 items (1) and (2), we deduce that $\langle \chi_i, \chi_j \rangle = \delta_{ij}$. This gives the result. ∎

**Corollary 10.51** *The inner product defined in (10.5) is non-singular.*

**Proof** Suppose that $f$ is an arbitrary elements of $Cf(G, \mathbb{F})^{\perp}$. Since $f \in Cf(G, \mathbb{F})$, it follows that $f = c_1\chi_1 + \cdots + c_h\chi_h$, for some $c_1, \ldots, c_h \in \mathbb{F}$. So, for each $1 \leq i \leq h$, we have

$$0 = \langle f, \chi_i \rangle = \langle c_1 \chi_1 + \cdots + c_h \chi_h, \chi_i \rangle = c_1 \langle \chi_1, \chi_i \rangle + \cdots + c_h \langle \chi_h, \chi_i \rangle = c_i,$$

which implies that $f = 0$. Hence, $\langle \, , \, \rangle$ is non-singular.     ■

**Theorem 10.52** *Let $G$ be a finite group and suppose that $\mathbb{F}$ is an algebraically closed field and its characteristic does not divide $|G|$. Suppose that $G$ has exactly $h$ conjugacy classes $C_1, \ldots, C_h$, and let $h_i = |C_i|$ and $x_i \in C_i$ (for $1 \leq i \leq h$). If $\chi_1, \ldots, \chi_h$ are irreducible characters of $G$, then*

$$\sum_{m=1}^{h} \chi_m(x_i) \chi_m(x_j^{-1}) = \frac{\delta_{ij} |G|}{h_j} |C_G(x_j)|.$$

**Proof** We define two $h \times h$ matrices $B = (b_{ij})$ and $C = (c_{ij})$ by $b_{ij} = h_j/|G| \chi_i(x_j^{-1})$ and $c_{ij} = \chi_j(x_i)$. By using Theorem 10.47 items (1) and (2), the $rs$th entry of $BC$ is

$$\sum_{k=1}^{h} b_{rk} c_{ks} = \sum_{k=1}^{h} \frac{h_k}{|G|} \chi_r(x_k^{-1}) \chi_s(x_k) = \delta_{rs}.$$

This shows that $BC = I_n$, and so $CB = I_n$. Consequently, $ij$th entry of $CB$ is

$$\delta_{ij} = \sum_{m=1}^{h} \chi_m(x_i) \frac{h_j}{|G|} \chi_m(x_j^{-1}).$$

This gives the result.     ■

If $\chi_1, \ldots, \chi_h$ are the irreducible characters and $\chi$ is an arbitrary character, then by Theorem 10.45, we can write $\chi = a_1 \chi_1 + \cdots a_h \chi_h$, for some nonnegative integers $a_1, \ldots, a_h$. The nonnegative integer $a_i$ is called the *multiplicity* of $\chi_i$ in $\chi$. It is clear that $a_i = \langle \chi, \chi_i \rangle$.

Now, we have the following two fundamental corollaries.

**Corollary 10.53** *We have $|G| = \displaystyle\sum_{i=1}^{h} \chi_i(e)^2$.*

**Proof** It is straightforward.     ■

**Corollary 10.54** *If $\langle \chi, \chi \rangle = 1$ and $\chi(e) > 0$, then $\chi$ is irreducible.*

**Proof** We can write

$$\langle \chi, \chi \rangle = \langle a_1 \chi_1 + \cdots a_h \chi_h, a_1 \chi_1 + \cdots a_h \chi_h \rangle = a_1^2 + \cdots + a_h^2 = 1.$$

This implies that $a_j = \pm 1$, for some $1 \leq j \leq h$, and so $\chi = \pm \chi_j$. Since $\chi(e) > 0$, it follows that $\chi = \chi_j$.     ■

**Table 10.1** Character table

| $x_i$ | $x_1$ | $\cdots$ | $x_h$ |
|---|---|---|---|
| $\lvert C_G(x_i)\rvert$ | $\lvert C_G(x_1)\rvert$ | $\cdots$ | $\lvert C_G(x_h)\rvert$ |
| $\chi_1$ | $\chi_1^{(1)}$ | $\cdots$ | $\chi_1^{(h)}$ |
| $\vdots$ | $\vdots$ | $\ddots$ | $\vdots$ |
| $\chi_h$ | $\chi_h^{(1)}$ | $\cdots$ | $\chi_h^{(h)}$ |

Assume that $\mathbb{F}$ is algebraically closed, $char\mathbb{F} \nmid \lvert G\rvert$, $G$ has exactly $h$ conjugacy classes $C_1, \ldots, C_h$, and $x_i \in C_i$ (for $1 \leq i \leq h$). The value of $\chi_i$ on $C_j$ will be denoted by $\chi_i^{(j)}$. The values of the characters can be conveniently displayed in the *character table* of $G$ as Table 10.1.

It is usual to number the irreducible characters and conjugacy classes of $G$, so that $\chi_1 = 1_G$, the trivial character, and $x_1 = e$, the identity element of $G$. Beyond this, the numbering is arbitrary. Note that in the character table, the rows are indexed by the irreducible characters of $G$ and the columns are indexed by the conjugacy classes (or, in practice, by conjugacy class representatives). We have already seen many uses for the relations $\langle \chi_r, \chi_s \rangle = \delta_{rs}$. These relations can be expressed in terms of the rows of the character table, by writing them as

$$\sum_{i=1}^{h} \frac{\chi_r(x_i)\chi_s(x_i^{-1})}{\lvert C_G(x_i)\rvert} = \delta_{rs}, \tag{10.6}$$

for all $1 \leq r, s \leq h$. The relations (10.6) are called the *row orthogonality relations*. Similar relations exist between the columns of the character table, and these are given by

$$\sum_{i=1}^{h} \chi_i(x_r)\chi_i(x_s^{-1}) = \delta_{rs}\lvert C_G(x_r)\rvert, \tag{10.7}$$

for all $1 \leq r, s \leq h$. The relations (10.7) are called the *column orthogonality relations*.

**Definition 10.55** If $\chi$ is a character of $G$, then the kernel of $\chi$, written $Ker\chi$, is defined by

$$Ker\chi = \{g \in G \mid \chi(g) = \chi(e)\}.$$

If $T$ is a representation of $G$ with character $\chi$, then $KerT = Ker\chi$. In particular, $Ker\chi \trianglelefteq G$. We call $\chi$ a *faithful character* if $Ker\chi = \{e\}$.

**Theorem 10.56** *Let $N$ be a normal subgroup of $G$ and $\mathbb{F}$ be a field. Suppose that $\widetilde{\chi}$ is a character of $G/N$. Define $\chi : G \to \mathbb{F}$ by $\chi(g) = \widetilde{\chi}(gN)$, for all $g \in G$. Then, $\chi$ is a character of $G$, and furthermore, $\chi$ and $\widetilde{\chi}$ have the same degree.*

**Proof** Let $\widetilde{T} : G/N \to GL_n(\mathbb{F})$ be a representation of $G/N$ with character $\widetilde{\chi}$. The function $T: G \to GL_n(\mathbb{F})$ which is given by $T(g) = \widetilde{T}(gN)$, for all $g \in G$, is a homomorphism and hence is a representation of $G$. The character $\chi$ of $T$ satisfies $\chi(g) = tr\big(T(g)\big) = tr\big(\widetilde{T}(gN)\big) = \widetilde{\chi}(gN)$, for all $g \in G$. In addition, $\chi(e) = \widetilde{\chi}(N)$. This means that $\chi$ and $\widetilde{\chi}$ have the same degree. ∎

**Definition 10.57** If $N$ is a normal subgroup of $G$ and $\widetilde{\chi}$ is a character of $G/N$, then the character $\chi$ of $G$ which is given by $\chi(g) = \widetilde{\chi}(gN)$, for all $g \in G$, is called the *lift* of $\widetilde{\chi}$.

**Theorem 10.58** *Suppose that $N$ is a normal subgroup of $G$. By associating each character of $G/N$ with its lift to $G$, we obtain a one to one correspondence between the set of characters of $G/N$ and the set of characters $\chi$ of $G$ which satisfy $N \leq Ker\chi$. Moreover, irreducible characters of $G/N$ correspond to irreducible characters of $G$ which have $N$ in their kernel.*

**Proof** If $\widetilde{\chi}$ is a character of $G/N$, and $\chi$ is the lift of $\widetilde{\chi}$ to $G$, then $\widetilde{\chi}(N) = \chi(e)$. Also, if $x \in N$, then $\chi(x) = \widetilde{\chi}(xN) = \widetilde{\chi}(N) = \chi(e)$, which implies that $x \in Ker\chi$. Hence, we deduce that $N \leq Ker\chi$. Now, let $\chi$ be a character of $G$ with $N \leq Ker\chi$. Assume that $T: G \to GL_n(\mathbb{F})$ is a representation of $G$ with character $\chi$. If $x, y \in G$ and $xN = yN$, then $x^{-1}y \in N$, and so $T(x^{-1}y) = I_n$. This gives $T(x) = T(y)$. Therefore, we may define a function $\widetilde{T} : G/N \to GL_n(\mathbb{F})$ by $\widetilde{T}(gN) = T(g)$, for all $g \in G$. Then, we have

$$\widetilde{T}\big((xN)(yN)\big) = \widetilde{T}(xyN) = T(xy) = T(x)\,T(y) = \widetilde{T}(xN)\,\widetilde{T}(yN),$$

for all $x, y \in G$. Thus, $\widetilde{T}$ is a representation of $G/N$. If $\widetilde{\chi}$ is the character of $\widetilde{T}$, then $\widetilde{\chi}(gN) = \chi(g)$, for all $g \in G$. Consequently, $\chi$ is the lift of $\widetilde{\chi}$. Now, we have established that the function which sends each character of $G/N$ to its lift to $G$ is a one to one correspondence between characters of $G/N$ and the set of characters of $G$ which have $N$ in their kernel.

Finally, we show that irreducible characters correspond to irreducible characters. In order to prove this, let $W$ be a subspace of $\mathbb{F}^n$. We have $T(g)w \in W$, for all $w \in W$, if and only if $\widetilde{T}(gN)w \in W$, for all $w \in W$. Then, $W$ is an $\mathbb{F}G$-submodule of $\mathbb{F}^n$ if and only if $W$ is an $\mathbb{F}(G/N)$-submodule of $\mathbb{F}^n$. Hence, the representation $T$ is irreducible if and only if the representation $\widetilde{T}$ is irreducible. Therefore, $\chi$ is irreducible if and only if $\widetilde{\chi}$ is irreducible. ∎

**Theorem 10.59** *If $N$ is a normal subgroup of $G$, then there are irreducible characters $\chi_1, \ldots, \chi_k$ of $G$ such that*

$$N = \bigcap_{i=1}^{k} Ker\chi_i.$$

**Proof** Suppose that $x$ belongs to the kernel of each irreducible character of $G$, then $\chi(x) = \chi(e)$, for all character $\chi$, and hence, $x = e$. Thus, the intersection of the kernels of all the irreducible characters of $G$ is $\{e\}$. Next, suppose that $\widetilde{\chi}_1, \ldots, \widetilde{\chi}_k$ are the irreducible characters of $G/N$. By the above observation,

$$\bigcap_{i=1}^{k} Ker\widetilde{\chi}_i = \{N\}.$$

For $1 \le i \le k$, suppose that $\chi_i$ is the lift to $G$ of $\widetilde{\chi}_i$. If $x \in Ker\chi_i$, then

$$\widetilde{\chi}_i(N) = \chi_i(e) = \chi_i(x) = \widetilde{\chi}_i(xN),$$

and so $xN \in Ker\widetilde{\chi}_i$. Therefore, if $x \in \bigcap_{i=1}^{k} Ker\chi_i$, then $xN \in \bigcap_{i=1}^{k} Ker\widetilde{\chi}_i = \{N\}$, and so $x \in N$. This completes the proof. ∎

**Exercises**

1. Show that a group $G$ is simple if and only if $Ker\chi = \{e\}$, for all irreducible characters $\chi$ of $G$.
2. Show that any normal $p$-subgroup of a group lies in the kernel of every irreducible representation over a field of characteristic $p$.
3. Show that the degree of an irreducible representation on the field $\mathbb{Q}$ need not divide the group order, and also $\langle \chi, \chi \rangle$ need not be 1 if $\chi$ is a irreducible character on $\mathbb{Q}$.
4. Show that a representation of a finite group $G$ over a field of positive characteristic $p$ is not in general determined by its character even if $p \nmid |G|$.
5. Suppose that a finite group $G$ has a faithful representation of degree $n$ over a field $\mathbb{F}$, where $n$ is less than the smallest prime divisor of $|G|$ and $\mathbb{F}$ is an algebraically closed field of characteristic 0. Prove that $G$ is abelian.
6. Let $\chi_1, \ldots, \chi_h$ be the distinct irreducible characters of a finite group $G$ over an algebraically closed field whose characteristic does not divide $|G|$. If $\chi_i$ has degree $m_i$, prove that $m_1\chi_1 + \cdots + m_h\chi_h$ is the character of the regular representation.

## 10.5 Algebraic Numbers and Algebraic Integers

We start by introducing two essential notions: number field and algebraic integer.

**Definition 10.60** A *number field* is a finite field extension $\mathbb{F}$ of $\mathbb{Q}$, i.e., a field which is a vector space of finite dimension on $\mathbb{Q}$. We call this dimension the degree of $\mathbb{F}$.

**Example 10.61** The field $\mathbb{Q}(\sqrt{2}) = \{a + b\sqrt{2} \mid a, b \in \mathbb{Q}\}$ is a number field.

**Example 10.62** Let $\lambda_n$ be a primitive root of unity. The field $\mathbb{Q}(\lambda_n)$ is a number field.

**Example 10.63** The fields $\mathbb{R}$ and $\mathbb{C}$ are not number field.

If $\alpha \in \mathbb{F}$, there must be a linear dependency among $\{1, \alpha, \ldots, \alpha^n\}$, because $\mathbb{F}$ is a vector space of dimension $n$ on $\mathbb{Q}$. In other words, there exists a polynomial $f(x) \in \mathbb{Q}[x]$ such that $f(\alpha) = 0$. We call $\alpha$ an *algebraic number*.

**Definition 10.64** An *algebraic integer* in a number field $\mathbb{F}$ is an element $\alpha \in \mathbb{F}$ which is a root of a monic polynomial with coefficients in $\mathbb{Z}$.

**Example 10.65** Since $x^2 - 2 = 0$, it follows that $\sqrt{2} \in \mathbb{Q}(\sqrt{2})$ is an algebraic integer. However, $a/b \in \mathbb{Q}$ is not algebraic integer, unless $b$ divides $a$.

**Definition 10.66** The polynomial $f(x) = a_0 + a_1 + \cdots + a_n x^n$, where $a_i$ (for $0 \le i \le n$) are integers, is said to be *primitive* if the greatest common divisor of $a_i$'s is 1.

**Lemma 10.67** *If $f(x)$ and $g(x)$ are primitive polynomials, then $f(x)g(x)$ is a primitive polynomial.*

**Proof** Suppose that $f(x) = a_0 + a_1 + \cdots + a_n x^n$ and $g(x) = b_0 + b_1 + \cdots + b_m x^m$. Let $d$ be the greatest common divisor of coefficients of $f(x)g(x)$. To obtain a contradiction, suppose that $d > 1$. Let $p$ be a prime dividing $d$. Let $j \ge 0$ be the smallest number such that $p \nmid a_j$ and let $k \ge 0$ be the smallest number such that $p \nmid b_k$. Then, in $f(x)g(x)$ the coefficient of $x^{j+k}$, $c_{j+k}$ is

$$c_{j+k} = a_j b_k + (a_{j+1}b_{k-1} + a_{j+2}b_{k-2} + \cdots + a_{j+k}b_0)$$
$$+(a_{j-1}b_{k+1} + a_{j-2}b_{k+2} + \cdots + a_0 b_{j+k}). \tag{10.8}$$

Now, by our choice of $b_k$, $p|b_{k-1}, b_{k-2}, \ldots, b_0$ so that $p|(a_{j+1}b_{k-1} + a_{j+2}b_{k-2} + \cdots + a_{j+k}b_0)$. Similarly, by our choice of $a_j$, $p|a_{j-1}, a_{j-2}, \ldots, a_0$ so that $p|(a_{j-1}b_{k+1} + a_{j-2}b_{k+2} + \cdots + a_0 b_{j+k})$. By assumption $p|c_{j+k}$. Hence, by (10.8), we obtain $p|a_j b_k$. This implies that $p|a_j$ or $p|b_k$, which is impossible. This completes the proof. $\blacksquare$

**Definition 10.68** The *content* the polynomial $f(x) = a_0 + a_1 + \cdots + a_n x^n$, where the $a_i$'s are integers, is the greatest common divisor of the integers $a_0, a_1, \ldots, a_n$.

It is clear that any polynomial $p(x)$ with integer coefficients it can be written as $p(x) = dq(x)$ where $d$ is the content of $p(x)$ and $q(x)$ is a primitive polynomial.

**Theorem 10.69 (Gauss' Lemma).** *If the primitive polynomial $f(x)$ can be factored as the product of two polynomials having rational coefficients, then it can be factored as the product of two polynomials having integer coefficients.*

**Proof** Let $f(x) \in \mathbb{Z}[x]$ be primitive, and suppose that $f(x) = u(x)v(x)$, where $u(x)$, $v(x) \in \mathbb{Q}[x]$. By clearing of denominators and taking out common factors, we can write $f(x) = (a/b)g(x)h(x)$, where $a, b$ are integers and $g(x), h(x) \in \mathbb{Z}[x]$. We can write

$$bf(x) = ag(x)h(x). \tag{10.9}$$

Since $f(x)$ is primitive, it follows that the content of the left side of (10.9) is $b$. On the other hand, since both $g(x)$ and $h(x)$ are primitive, by Lemma 10.67, $f(x)g(x)$ is primitive. Hence, the content of the right side of (10.9) is $a$. Thus, we must have $a = b$. This shows that $f(x) = g(x)h(x)$, where $f(x)$ and $g(x)$ have integers coefficients. ∎

**Definition 10.70** A polynomial is said to be *integer monic* if all its coefficients are integers and its highest coefficient is 1.

**Corollary 10.71** *If an integer monic polynomial factors as the product of two non-constant polynomials having rational coefficients, then it factors as the product of two integer monic polynomials.*

*Proof* We leave the proof of the corollary as an exercise for the reader.                  ∎

**Definition 10.72** Let $\mathbb{F}$ be a field. A polynomial $f(x) \in \mathbb{F}[x]$ is said to be *irreducible* over $\mathbb{F}$ if whenever $f(x) = g(x)h(x)$ with $g(x), h(x) \in \mathbb{F}[x]$, then one of $g(x)$ or $h(x)$ is a constant.

**Definition 10.73** The *minimal polynomial $f$* of an algebraic number $\alpha$ is the monic polynomial in $\mathbb{Q}[x]$ of smallest degree such that $f(\alpha) = 0$.

**Theorem 10.74** *The minimal polynomial of $\alpha$ has integer coefficients if and only if $\alpha$ is an algebraic integer.*

*Proof* If the minimal polynomial of $\alpha$ has integer coefficients, then by Definition 10.64, $\alpha$ is an algebraic integer.

Conversely, let us assume that $\alpha$ is an algebraic integer. This means by the definition that there exists a monic polynomial $f \in \mathbb{Z}[x]$ such that $f(\alpha) = 0$. Let $g(x) \in \mathbb{Q}[x]$ be the minimal polynomial of $\alpha$. Then, there exists a monic polynomial $h(x) \in \mathbb{Q}[x]$ such that $f(x) = g(x)h(x)$. Now, by Gauss' lemma, we conclude that the coefficients of $g(x)$ actually belongs to $\mathbb{Z}$.                                    ∎

**Corollary 10.75** *The set of algebraic integers of $\mathbb{Q}$ is $\mathbb{Z}$.*

*Proof* Let $a/b \in \mathbb{Q}$. Its minimal polynomial is $x - (a/b) = 0$. By Theorem 10.74, $a/b$ is an algebraic integer if and only $b = \pm 1$.                              ∎

**Theorem 10.76** *Let $\mathbb{F}$ be a finite extension field of the rational number $\mathbb{Q}$, and let $\alpha \in \mathbb{F}$. Then, the following are equivalent:*

*(1)  $\alpha$ is an algebraic integers.*
*(2)  If $f(x) \in \mathbb{Q}[x]$ is the monic polynomial irreducible over $\mathbb{Q}$ such that $f(\alpha) = 0$, then $f(x) \in \mathbb{Z}[x]$.*
*(3)  $\mathbb{Z}[\alpha]$, the subring of $\mathbb{F}$ generated by $\mathbb{Z}$ and $\alpha$, is a finitely generated $\mathbb{Z}$-module.*

*Proof* $(1 \Rightarrow 2)$: It follows from Theorem 10.74.

$(2 \Rightarrow 3)$: Assume that $f(x) = a_0 + a_1 x + \cdots + a_{n-1}x^{n-1} + x^n$, where $a_i \in \mathbb{Z}$ (for all $0 \le i \le n-1$) such that $f(\alpha) = 0$. So, we have $\alpha^n = -a_0 - a_1\alpha - \ldots - a_{n-1}\alpha^{n-1}$. Therefore, we conclude that $\mathbb{Z}[\alpha] = \mathbb{Z} \oplus \mathbb{Z}\alpha \oplus \cdots \oplus \mathbb{Z}\alpha^{n-1}$.

$(3 \Rightarrow 1)$: Suppose that $\mathbb{Z}[\alpha] = \mathbb{Z}f_1(\alpha) + \cdots + \mathbb{Z}f_k(\alpha)$, where $f_i(\alpha) \in \mathbb{Z}[\alpha]$. Choose an integer $m$ such that $m > \max\{degf_1(x), \ldots, degf_k(x)\}$ and $\alpha^m \in \mathbb{Z}[\alpha]$. Hence, we have

$$\alpha^m = a_1 f_1(\alpha) + \cdots + a_k f_k(\alpha),$$

where $a_i \in \mathbb{Z}$ $(1 \le i \le k)$. Consequently, there is the monic polynomial $f(x) = x^m - a_1 f_1(x) - \cdots - a_k f_k(x)$ over $\mathbb{Z}$ such that $f(\alpha) = 0$. Thus, $\alpha$ is an algebraic integer. ∎

**Example 10.77** We have that

$$\mathbb{Z}[1/2] = \{a/b \mid b \text{ is a power of } 2\}$$

is not finitely generated, since $1/2$ is not an algebraic integer. Its minimal polynomial is $x - (1/2)$.

**Corollary 10.78** *The algebraic integers in an algebraic number field $\mathbb{F}$ form a subring of $\mathbb{F}$.*

**Proof** Suppose that $\alpha$, $\beta \in \mathbb{F}$ are algebraic integers. Theorem 10.76 tells us that $\mathbb{Z}[\alpha]$ and $\mathbb{Z}[\beta]$ are finitely generated. Let $\mathbb{Z}[\alpha]$ is generated by $1, \alpha, \ldots, \alpha^{m-1}$ and $\mathbb{Z}[\beta]$ is generated by $1, \beta, \ldots, \beta^{n-1}$. Then, $\mathbb{Z}[\alpha, \beta]$ is generated by $\{\alpha^i \beta^j \mid i < n, \ j < n\}$. Therefore, the submodules $\mathbb{Z}[\alpha \pm \beta]$ and $\mathbb{Z}[\alpha\beta]$ are finitely generated. By invoking again Theorem 10.76, $\alpha \pm \beta$ and $\alpha\beta$ are algebraic integers. ∎

**Theorem 10.79** *Let $G$ be a finite group and $\mathbb{F}$ be an algebraic closed field of characteristic $0$. If*

*(1) $\mathcal{C}_1, \ldots, \mathcal{C}_h$ are conjugacy classes of $G$ such that $h_i = |\mathcal{C}_i|$ and choose representatives $x_i \in \mathcal{C}_i$, for $1 \le i \le h$,*

*(2) $\chi_1, \ldots, \chi_h$ are irreducible characters of $G$ such that the degree of $\chi_i$ is $n_i$,*

*then the numbers*

$$\frac{h_i \chi_k(x_i)}{n_k}, \ i, k = 1, \ldots, h,$$

*are algebraic integers.*

**Proof** Suppose that $\overline{\mathcal{C}_i} = \sum_{x \in \mathcal{C}_i} x \in \mathbb{F}G$. In Theorem 10.30, we saw that $\{\overline{\mathcal{C}_i}, \ldots, \overline{\mathcal{C}_h}\}$ is a basis for the center of $\mathbb{F}G$. So, there are integers $a_{ijm}$ such that

$$\overline{\mathcal{C}_i} \, \overline{\mathcal{C}_j} = \sum_{m=1}^{h} a_{ijm} \overline{\mathcal{C}_m}. \tag{10.10}$$

Let $\chi_k$ be the character of the representation $T_k$, and $V_k$ be the corresponding irreducible $\mathbb{F}G$-module. We extend $T_k : G \to End_{\mathbb{F}G}(V_k)$ to all of $\mathbb{F}G$-modules by linearity on $\mathbb{F}$. Then, we have

$$T_k\left(\overline{C_i}\right)T_k\left(\overline{C_j}\right) = \sum_{m=1}^{h} a_{ijm}T_k\left(\overline{C_m}\right).$$

Since $\overline{C_i} \in Z(\mathbb{F}G)$, it follows that $T_k\left(\overline{C_i}\right)$ commutes with $T_k(x)$, for all $x \in \mathbb{F}G$. For any $v \in V_i$, this means that

$$T_k\left(\overline{C_i}\right) \cdot xv = T_k\left(\overline{C_i}\right)T_k(x)v = T_k(x)T_k\left(\overline{C_i}\right)v = x \cdot T_k\left(\overline{C_i}\right)v,$$

and so $T_k\left(\overline{C_i}\right) \in End_{\mathbb{F}G}(V_k) \cong \mathbb{F}$. This implies that

$$T_k\left(\overline{C_i}\right) = \alpha_{ik}I_{n_k}, \tag{10.11}$$

for some $\alpha_{ik} \in \mathbb{F}$. Taking traces in both sides of (10.11), we obtain

$$tr\left(T_k\left(\overline{C_i}\right)\right) = tr\left(\alpha_{ik}I_{n_k}\right).$$

This yields that

$$\sum_{x_i \in C_i} tr\left(T_k(x_i)\right) = \alpha_{ik}n_k,$$

or equivalently $h_i\chi_k(x_i) = \alpha_{ik}n_k$. Hence, we conclude that

$$\alpha_{ik} = \frac{h_i\chi_k(x_i)}{n_k}.$$

Now, the equation (10.11) becomes

$$\alpha_{ik}\alpha_{jk} = \sum_{m=1}^{h} a_{ijm}\alpha_{mk},$$

which implies that

$$0 = \sum_{m=1}^{h} \left(a_{ijm} - \delta_{jm}\alpha_{ik}\right)\alpha_{mk}.$$

If we denote $C_1 = \{e\}$, then $h_1 = 1$ and $\alpha_{ik} = 1 \cdot \chi_k(e)/n_k = 1$. Now, for fixed $i$ and $k$, since $\alpha_{1k} = 1 \neq 0$, it follows that the system of equations

$$0 = \sum_{m=1}^{h} \left(a_{ijm} - \delta_{jm}\alpha_{ik}\right)X_m$$

has a non-trivial solution. Therefore, the determinant of the $h \times h$ matrix

$$\begin{bmatrix} a_{i11} - \delta_{11}\alpha_{ik} & a_{i12} & \cdots & a_{i1h} \\ a_{i21} & a_{i22} - \delta_{22}\alpha_{ik} & \cdots & a_{i2h} \\ \vdots & \vdots & \ddots & \vdots \\ a_{ih1} & a_{ih2} & \cdots & a_{ihh} - \delta_{hh}\alpha_{ik} \end{bmatrix}$$

must vanish. This shows that $\alpha_{ik}$'s are eigenvalues of the matrix $\left(a_{ijm}\right)_{h\times h}$. The entries of this matrix are integers. Hence, $\alpha_{ik}$'s satisfy a monic polynomial over the integers, and so, they are algebraic integers. ∎

**Theorem 10.80** *Let $G$ be a finite group and $\mathbb{F}$ be an algebraic closed field of characteristic $0$. Then, the degree of the irreducible representations of $G$ divide the order of group.*

***Proof*** Suppose that $T_k$ is an irreducible representation of $G$ and $\chi_k$ be its character. We employ the notation of Theorem 10.79. Suppose that $x_i \in C_i$. Then, we can write

$$\sum_{i=1}^{h} \alpha_{ik}\chi_k(x_i^{-1}) = \sum_{i=1}^{h} \frac{h_i\chi_k(x_i)}{\chi_k(e)}\chi_k(x_i^{-1}) = \frac{1}{\chi_k(e)}\sum_{i=1}^{h}\chi_k(x_i)\chi_k(x_i^{-1})$$
$$= \frac{1}{\chi_k(e)}\sum_{x\in G}\chi_k(x)\chi_k(x^{-1}) = \frac{|G|}{\chi_k(e)}.$$

Consequently, $|G|/\chi_k(e)$ is an algebraic integer. Finally, Corollary 10.75 shows that $|G|/\chi_k(e)$ is an integer. This yields that $\chi_k(e)\big||G|$. ∎

**Exercises**

1. Show that $\sqrt{2} + \sqrt[3]{5}$ is algebraic over $\mathbb{Q}$.
2. If $\alpha$ is an algebraic number, show that there is a positive integer $n$ such that $n\alpha$ is an algebraic integer.
3. Prove that a complex number $\lambda$ is an algebraic integer if and only if $\lambda$ is a root of a polynomial of the form

$$a_0 + a_1 x + \cdots + a_{n-1}x^{n-1} + x^n,$$

   where each $a_k$ $(0 \le k \le n-1)$ is an integer.
4. Let $G$ be a group and $\chi$ be a character of $G$. If $|\chi(g)| = 1$, show that $\chi(g)$ is a root of unity.
5. Let $\chi$ be a (possibly reducible) character which is constant on $G - \{e\}$.

   (a) Show that $\chi = a\,1 + b\rho_G$, where $a, b \in \mathbb{Z}$ and $\rho_G$ is the regular character of $G$;
   (b) Show that $a + b$ and $a + b|G|$ are integers;
   (c) If $\chi$ is a non-trivial irreducible character of $G$, then $b\chi(e)$ is an integer;
   (d) Deduce that both $a$ and $b$ are integers.

## 10.6   Complex Characters

In this section we consider the field $\mathbb{C}$.

**Definition 10.81** If $\chi$ is the character of the $\mathbb{C}G$-module $V$, then the dimension of $V$ is called the *degree* of $\chi$.

The next result gives information about the complex numbers $\chi(g)$, where $\chi$ is a character of $G$ and $x \in G$.

**Theorem 10.82** *Let $\chi$ be the character of a $\mathbb{C}G$-module $V$. Suppose that $x$ is an element of $G$ of order $M$. Then,*

*(1) $\chi(x)$ is a sum of $|G|$th roots of unity;*
*(2) $\chi(x^{-1}) = \overline{\chi(x)}$;*
*(3) $\chi(x)$ is real number if $x$ is conjugate to $x^{-1}$.*

**Proof** Suppose that $n = dim V$, and $T: G \to GL_n(\mathbb{F})$ is a representation of $G$ such that $\chi(x) = tr(T(x))$, for all $x \in G$.

(1) Note that $T(x)$ is similar to an upper triangular matrix $A$ such that

$$A = \begin{bmatrix} \lambda_1 & & & \\ & \lambda_2 & \bigstar & \\ & & \ddots & \\ \mathbf{0} & & & \lambda_n \end{bmatrix},$$

where $\lambda_i$'s are eigenvalues of $A$. Since similar matrices have the same eigenvalues, we conclude that $\chi(x) = tr(T(x)) = \sum_{i=1}^{n} \lambda_i$. Now, we have

$$T(x)^{|G|} = T\left(x^{|G|}\right) = T(e) = I_n.$$

For each $1 \le i \le n$, since $\lambda_i$ is an eigenvalue of $T(x)$, it follows that there exists a column vector $0 \ne X_i$ such that $T(x)X_i = \lambda_i X_i$. So, we deduce that

$$\left(T(x)\right)^{|G|} X_i = \lambda_i^{|G|} X_i,$$

and so $I_n X_i = \lambda_i X_i$. This implies that $\lambda_i^{|G|} = 1$. Therefore, $\lambda_i$'s are $|G|$th roots of 1.
(2) If $\lambda$ is an eigenvalue of $T(x)$, then $\lambda^{-1}$ is an eigenvalue of $T(x^{-1})$. Hence, we can write $\chi(x^{-1}) = tr(T(x^{-1})) = \sum_{i=1}^{n} \lambda_i^{-1}$. Since $\lambda_i^n = 1$, it follows that

$$\lambda_i = e^{2k\pi i/n} = \cos\left(\frac{2k\pi}{n}\right) + i\sin\left(\frac{2k\pi}{n}\right).$$

This yields that $\lambda_i^{-1} = \overline{\lambda_i}$. Thus, we obtain

$$\chi(x^{-1}) = \sum_{i=1}^{n} \lambda_i^{-1} = \sum_{i=1}^{n} \overline{\lambda_i} = \overline{\sum_{i=1}^{n} \lambda_i} = \overline{\chi(x)}.$$

(3) Suppose that $x$ and $x^{-1}$ are conjugate. Then, we have $\chi(x) = \chi(x^{-1})$. Now, by (2), since $\chi(x^{-1}) = \overline{\chi(x)}$, we conclude that $\chi(x) = \overline{\chi(x)}$. This means that $\chi(x)$ is a real number.    ∎

**Theorem 10.83** *Let* $T: G \to GL_n(\mathbb{C})$ *be a representation affording the character* $\chi$. *Then, for any* $g \in G$,

$$|\chi(g)| \le n; \tag{10.12}$$

*In (10.12), equality holds if and only if* $T(g) = \lambda I_n$, *for some* $\lambda \in \mathbb{C}$.

**Proof** Suppose that $T(g)$ have eigenvalues $\lambda_1, \dots, \lambda_n$. Hence, we have $\chi(g) = \lambda_1 + \cdots + \lambda_n$, which implies that

$$|\chi(g)| = \left| \sum_{i=1}^{n} \lambda_i \right| \le \sum_{i=1}^{n} |\lambda_i| = \sum_{i=1}^{n} 1 = n.$$

So, if $\lambda_1 = \dots = \lambda_n$, then equality holds in (10.12).

Conversely, if equality holds in (10.12), then the characteristic equation of $T(g)$ is $(x - \lambda)^n = 0$. On the other hand, $T(g)$ satisfies $x^{|G|} - 1 = 0$. Consequently, $T(g)$ is a zero of $\left( (x - \lambda)^n, x^{|G|} - 1 \right) = x - \lambda$.    ∎

**Theorem 10.84** *Let* $G$ *be a finite group and* $\chi_1, \dots, \chi_h$ *be the irreducible* $\mathbb{C}G$-*characters. If we define* $\overline{\chi_i}$ *by* $\overline{\chi_i}(x) = \overline{\chi_i(x)}$, *for all* $x \in G$, *then* $\overline{\chi_i}$ *is one of the* $\chi_j$'s.

**Proof** Suppose that $T_i : G \to GL_{n_i}(\mathbb{C})$ is matrix representation with $\chi_i(x) = tr\left( T_i(x) \right)$, for all $x \in G$. For any $1 \le i \le h$, we define $\overline{T_i} : G \to GL_{n_i}(\mathbb{C})$ by $\overline{T_i}(x) = \left( T_i(x^{-1}) \right)^t$, the transpose of the matrix $T_i(x^{-1})$. Then, we have

$$\overline{T_i}(xy) = \left( T_i(xy)^{-1} \right)^t = \left( T_i(y^{-1}x^{-1}) \right)^t = \left( T_i(y^{-1}) T_i(x^{-1}) \right)^t$$
$$= \left( T_i(x^{-1}) \right)^t \left( T_i(y^{-1}) \right)^t = \overline{T_i}(x) \overline{T_i}(y).$$

Hence, $\overline{T_i}$ is a homomorphism, and so is a representation of $G$. Now, we determine $\overline{\chi_i}$, we have

$$tr\left( \overline{T_i}(x) \right) = tr\left( T_i(x^{-1}) \right)^t = tr\left( T_i(x^{-1}) \right) = \chi_i(x^{-1}) = \overline{\chi_i(x)} = \overline{\chi_i}(x).$$

Thus, we conclude that

$$\langle \overline{\chi_i}, \overline{\chi_i} \rangle = \frac{1}{|G|} \sum_{x \in G} \overline{\chi_i}(x) \, \overline{\chi_i}(x^{-1}) = \frac{1}{|G|} \sum_{x \in G} \overline{\chi_i(x)} \, \overline{\chi_i(x^{-1})}$$

$$= \frac{1}{|G|} \sum_{x \in G} \overline{\chi_i(x)} \, \overline{\overline{\chi_i(x)}} = \frac{1}{|G|} \sum_{x \in G} \overline{\chi_i(x)} \, \chi_i(x)$$

$$= \langle \chi_i, \chi_i \rangle = 1.$$

This shows that $\overline{\chi_i}$ is irreducible, and so it must be one of the $\chi_j$'s. ∎

Now, we present three theorems which we shall use in the construction of our examples of character tables. The first is an easy refinement of Theorem 10.82.

**Theorem 10.85** *Let $G$ be a finite group, $x \in G$, and $\chi$ be an irreducible character over $\mathbb{C}$. Then,*

*(1) If $x^2 = e$, then $\chi(x) \in \mathbb{Z}$;*
*(2) If $x$ is conjugate to its inverse, then $\chi(x) \in \mathbb{R}$;*
*(3) If $x$ is an element of order 3 which is conjugate to its inverse, then $\chi(x) \in \mathbb{Z}$.*

**Proof** This follows immediately from the consideration of possible eigenvalues in the matrix representation of a given element. ∎

**Theorem 10.86** *Let $G$ be a finite group. If $\chi$ is a complex character of $G$ and $\theta$ is a linear character of $G$, then the product $\theta\chi$, defined by $\theta\chi(x) = \theta(x)\,\chi(x)$, for all $x \in G$, is a character of $G$. In addition, if $\chi$ is irreducible, then so is $\theta\chi$.*

**Proof** Suppose that $T: G \to GL_n(\mathbb{C})$ is a representation with character $\chi$. We define $\theta T: G \to GL_n(\mathbb{C})$ by $\theta T(x) = \theta(x)\,T(x)$, for all $x \in G$. Hence, $\theta T(x)$ is the matrix $T(x)$ multiplied by the complex number $\theta(x)$. Since $T$ and $\theta$ are homomorphisms, it follows that $\theta T$ is a homomorphism too. Consequently, $\theta T$ is a representation of $G$ with character $\theta\chi$.

Now, for any $x \in G$, the complex number $\theta(x)$ is a root of unity, so $\theta(x)\,\overline{\theta(x)} = 1$. Therefore, we obtain

$$\langle \theta\chi, \theta\chi \rangle = \frac{1}{|G|} \sum_{g \in G} \theta(x)\,\chi(x)\,\overline{\theta(x)}\,\overline{\chi(x)} = \frac{1}{|G|} \sum_{g \in G} \chi(x)\,\overline{\chi(x)} = \langle \chi, \chi \rangle.$$

Finally, by Corollary 10.54, we conclude that $\theta\chi$ is irreducible if and only if $\chi$ is irreducible. ∎

**Definition 10.87** Let $G$ be a permutation group on $X = \{1, \ldots, n\}$, and $V$ be a vector space on $\mathbb{C}$ with basis $\{v_1, \ldots, v_n\}$. Then, $V$ becomes a $\mathbb{C}G$-module by setting $gv_i = v_{ig}$, for all $g \in G$ and $1 \le i \le n$. The character $\theta$ of this representation is the *permutation character* of $G$. It is clear that $\theta(g)$ is the number of fixed points of $g$ on $X$.

**Theorem 10.88** *Let $G$ be a permutation group on $X = \{1, \ldots, n\}$ and $\theta$ be the permutation character of $G$. Then,*

*(1)  If G has k orbits on X, then*

$$\sum_{g \in G} \theta(g) = k|G|,$$

*i.e.,* $\langle \theta, 1 \rangle = k$;

*(2)  If G is transitive on X and* $\mathrm{Stab}_G(x)$ *has r orbits on X, then*

$$r = \langle \theta, \theta \rangle = \sum_{g \in G} \theta(g)^2;$$

*(3)  Let G be transitive. Then, G is doubly transitive if and only if* $\theta = 1 + \chi$, *where* $\chi$ *is an irreducible character of G.*

**Proof**  (1) First, we assume that $k = 1$. Then, $G$ is transitive and we have

$$\langle \theta, 1 \rangle = \frac{1}{|G|} \sum_{g \in G} \theta(g) \, 1(g^{-1}) = \sum_{g \in G} \theta(g)$$

$$= \frac{1}{|G|} \sum_{g \in G} |\{i \in X \mid i^g = i\}| = \frac{1}{|G|} \sum_{i=1}^{n} |\{g \in G \mid i^g = i\}|$$

$$= \frac{1}{|G|} \sum_{i=1}^{n} |\mathrm{Stab}_G(i)| = \frac{1}{|G|} n |\mathrm{Stab}_G(i)| = 1.$$

In general case, assume that $\mathrm{Orb}_G(x_1), \ldots, \mathrm{Orb}_G(x_k)$ are disjoint orbits of $G$. We know that the action of $G$ on each orbit is transitive. We denote the permutation character of $G$ on $\mathrm{Orb}_G(x_i)$ by $\theta_i$, for all $1 \le i \le k$. Let $\theta_i(g)$ be the number of fixed points of $G$ on $\mathrm{Orb}_G(x_i)$, for all $1 \le i \le k$. Then, we have

$$\langle \theta, 1 \rangle = \frac{1}{|G|} \sum_{g \in G} \theta(g) = \frac{1}{|G|} \sum_{g \in G} \theta_1(g) + \cdots + \frac{1}{|G|} \sum_{g \in G} \theta_k(g)$$

$$= \underbrace{1 + \cdots + 1}_{k \text{ times}} = k.$$

(2) All $\mathrm{Stab}_G(i)$'s are conjugate, and so each has exactly $r$ orbits on $X$. Thus, we have

$$\langle \theta, \theta \rangle = \frac{1}{|G|} \sum_{g \in G} \theta(g) \, \theta(g^{-1}) = \frac{1}{|G|} \sum_{g \in G} \theta(g) \, \overline{\theta(g)}$$

$$= \frac{1}{|G|} \sum_{g \in G} \theta(g) \, \theta(g) = \frac{1}{|G|} \sum_{g \in G} \theta(g)^2 = \frac{1}{|G|} \sum_{i=1}^{n} \sum_{g \in \mathrm{Stab}_G(i)} \theta(g)$$

$$= \frac{1}{|G|} \sum_{i=1}^{n} r |\mathrm{Stab}_G(i)| = \frac{1}{|G|} rn |\mathrm{Stab}_G(i)| = r.$$

(3) Suppose that $\chi_1, \ldots, \chi_h$ are irreducible characters of $G$. Then, we have $\theta = a_1\chi_1 + \cdots + a_h\chi_h$, for some nonnegative integers $a_1, \ldots, a_h$. Since $\chi_1 = 1$, then by (1), we have $a_1 = \langle \theta, 1 \rangle = 1$. Hence, $r = 1 + a_2^2 + \cdots + a_h^2$. The group $G$ is doubly transitive if and only if $\mathrm{Stab}_G(i)$ is transitive on $G - \{i\}$, that is , if and only if $r = 2$. This occurs when exactly one $a_i$ equals 1 and $a_j = 0$ if $j \geq 2$ and $j \neq i$. Consequently, we obtain $\theta = 1 + \chi_i$, for some $1 \leq i \leq h$.                                 ∎

**Example 10.89** Let $G$ be the symmetric group $S_3$. Then, $G$ has three conjugacy classes with representatives $id$, $(1\,2)$ and $(1\,2\,3)$. From $6 = n_1^2 + n_2^2 + n_3^2$, we conclude that the degrees of irreducible characters are $1, 1$, and $2$. Together with Theorem 10.85 and orthogonality relations now determine the character table of $G$ in Table 10.2.

Notice that these characters may be checked by considering the complete irreducible characters of $S_3$.

**Example 10.90** Let $G$ be the symmetric group $S_4$. There are 5 conjugacy classes, and so 5 irreducible characters. We know two of them, say $\chi_1 = 1$ and $\chi_2$, the trivial and sign characters, have degree 1. If we let $S_4$ act on $\{1,\ 2,\ 3,\ 4\}$, then the characters values, call then $\theta(x)$ are given counting fixed points. Hence, we obtain $\theta(id) = 4$, $\theta((1\,2)) = 2, \theta((1\,2\,3)) = 1, \theta((1\,2\,3\,4)) = 0$ and $\theta((1\,2)(3\,4)) = 0$. Now, we have $\theta = \chi_3 + 1$ or $\chi_3 = \theta - 1$, where $\chi_3$ is irreducible. So, we found three irreducible characters. By Theorem 10.86, $\chi_2\chi_3$ is also irreducible. Since the sum of squares of the degrees of the irreducible characters is 24, it follows that the last irreducible character must have degree 2. Then, we use the column orthogonality to fill the last row of the character table (Table 10.3).

**Exercises**

1. An element of a group is said to be *real* if it is conjugate to its inverse.

   (a) Let $g$ be a real element of a finite group $G$. Prove that $\chi(g) \in \mathbb{R}$, for all characters $\chi$ of $G$;
   (b) Prove that all elements of the symmetric group $S_n$ are real;
   (c) Find all real elements in the alternating group $A_n$.

2. Let $T\colon G \to GL_n(\mathbb{C})$ be a representation of a finite group $G$ with character $\chi$ such that the elements of $T(G)$ are linearly independent and $T$ is injective. Prove that $\langle \chi, \psi \rangle > 0$, for all irreducible characters $\psi$ of $G$.

**Table 10.2**  Character table of $S_3$

| $x_i$ | $id$ | $(1\,2)$ | $(1\,2\,3)$ |
|---|---|---|---|
| $|C_G(x_i)|$ | 6 | 2 | 3 |
| $\chi_1$ | 1 | 1 | 1 |
| $\chi_1$ | 1 | $-1$ | 1 |
| $\chi_3$ | 2 | 0 | $-1$ |

**Table 10.3**  Character table of $S_4$

| $x_i$ | id | (1 2) | (1 2 3) | (1 2 3 4) | (1 2)(3 4) |
|-------|----|----|----|----|----|
| $|C_G(x_i)|$ | 24 | 4 | 3 | 4 | 8 |
| $\chi_1$ | 1 | 1 | 1 | 1 | 1 |
| $\chi_2$ | 1 | −1 | 1 | −1 | 1 |
| $\chi_3$ | 3 | 1 | 0 | −1 | −1 |
| $\chi_4$ | 3 | −1 | 0 | 1 | −1 |
| $\chi_5$ | 2 | 0 | −1 | 0 | 2 |

3. Let $T: G \to GL_n(\mathbb{C})$ be a representation of a finite group $G$. Prove that $|\det (T(x))| = 1$, for all $x \in G$.
4. Prove that $Q_8$ and $D_4$ are non-isomorphic groups with the same character table.
5. Let $\{v_1, \ldots, v_n\}$ be a basis of a complex vector space $V$ with $n \geq 2$. For $\sigma \in S_n$) and $a_1, \cdots, a_n \in \mathbb{C}$ we define

$$\sigma\left(\sum_{i=1}^{n} a_i v_i\right) := \sum_{i=1}^{n} a_i v_{i\sigma}.$$

(a) Prove that this makes $V$ into a $\mathbb{C}S_n$-module. It is called the *permutation module*;
(b) Let $X = \left\{ \sum_{i=1}^{n} a_i v_i \mid a_1 + \cdots + a_n = 0 \right\}$. Prove that $X$ is a submodule of $V$;
(c) In order to prove that $X$ is simple, assume from now on that $Y$ is a nonzero submodule of $X$. Prove that there exists $y \in Y$ of the form $y = \sum a_i v_i$ with $a_n = 1$;
(d) Put $w = v_1 + \cdots + v_n$ and $S_{n-1} = \{g \in S_n \mid n^g = n\}$. Define

$$z = \frac{1}{(n-2)!} \sum_{g \in S_{n-1}} yg.$$

Prove that $zg = z$ for all $g \in S_{n-1}$. Prove that $z = nv_n - w$;
(e) Prove that $nv_i - w \in Y$, for all $i$;
(f) Prove that $v_n - v_i \in Y$, for all $i$. Prove $Y = X$, i.e., $X$ is simple;
(g) Find an explicit submodule $W$ of $V$ such that $V = W \oplus X$. Prove that $W$ is simple.

6. Let $0 \leq k \leq n$ and write $A_n = \{1, \ldots, n\}$. Let $V$ be an $\binom{n}{k}$ dimensional vector space with basis $\{v(I) \mid I \subseteq A_n, |I| = k\}$. Let $G = S_n$ act on $V$ by $gv(I) = v(gI)$, where $gI = \{gi \mid i \in I\}$. Let $\chi$ be the character of the $\mathbb{C}G$-module $V$. Prove that

$$\langle \chi, \chi \rangle_G = \min(k, n - k) + 1.$$

7. We define a group

$$\langle x, y, z \mid x^2 = z, \ y^2 = z, \ (xy)^2 = z, \ z^2 = e \rangle$$

and consider $x, y, z$ as elements of $G$. We write $(g_1, \ldots, g_5) = (e, x, y, xy, z)$.

(a) Prove that there are precisely 4 linear characters of $G$, written $\chi_1, \ldots, \chi_4$. Give a table with the values $\chi_i(g_j)$ whenever $1 \le i \le 4$ and $1 \le j \le 5$;

(b) Prove that there exists a unique representation $T$ of $G$ such that $T(x) = A$ and $T(y) = B$, where we write

$$\begin{bmatrix} i & 0 \\ 0 & -i \end{bmatrix} \text{ and } \begin{bmatrix} 0 & 1 \\ -1 & 0 \end{bmatrix}$$

(c) Prove that $T$ is irreducible;

(d) Prove that $\langle z \rangle$ is a normal subgroup of $G$. Give an explicit presentation of the quotient group $G/\langle z \rangle$;

(e) Prove that $|G| = 8$.

## 10.7  Tensor Products and Induced Characters

Let $V$ and $W$ be vector spaces over a field $\mathbb{F}$. The *tensor product* $V \otimes W$ is the vector space linearly spanned over $\mathbb{F}$ by symbols $v \otimes w$, for $v \in V$ and $w \in W$, which satisfy the relations

$$\lambda(v \otimes w) = (\lambda v) \otimes w = v \otimes (\lambda w),$$
$$(v + v') \otimes w = (v \otimes w) + (v' \otimes w),$$
$$v \otimes (w + w') = (v \otimes w) + (v \otimes w'),$$

for all $\lambda \in \mathbb{F}$, $v, v' \in V$ and $w, w' \in W$. If $\{v_1, \ldots, v_n\}$ is a basis of $V$ and $\{w_1, \ldots, w_m\}$ is a basis of $W$, then $V \otimes W$ is a vector space of dimension $mn$ over $\mathbb{F}$ with a basis given by

$$\{v_i \otimes w_j \mid 1 \le i \le n, \ 1 \le j \le m\}.$$

Therefore, $V \otimes W$ consists of all expressions of the form

$$\sum_{i,j} \lambda_{ij}(v_i \otimes w_j),$$

where $\lambda_{ij} \in \mathbb{F}$. For $v \in V$ and $w \in W$ with $v = \sum_{i=1}^{n} \lambda_i v_i$ and $w = \sum_{j=1}^{m} \mu_j w_j$, where $\lambda_i, \mu_j \in \mathbb{F}$, we define

$$v \otimes w = \sum_{i,j} \lambda_i \mu_j (v_i \otimes w_j).$$

Now, we define the tensor product of two $\mathbb{F}G$-modules. Let $G$ be a finite group and let $V$ and $W$ be two $\mathbb{F}G$-modules with basis $\{v_1, \ldots, v_n\}$ and $\{w_1, \ldots, w_m\}$, respectively. We know that the elements $v_i \otimes w_j$, for $1 \leq i \leq n$ and $1 \leq j \leq m$, give a basis for $V \otimes W$. The multiplication of $v_i \otimes w_j$ by an element of $G$ is defined by

$$(v_i \otimes w_j)g = v_i g \otimes w_j g,$$

and more generally, let

$$\left( \sum_{i,j} \lambda_{ij} (v_i \otimes w_j) \right) g = \sum_{i,j} \lambda_{ij} (v_i g \otimes w_j g),$$

for arbitrary $\lambda_{ij} \in \mathbb{F}$. It is easy to check that $(v \otimes w)g = vg \otimes wg$, for all $v \in V$, $w \in W$ and $g \in G$. The rule for multiplying an element of $V \otimes W$ by an element of $G$ makes the vector space $V \otimes W$ into an $\mathbb{F}G$-module.

**Definition 10.91** Let $T \colon G \to GL(V, \mathbb{F})$ and $T' \colon G \to GL(W, \mathbb{F})$ be representations of $G$ over $\mathbb{F}$. Then, $T \otimes T' \colon G \to GL(V \otimes W, \mathbb{F})$ is the representations defined by

$$(T \otimes T')(g)(v \otimes w) = T(g)v \otimes T'(g)w.$$

Let $g$ be represented by the matrix $A = \left( a_{ij} \right)_{n \times n}$ with respect to a basis $\{v_1, \ldots, v_n\}$ of $V$ and by the matrix $B = \left( b_{kl} \right)_{m \times m}$ with respect to a basis $\{w_1, \ldots, w_m\}$ of $W$. Let us work out the matrix $g$ on the basis

$$v_1 \otimes w_1, \ \ldots, \ v_1 \otimes w_m, \ v_2 \otimes w_1, \ \ldots, \ v_2 \otimes w_m, \ \ldots, \ v_n \otimes w_1, \ \ldots, \ v_n \otimes w_m.$$

We have

$$g(v_i \otimes w_k) = gv_i \otimes gw_k = (a_{i1}v_1 + \cdots + a_{in}v_n) \otimes (b_{k1}w_1 + \cdots + b_{km}w_m)$$
$$= \sum_{j,l} a_{ij} b_{kl} v_j w_l.$$

From here, it immediately follows that the matrix of $g$ with respect to the above basis on $V \otimes W$ is the $nm \times nm$ block matrix

$$A \otimes B = \begin{bmatrix} a_{11}B & a_{12}B & \ldots & a_{1n}B \\ a_{21}B & a_{22}B & \ldots & a_{2n}B \\ \vdots & \vdots & \ddots & \vdots \\ a_{n1}B & a_{n2}B & \ldots & a_{nn}B \end{bmatrix}. \tag{10.13}$$

This calculation has a very useful consequence.

**Theorem 10.92** *Let $T$ and $T'$ be representations of $G$ with characters $\chi$ and $\psi$, respectively. Then, the character of $T \otimes T'$ is the product character $\chi\psi$, where $\chi\psi(g) = \chi(g)\psi(g)$, for all $g \in G$.*

**Proof** It follows from (10.13). ∎

Theorem 10.92 yields that if $\chi$ is a character of $G$, then so is $\chi^2$, where $\chi^2 = \chi\chi$, the product of $\chi$ with itself. More generally, for every nonnegative integer $n$, we define $\chi^n$ by $\chi^n(g) = (\chi(g))^n$, for all $g \in G$. An induction proof using Theorem 10.92 shows that $\chi^n$ is a character of $G$.

If $H$ is a subgroup of a group $G$ and $T$ is a representation of $G$ over a field $\mathbb{F}$, then a representation of $H$ is obtained by simply restricting $T$ to $H$. A less trivial problem is to construct a representation of $G$ starting with a representation of $H$. This leads to the important concept of an *induced representation*, which is due to Frobenius.

Let $G$ be a group with subgroup $H$, $\mathbb{F}$ be a field, and $V$ an $\mathbb{F}G$-module. Then, $\mathbb{F}H \subseteq \mathbb{F}G$, and so $V$ is also an $\mathbb{F}H$-module; when so considered, we denote $V$ by $V_H$, the restriction of $V$ to $\mathbb{F}H$.

**Definition 10.93** Let $G$ be a finite group with subgroup $H$, $\mathbb{F}$ be a field, and $W$ be an $\mathbb{F}H$-module. Let $\{x_1, \ldots, x_k\}$ be a set of representatives of left cosets of $H$ in $G$. Then, we have $G = x_1 H \cup \cdots \cup x_k H$. The group algebra $\mathbb{F}G$ has the structure

$$\mathbb{F}G = \oplus \sum_{i=1}^{k} \mathbb{F}x_i H.$$

Considering $\mathbb{F}G$ as an $\mathbb{F}H$-module, this expresses $\mathbb{F}G$ as a direct sum of $\mathbb{F}H$-modules. Define an $\mathbb{F}G$-module $W^G$ by forming the tensor product

$$W^G = \mathbb{F}G \otimes_{\mathbb{F}G} W,$$

wherein $\mathbb{F}G$ is to be regarded as an $\mathbb{F}H$-module. The module $W^G$ is said to have been *induced* by $W$, or *induced* from the subgroup $H$, and the representation $T^G$ that $W^G$ affords is called *induced representation*. The construction above does not depend on the particular representative set taken.

Now, we can write

$$\mathbb{F}G = \oplus \sum_{i=1}^{k} \mathbb{F}x_i H = \oplus \sum_{i=1}^{k} x_i \mathbb{F}H,$$

and so

$$W^G = \oplus \sum_{i=1}^{k} x_i \mathbb{F}H \otimes W = \oplus \sum_{i=1}^{k} x_i \otimes W,$$

at least a direct sum as $\mathbb{F}$-modules.

**Lemma 10.94**  *With the above notation, let*

$$W_i = x_i \otimes W = \{x_i \otimes w \mid w \in W\},$$

*for $i = 1, \ldots k$. Then, $W^G = W_1 \oplus \cdots \oplus W_k$ as a vector space over $\mathbb{F}$. If $x \in G$, $x$ acts on $W^G$ as follows: $xx_i = x_jh$, for some $h \in H$. Then, $xW_i \subseteq W_j$ and $x(x_i \otimes w) = x_j \otimes hw$.*

**Proof**  The first statement is obvious. Now, we have

$$x(x_i \otimes w) = xx_i \otimes w = x_jh \otimes w = x_j \otimes hw \in x_j \otimes W = W_j,$$

and the inclusion follows.                                                        ∎

Since an element $x$ acts as a non-singular linear transformation on $W^G$ and the subspaces $W_i$ have the same dimension, it follows that the inclusion of Lemma 10.94 may be replaced by equality. Thus, the action of an element $x$ on $W^G$ may be viewed as first inducing the same permutation of the subspace $W_i$ as on the left cosets of $H$, and then an action by elements of $H$ on each.

If $\{w_1, \ldots, w_m\}$ is a basis of $W$ over $\mathbb{F}$, the elements $x_i \otimes w_j$, where $i = 1, \ldots, k$ and $j = 1, \ldots, m$, form a basis for $W^G$ over $\mathbb{F}$. In particular, we have

$$\text{degree } T^G = (\text{degree } T) \cdot [G : H].$$

**Theorem 10.95**  *Let $G$ be a finite group with subgroup $H$, $\mathbb{F}$ be a field and $W$ be an $\mathbb{F}H$-module with basis $\{w_1, \ldots, w_m\}$. Let $T$ be the matrix representation of $H$ afforded by $W$ in this basis. Define $\widehat{T}$ on all $G$ by*

$$\widehat{T}(g) = \begin{cases} T(g) & \text{if } g \in H \\ 0, \text{ the } m \times m \text{ zero matrix,} & \text{if } g \in H. \end{cases}$$

*Let $x_1, \ldots, x_k$ be a representative set of left cosets of $H$ in $G$, so $[G : H] = k$. Then, $W^G$ has a basis $\{x_i \otimes w_j \mid 1 \leq i \leq k \text{ and } 1 \leq j \leq m\}$. By arranging the elements of this basis in the order*

$$x_1 \otimes w_1, \ \ldots, \ x_1 \otimes w_m, \ x_2 \otimes w_1, \ \ldots, \ x_2 \otimes w_m, \ \ldots, \ x_k \otimes w_1, \ \ldots, \ x_k \otimes w_m.$$

*Then, the matrix representation $T^G$ of $G$ afforded by $W^G$ is*

$$T^G(g) = \begin{bmatrix} \widehat{T}(x_1^{-1}gx_1) & \widehat{T}(x_1^{-1}gx_2) & \cdots & \widehat{T}(x_1^{-1}gx_k) \\ \widehat{T}(x_2^{-1}gx_1) & \widehat{T}(x_2^{-1}gx_2) & \cdots & \widehat{T}(x_2^{-1}gx_k) \\ \vdots & \vdots & \ddots & \vdots \\ \widehat{T}(x_k^{-1}gx_1) & \widehat{T}(x_k^{-1}gx_2) & \cdots & \widehat{T}(x_k^{-1}gx_k) \end{bmatrix}$$

*In this block array, only one block in each row and each column is a nonzero block.*

**Proof**  Suppose that $T(h) = \big(a_{ij}(h)\big)_{m \times m}$, where $hw_j = \sum_{i=1}^{m} a_{ij}(h)w_i$. Let $g \in G$ and $i, j$ be fixed. If $gx_i = x_l h$, for some $h \in H$, then

$$g(x_i \otimes w_j) = gx_i \otimes w_j = x_l h \otimes w_j = x_l \otimes hw_j = x_l \otimes \sum_{i=1}^{m} a_{ij}(h)w_i$$

$$= \sum_{i=1}^{m} a_{sj}(h)x_l \otimes w_s = \sum_{i=1}^{m} a_{sj}(x_l^{-1}gx_i)x_l \otimes w_s.$$

This gives the result.  ∎

On the basis of Theorem 10.95 the induced character $\chi^G$ can be calculated.

**Theorem 10.96**  *Let $G$ be a finite group with a subgroup $H$, and $\mathbb{F}$ be a field whose characteristic does not divided the order of $H$. If $\chi$ is a character of $H$ over $\mathbb{F}$, then the value of the induced character is given by*

$$\chi^G(g) = \frac{1}{|H|} \sum_{x \in G} \widehat{\chi}(x^{-1}gx),$$

*where*

$$\widehat{\chi}(y) = \begin{cases} \chi(y) & \text{if } y \in H \\ 0 & \text{if } y \notin H. \end{cases}$$

**Proof**  Suppose that $\{x_1, \ldots, x_k\}$ is a set of representatives of left cosets of $H$ in $G$. Taking trace in Theorem 10.95, we obtain

$$\chi^G(g) = \sum_{i=1}^{k} \widehat{\chi}(x_i^{-1}gx_i),$$

For any $h \in H$ and $g \in G$, we have $\widehat{\chi}(h^{-1}gh) = \widehat{\chi}(g)$, whether $g \in H$ or $g \notin H$. Consequently, we have

$$\chi^G(x_i^{-1}gx_i) = \frac{1}{|H|} \sum_{h \in H} \widehat{\chi}(h^{-1}x_i^{-1}gx_ih),$$

and

$$\chi^G(g) = \sum_{i=1}^{k} \widehat{\chi}(x_i^{-1}gx_i) = \frac{1}{|H|} \sum_{i=1}^{k} \sum_{h \in H} \widehat{\chi}(h^{-1}x_i^{-1}gx_ih)$$

$$= \frac{1}{|H|} \sum_{x \in G} \widehat{\chi}(x^{-1}gx).$$

∎

**Definition 10.97** If $\chi$ is a character of a subgroup $H$ of $G$, then $\chi^G$ is called the *induced character*.

For the practical propose, a formula for the values of induced characters different from that given in Theorem 10.96 is more useful, and we derive this next.

**Theorem 10.98** *Let $\chi$ be a character of subgroup $H$ of $G$, and suppose that $g \in G$. Let $C_G(g) \cap H = \bigcup_{i=1}^{m} C_H(x_i)$ (disjoint union), where $x_i$'s are representatives of the $m$ conjugacy classes of elements of $H$ conjugate to $g$. Then, we have*

$$\chi^G(g) = |C_G(g)| \sum_{i=1}^{m} \frac{\chi(x_i)}{|C_H(x_i)|}.$$

***Proof*** We can write

$$\chi^G(g) = \frac{1}{|H|} \sum_{x \in G} \widehat{\chi}(x^{-1}gx) = \frac{1}{|H|} \sum_{i=1}^{m} |C_G(g)| \, [H : C_H(x_i)] \, \chi(x_i),$$

and this gives the result.              ∎

The following theorem is used quite frequently in computations with induced characters.

**Theorem 10.99 (Frobenius Reciprocity Theorem).** *Let $G$ be a finite group, and let $\mathbb{F}$ be a field whose characteristic does not divide the order of $G$. Assume that $\psi$ and $\chi$ are characters of $H$ and $G$ over $\mathbb{F}$, respectively. Then, we have*

$$\langle \psi^G, \chi \rangle_G = \langle \psi, \chi_H \rangle_H,$$

*where $\chi_H$ denotes the restriction of $\chi$ to $H$.*

***Proof*** We have

$$\langle \psi^G, \chi \rangle_G = \frac{1}{|G|} \sum_{x \in G} \psi^G(x) \, \chi(x^{-1}) = \frac{1}{|G|} \sum_{x \in G} \left( \frac{1}{|H|} \sum_{y \in G} \widehat{\psi}(y^{-1}xy) \right) \chi(x^{-1})$$

$$= \frac{1}{|G| \, |H|} \sum_{x \in G} \sum_{y \in G} \widehat{\psi}(y^{-1}xy) \, \chi(x^{-1})$$

$$= \frac{1}{|G| \, |H|} \sum_{x \in G} \sum_{y \in G} \widehat{\psi}(y^{-1}xy) \, \chi(y^{-1}x^{-1}y)$$

$$= \frac{1}{|G| \, |H|} \sum_{t \in G} \widehat{\psi}(t) \, \chi(t^{-1}) \, |G| = \frac{1}{|H|} \sum_{t \in G} \widehat{\psi}(t) \, \chi(t^{-1})$$

$$= \frac{1}{|H|} \sum_{t \in G} \psi(t) \, \chi(t^{-1}) = \langle \psi, \chi_H \rangle_H,$$

and we are done.              ∎

**Exercises**

1. Let $\chi$ and $\psi$ be irreducible characters of $G$ over $\mathbb{C}$. Prove that

$$\langle \chi\psi, 1\rangle = \begin{cases} 1 & \text{if } \chi = \overline{\psi} \\ 0 & \text{if } \chi \neq \overline{\psi}. \end{cases}$$

2. Let $H$ and $K$ be a subgroup of a group $G$ with $K \subseteq H$, and let $U$ and $W$ be $\mathbb{F}K$-modules. Prove that

   (a) $(U \oplus W)^G \cong U^G \oplus W^G$;
   (b) $(W^H)^G \cong W^G$.

3. Let $G$ be a dihedral group of order $2n$, $n$ odd. Show that every two-dimensional irreducible representation of $G$ over $\mathbb{C}$ is induced from a representation of the cyclic subgroup of index 2. Obtain the same result if $n$ is even and $n > 2$.

4. If $\chi$ is a faithful complex character of the subgroup $H$ of $G$, show that $\chi^G$ is a faithful character of $G$.

5. Suppose that $G$ is a group with a subgroup $H$ of index 3, and let $\chi$ be an irreducible character of $G$. Prove that

$$\langle \chi_H, \chi_H \rangle = 1, 2 \text{ or } 3.$$

   Give examples to show that each possibility can occur.

6. Let $H$ and $K$ be subgroups of a finite group $G$. Let $\chi$ and $\psi$ denote the trivial characters of $H$ and $K$ over an algebraically closed field of characteristic 0. Prove that $\langle \chi^G, \psi^G \rangle_G$ equals to the number of $(H, K)$-double cosets.

7. Suppose that $H$ is a subgroup of $G$, and let $\chi_1, \ldots, \chi_h$ be the irreducible characters of $G$. Let $\psi$ be an irreducible character of $H$. Show that the integers $a_1, \ldots, a_h$, which are given by $\psi^G = a_1\chi_1 + \cdots + a_h\chi_h$, satisfy

$$\sum_{i=1}^{h} a_i^2 \leq [G:H].$$

## 10.8 Representations of Abelian Groups

We shall now investigate the representations of the abelian groups.

**Theorem 10.100** *If $G$ is a finite group, then $G$ has exactly $[G:G']$ linear representation on $\mathbb{C}$.*

**Proof** Suppose that $T: G \to \mathbb{C}$ is a linear representation, then $ImT$ is a multiplicative subgroup of $\mathbb{C}$, and so it is abelian. Since $G/KetT \cong ImT$, it follows that $G/KerT$ is abelian. Hence, we have $G' \subseteq KerT$. Note that any homomorphism $G \to \mathbb{C}$ with kernel containing $G'$ induces a homomorphism $G/G' \to \mathbb{C}$, and conversely. ∎

**Theorem 10.101** *All irreducible characters of a finite abelian group are linear.*

**Proof** Let $G$ be a finite abelian group of order $n$. If $x \in G$ is an arbitrary element, then the conjugacy class of $x$ is $\{x\}$. Hence, $G$ consists $n$ conjugacy classes, which implies that there are $n$ irreducible characters, say $\chi_1, \ldots, \chi_n$. Now, we must have

$$\sum_{i=1}^{n} \chi_i^2(e) = |G| = n.$$

This yields that $\chi_i(e) = 1$, for all $i = 1, \ldots, n$.                                  ∎

By the fundamental theorem of finite abelian groups (Theorem 4.45), we know that each finite abelian group is a direct product of cyclic groups. Hence, in order to determine the irreducible characters of finite abelian groups, it is necessary to have information about the irreducible characters of the following groups:

(1) Cyclic groups;
(2) Direct product of groups.

**Theorem 10.102** *Let $G = \langle x \rangle$ be a cyclic group of order $n$ and let $\omega = 2\pi i/n \in \mathbb{C}$ be a primitive nth root of unity. Then, the irreducible complex characters of $G$ are the n functions $\chi_{k-1}(x^m) = \omega^{(k-1)m}$, where $1 \leq k \leq n$.*

**Proof** We merely observe that the mapping $x^m \mapsto \omega^{km}$ is a homomorphism $G \to GL_1(\mathbb{C}) \cong \mathbb{C}^*$ giving a one-dimensional representation with character $\chi_k$, that must necessary be irreducible. These characters are all distinct, and since the number of them equals the group order we have them all.                                  ∎

The character table for $G$ is given in Table 10.4.

Let $G$ and $H$ be two finite groups. Suppose that $\mathcal{C}_1, \ldots, \mathcal{C}_h$ are distinct conjugacy classes of $G$ and $\mathcal{D}_1, \ldots, \mathcal{D}_k$ are distinct conjugacy classes of $H$. If $x_i \in \mathcal{C}_i$ and $y_j \in \mathcal{D}$, then $(x_i, y_j) \in \mathcal{C}_i \times \mathcal{D}_j$, and we have

**Table 10.4** Character table of a cyclic group of order $n$

| $x_i$ | $e$ | $x$ | $x^2$ | $\cdots$ | $x^{n-2}$ | $x^{n-1}$ |
|---|---|---|---|---|---|---|
| $|C_G(x_i)|$ | $n$ | $n$ | $n$ | $\cdots$ | $n$ | $n$ |
| $\chi_1$ | $1$ | $1$ | $1$ | $\cdots$ | $1$ | $1$ |
| $chi_2$ | $1$ | $\omega$ | $\omega^2$ | $\cdots$ | $\omega^{n-2}$ | $\omega^{n-1}$ |
| $\chi_3$ | $1$ | $\omega^2$ | $\omega^4$ | $\cdots$ | $\omega^{n-4}$ | $\omega^{n-2}$ |
| $\vdots$ | $\vdots$ | $\vdots$ | $\vdots$ | $\ddots$ | $\vdots$ | $\vdots$ |
| $\chi_{n-1}$ | $1$ | $\omega^{n-2}$ | $\omega^{n-4}$ | $\cdots$ | $\omega^4$ | $\omega^2$ |
| $\chi_n$ | $1$ | $\omega^{n-1}$ | $\omega^{n-2}$ | $\cdots$ | $\omega^2$ | $\omega$ |

$$C_{G \times H}(x_i, y_j) = C_G(x_i) \times C_H(y_j).$$

Therefore, the number of irreducible characters of $G \times H$ equals to $hk$.

**Theorem 10.103** *If $\chi_1, \ldots, \chi_h$ are irreducible characters of $G$ and $\psi_1, \ldots, \psi_k$ are irreducible characters of $H$, then the irreducible characters of direct product $G \times H$ are $\chi_i \times \psi_j$, for $1 \le i \le h$ and $1 \le j \le k$, where*

$$\chi_i \times \psi_j(x, y) = \chi_i(x) \, \psi_j(y),$$

*for all $x \in G$ and $y \in G$.*

**Proof** It is clear that each $\chi_i \times \psi_j$ is a character of $G \times H$. Moreover, we have

$$
\begin{aligned}
\langle \chi_i \times \psi_j, \chi_i \times \psi_j \rangle &= \frac{1}{|G \times H|} \sum_{(x,y) \in G \times H} \chi_i \times \psi_j(x,y) \, \chi_i \times \psi_j(x^{-1}, y^{-1}) \\
&= \frac{1}{|G \times H|} \sum_{(x,y) \in G \times H} \chi_i(x) \, \psi_j(y) \, \chi_i(x^{-1}) \, \psi_j(y^{-1}) \\
&= \Big( \frac{1}{|G|} \sum_{x \in G} \chi_i(x) \, \chi_i(x^{-1}) \Big) \Big( \frac{1}{|H|} \sum_{y \in H} \psi_j(y) \psi_j(y^{-1}) \Big) \\
&= 1 \cdot 1 = 1.
\end{aligned}
$$

This gives the result.                                                         ∎

**Example 10.104** Let $G = S_3$ and $H = C_2$, and consider the direct product $G \times H$. The character table of $S_3$ is given in Example 10.89, and the character table of cyclic group $C_2$ is given in Table 10.5

The conjugacy classes of $S_3 \times C_2$ are represented by $(id, 1), \big((1\,2), 1\big), \big((1\,2\,3), 1\big)$, $(id, -1)$, $\big((1\ 2), -1\big)$, and $\big((1\ 2\ 3), -1\big)$. Now, by Theorem 10.103, we determine the character table of $S_3 \times C_2$ as given in Table 10.6.

**Exercises**

1. What is the orthogonality relation for the characters of the cyclic group of order n?

**Table 10.5** Character table of $C_2$

| $y_j$ | 1 | $-1$ |
|---|---|---|
| $|C_H(y_j)|$ | 2 | 2 |
| $\psi_1$ | 1 | 1 |
| $\psi_2$ | 1 | $-1$ |

**Table 10.6**  Character table of $S_3 \times C_2$

| $(x_i, y_j)$ | $(id, 1)$ | $((1\,2), 1)$ | $((1\,2\,3), 1)$ | $(id, -1)$ | $((1\,2), -1)$ | $((1\,2\,3), -1)$ |
|---|---|---|---|---|---|---|
| $|C_{G\times H}(x_i, y_j)|$ | 12 | 4 | 6 | 12 | 4 | 6 |
| $\chi_1 \times \psi_1$ | 1 | 1 | 1 | 1 | 1 | 1 |
| $\chi_2 \times \psi_1$ | 1 | $-1$ | 1 | 1 | $-1$ | 1 |
| $\chi_3 \times \psi_1$ | 2 | 0 | $-1$ | 2 | 0 | $-1$ |
| $\chi_1 \times \psi_2$ | 1 | 1 | 1 | $-1$ | $-1$ | $-1$ |
| $\chi_2 \times \psi_2$ | 1 | $-1$ | 1 | $-1$ | 1 | $-1$ |
| $\chi_3 \times \psi_2$ | 2 | 0 | $-1$ | $-2$ | 0 | 1 |

2. Let $G$ be a finite group of order $p^2$, where $p$ is a prime number. By considering the possible dimensions of the irreducible representations of $G$, prove that $G$ is abelian.

3. Determine the character table of $D_4 \times C_3$.

4. Let $\chi$ be a character of a cyclic group $G$. Show that $\prod_{g \in X} \chi(g)$ is rational, where $X = \{g \in G \mid G = \langle g \rangle\}$. Deduce that if $\chi(g) \neq 0$, then

$$\prod_{g \in X} |\chi(g)|^2 \geq 1,$$

and hence that

$$\sum_{g \in X} |\chi(g)|^2 \geq |X|.$$

5. Let $\chi$ be a character of an abelian group $G$. Show that

$$\sum_{g \in G} |\chi(g)|^2 \geq |G|\chi(e).$$

6. Let $G$ be a group and let $\chi$ be a nonlinear irreducible character of $G$. Show that $\chi(g) = 0$, for some $g \in G$.
   *Hint:* First, show that an eqivalence relation of $G$ can be defined by $x \sim y$ whenever $\langle x \rangle = \langle y \rangle$. Then, apply Exercise 4.

7. Suppose that $G = \bigcup_{i=1}^n H_i$, where $H_i$'s are abelian subgroups of $G$ and $H_i \cap H_j = \{e\}$ if $i \neq j$. Let $\chi$ be an irreducible character of $G$. Show that

   (a) If $\chi(e) > 1$, then $\chi(e) \geq |G|/(n-1)$;
   (b) If $G$ is non-abelian, then $|H_i| \leq n - 1$, for each $i$, and $n - 1 \geq |G|^{1/2}$.

8. Let $G$ be a finite group. Suppose that every irreducible character of $G$ with $\chi(e) > 1$ vanishes off of $G'$. Show that each non-identity coset of $G'$ is a conjugacy class.

## 10.9  Burnside's $p^a q^b$ Theorem

We shall devote this section to the proof of the classical result of Burnside. We start with a number theoretic lemma.

**Lemma 10.105** *If $\alpha$ is a sum of $n$ roots of unity and $\alpha/n$ is an algebraic integer, then either $\alpha/n = 0$ or $\alpha/n$ is a root of unity.*

**Proof** Suppose that $\alpha/n$ have minimal polynomial $f(x)$ over $\mathbb{Q}$. Then, $f(x)$ is a monic integer polynomial, say $f(x) = a_0 + a_1 x + \cdots + a_{n-1} x^{n-1} + x^n$, where $a_i \in \mathbb{Z}$, for all $i$. In particular, the constant term $a_0$ is an integer.

If $\alpha/n = (\lambda_1 + \cdots + \lambda_n)/n$ with each $\lambda_j$ a root of unity, then

$$\left| \frac{\alpha}{n} \right| = \left| \frac{\lambda_1 + \cdots + \lambda_n}{n} \right| \leq \frac{|\lambda_1| + \cdots + |\lambda_n|}{n} \leq 1.$$

Let $\alpha_1, \ldots, \alpha_k$ are algebraic conjugate of $\alpha$. Each algebraic conjugate of $\alpha$ is of the form $(\lambda_1' + \cdots + \lambda_n')/n$, where $\lambda_1', \ldots, \lambda_n'$ are roots of unity. It follows that if $\xi$ is the product of all algebraic conjugate of $\alpha$, then $|\xi| \leq 1$. But the algebraic conjugate of $\alpha$ are, by definition, the roots of polynomial $f(x)$, and the product of these roots is equal to $\pm a_0$. Thus, we have $\xi = \pm a_0$. If any $|\alpha_i| < 1$, then $|\xi| < 1$, so that $\xi = 0$ and $\alpha/n = 0$. Otherwise $|\alpha/n| = 1$, and the proof completes. ∎

The following lemma and theorem are general results.

**Lemma 10.106** *Let $\chi$ be an irreducible character of a finite group $G$ over $\mathbb{C}$, and suppose that $(|C|, \chi(e)) = 1$, for some conjugate class $C$ of $G$. Then, either $\chi(x) = 0$ or $\chi(x) = \lambda \chi(e)$, for some root $\lambda$ of unity.*

**Proof** There exist integers $r$ and $s$ such that $r\chi(e) + s|C| = 1$. This implies that

$$r\chi(x) + s \frac{|C| \chi(x)}{\chi(e)} = \frac{\chi(x)}{\chi(e)}.$$

By Theorem 10.79, $\frac{|C| \chi(x)}{\chi(e)}$ is an algebraic integer, and so $\frac{\chi(x)}{\chi(e)}$ is an algebraic integer. Now, the result follows from Lemma 10.105. ∎

**Theorem 10.107** *If $G$ has a conjugacy class $C$ such that $|C|$ is positive power $p$, where $p$ is a prime number, then $G$ is not simple.*

**Proof** Let $|C| = p^\alpha$. If $\alpha = 0$, then there is a non-identity element $x \in Z(G)$, and so $G$ is not simple. Now, suppose that $x \in C$ and $|C| = p^\alpha > 1$, so that $G$ is not abelian and $x \neq e$. Let $\chi_1, \ldots, \chi_h$ are irreducible characters of $G$ with $\chi_1 = 1$. From the column orthogonality relation, we have

$$0 = \sum_{i=1}^{h} \chi_i(x) \chi_i(e) = 1 + \sum_{i=2}^{h} \chi_i(x) \chi_i(e). \tag{10.14}$$

Then, we can write

$$\sum_{i=2}^{h} \chi_i(x) \frac{\chi_i(e)}{p} = -\frac{1}{p}.$$

Since $-1/p$ is not algebraic integer, it follows that $\chi_i(x)\chi_i(e)/p$ is not an algebraic integer, for some $i \geq 2$. Since $\chi_i(x)$ is an algebraic integer, it follows that $\chi_i(e)/p$ is not an algebraic integer, in other words $p \nmid \chi_i(e)$. We may arrange $\chi_i$'s such that $p \nmid \chi_i(e)$, for all $2 \leq i \leq k$; and $p|\chi_i(e)$, for all $k+1 \leq i \leq h$.

If $2 \leq i \leq k$, then $(p^\alpha, \chi_i(e)) = 1$, and so by Lemma 10.106, we conclude that

$$\chi_i(x) = 0 \text{ or } |\chi_i(x)| = \chi_i(e).$$

Suppose that $\chi_i(x) = 0$, for $2 \leq i \leq k$. For $i > k$, we have $p|\chi_i(e)$. This means that there is an integer $m_i$ such that $\chi_i(e) = pm_i$. Hence, the orthogonality relation (10.14) becomes

$$0 = 1 + \sum_{i=k+1}^{h} pm_i\chi_i(x) = 1 + p\sum_{i=k+1}^{h} m_i\chi_i(x).$$

If $\omega = \sum_{i=k+1}^{h} m_i\chi_i(x)$, then $\omega$ is an algebraic integer. Now, $\omega = -1/p$ is impossible.

Now, let $|\chi_i(x)| = \chi_i(e)$. Suppose that $T$ is a representation of $G$ with character $\chi_i$. By Theorem 10.83, there is $\lambda \in \mathbb{C}$ such that $T(x) = \lambda I$. Let $K = KerT$, so that $K \lhd G$. Since $\chi_i$ is not the trivial character, it follows that $K \neq G$. If $K \neq \{e\}$, then $G$ is not simple, as required. So, assume that $K = \{e\}$, i.e., $T$ is a faithful representation. Since $T(x)$ is a scalar multiple of identity, it follows that $T(x)$ commutes with $T(y)$, for all $y \in G$. As $T$ is faithful, we conclude that $x$ commutes with all $y \in G$. This yields that $x \in Z(G)$, and so $Z(G) \neq \{e\}$. Since $Z(G)$ is a proper subgroup of $G$, it follows that $G$ is not simple.  ∎

**Theorem 10.108  (Burnside's $p^a q^b$ Theorem).** *Let $p$ and $q$ be prime numbers, and suppose that $a$ and $b$ are nonnegative integers with $a + b \geq 2$. If $|G| = p^a q^b$, then $G$ is not simple.*

**Proof** If $a = 0$ or $b = 0$, then the order of $G$ is a power of a prime number, and so $Z(G) \neq \{e\}$. Hence, there is a non-identity element $x \in Z(G)$ of prime order. Since the subgroup generated by $x$ is a non-trivial proper normal subgroup of $G$, it follows that $G$ is not simple.

Now, assume that $a > 0$ and $b > 0$. By Sylow's theorem, $G$ has a subgroup $H$ of order $q^b$. Again, we have $Z(H) \neq \{e\}$. Let $x$ be a non-identity element of $Z(H)$, which lies to a conjugacy class $\mathcal{C}$. Then, we have $H \leq C_G(x)$, and consequently $|\mathcal{C}| = [G : C_G(x)] = p^m$, for some integer $m$. If $p^m = e$, then $x \in Z(G)$, and hence $Z(G) \neq \{e\}$, which implies that $G$ is not simple. If $p^m > 1$, then by Theorem 10.107, $G$ is not simple.  ∎

Indeed, Burnside's $p^a q^b$ theorem leads to a somewhat more informative result about groups of order $p^a q^b$:

**Theorem 10.109** *A finite group whose order is divisible by at most two distinct primes is solvable.*

**Proof** Suppose that $|G| = p^a q^b$. We use mathematical induction on $a + b$. The result is clear if $a + b \leq 1$. So, assume that $a + b \geq 2$. By Burnside's theorem, $G$ has a non-trivial proper normal $N$. Both $N$ and the factor group $G/N$ have order equal to a product of powers of $p$ and $q$, and these orders are less than $p^a q^b$. Thus, by induction hypothesis, $N$ and $G/N$ are solvable. Consequently, there exist solvable series

$$\{e\} = H_0 \subseteq H_1 \subseteq \cdots \subseteq H_m = N$$
$$\{N\} = H_m/N \subseteq H_{m+1}/N \subseteq \cdots \subseteq H_n/N = G/N,$$

with all factor groups $H_i/H_{i-1}$ of prime order. Then, the solvable series

$$\{e\} = H_0 \subseteq H_1 \subseteq \cdots \subseteq H_n = G$$

yields that $G$ is solvable. ∎

### Exercises

1. Let $G$ be a group having a nonlinear faithful irreducible character $\chi$ of degree $p^a$, where $p$ is a prime number, and suppose that $Z(G) = \{e\}$. Show that $\chi$ vanishes on the non-identity elements in the center of a Sylow $p$-subgroup of $G$.
2. Show that a non-abelian simple group can not have an abelian subgroup of prime power index.
3. Show that a non-abelian simple group can not have a nilpotent subgroup of prime power index.
4. Prove that if $G$ is a non-abelian simple group of order less than 80, then $|G| = 60$.

## 10.10 Worked-Out Problems

**Problem 10.110** Show that if $\chi$ is an irreducible complex character of $G$, and $x \in G$ has order 2, then $\chi(x)$ is an ordinary integer and

$$\chi(x) \equiv \chi(1)(\text{mod } 2).$$

*Solution* By Theorem 10.82, we have $\chi(x) = \lambda_1 + \cdots + \lambda_n$, where $n = \chi(e)$ and each $\lambda_i$ is a square root of unity. So, each $\lambda_i$ is 1 or $-1$. Assume that $r$ of them are 1, and $s$ of them are $-1$. Hence, we have $\chi(x) = r - s$ and $\chi(e) = r + s$. It is clear that $\chi(x)$ is an integer. Since $r - s = r + s - 2s \equiv r + s(\text{mod } 2)$, it follows that $\chi(x) \equiv \chi(1)(\text{mod } 2)$. ∎

**Problem 10.111** Let $G$ be a group of odd order, and write $G - \{e\}$ as the union of $(|G| - 1)/2$ pairs $\{x_i, x_i^{-1}\}$. Let $\chi \neq 1$ be an irreducible complex character of $G$, and show that $\chi \neq \overline{\chi}$.

*Solution* Since $\chi \neq 1$, it follows that $\langle \chi, 1 \rangle = 0$, and so $\sum_{g \in G} \chi(g) = 0$. Now, we can write

$$0 = \chi(e) + \sum_{i=1}^{n} \left( \chi(x_i) + \chi(x_i^{-1}) \right) = \chi(e) + \sum_{i=1}^{n} \left( \chi(x_i) + \overline{\chi(x_i)} \right)$$

$$= \chi(e) + \sum_{i=1}^{n} \left( \chi(x_i) + \overline{\chi}(x_i) \right).$$

Since $\chi(e) \big| |G|$, it follows that $\chi(e)$ is an odd integer. If $\chi = \overline{\chi}$, then we obtain

$$\sum_{i=1}^{n} \chi(x_i) = -\frac{\chi(e)}{2}.$$

Since $\sum_{i=1}^{n} \chi(x_i)$ is an algebraic integer, it follows that $-\chi(e)/2$ is an algebraic integer. As $-\chi(e)/2 \in \mathbb{Q}$ we must have $-\chi(e)/2 \in \mathbb{Z}$. This is impossible, because $\chi(e)$ is an odd number.  ∎

**Problem 10.112** Let $\chi_1, \dots, \chi_h$ be the irreducible characters of $G$. Show that

$$Z(G) = \left\{ x \in G \mid \sum_{i=1}^{k} \chi_i(x) \, \overline{\chi_i(x)} = |G| \right\}.$$

*Solution* Suppose that

$$A = \left\{ x \in G \mid \sum_{i=1}^{k} \chi_i(x) \, \overline{\chi_i(x)} = |G| \right\}.$$

By the column orthogonality relation, we have

$$\sum_{i=1}^{k} \chi_i(x) \, \overline{\chi_i(x)} = |C_G(x)|.$$

Hence, we can write

$$x \in A \Leftrightarrow \sum_{i=1}^{k} \chi_i(x) \, \overline{\chi_i(x)} = |G| \Leftrightarrow |C_G(x)| = |G| \Leftrightarrow x \in Z(G).$$

This shows that $A = Z(G)$, as required.  ∎

**Problem 10.113** Let $G$ be a finite group and $\mathbb{F}$ be a field. Let $T$ be a non-identity irreducible representation of $G$ affording the character $\chi$.

(1)  Show that $A = \sum_{g \in G} T(g) = 0$;

(2) Suppose that $H \leq G$ and $g \in G$ such that $x$ is conjugate to $y$ in $G$, for all $x, y \in Hg$. If $\langle \chi_H, 1 \rangle_H = 0$, show that $\chi(g) = 0$.

*Solution* (1) Since $T$ is non-identity, it follows that there is $x \in G$ such that $T(x) \neq I$. Take $B = T(x) - I$; Since $B \neq 0$, by Schur's lemma, we deduce that $B$ is invertible. Moreover, we have

$$AB = \left( \sum_{g \in G} T(g) \right)(T(x) - I) = \sum_{g \in G} T(gx) - \sum_{g \in G} T(g) = 0.$$

Since $AB = 0$ and $B$ is invertible, it follows that $A = 0$.
(2) $A = \sum_{h \in H} T(hg)$. If we take the trace, then we obtain

$$tr(A) = tr\left( \sum_{h \in H} T(hg) \right) = \sum_{h \in H} tr(T(hg)) = \sum_{h \in H} tr(T(g))$$

This yields that

$$tr(A) = |H| \, \chi(g). \tag{10.15}$$

On the other hand, we can write $\chi_H = a_1 \chi_1 + \cdots + a_k \chi_k$, where $\chi_i$'s are irreducible characters of $H$ and $a_i$'s are integers. Now, let $T_1, \ldots, T_k$ are irreducible representations of $G$ such that $tr(T_i(x)) = \chi_i(x)$, for all $1 \leq i \leq k$. Then, we can write $\chi_H(g) = tr(T_H(g)) = a_1(T_1(g)) + \cdots + a_n(T_k(g))$. Hence, we obtain

$$A = \sum_{h \in H} T(h)T(g) = T(g) \sum_{h \in H} T(h) = T(g) \sum_{h \in H} T_H(h)$$
$$= T(g) \sum_{h \in H} (a_1 T_1(h) + \cdots + a_k T_k(h))$$
$$= T(g)\left(a_1 \sum_{h \in H} T_1(h) + \cdots + a_k \sum_{h \in H} T_k(h)\right).$$

Now, by previous part, each term in the above left sum is zero, and so $A = 0$. This implies that $tr(A) = 0$. This together with (10.15) gives $\chi(g) = 0$. ∎

**Problem 10.114** Prove that a group $G$ is not simple if and only if $\chi(g) = \chi(e)$, for some non-trivial irreducible character $\chi$ of $G$ and some non-identity element $g \in G$.

*Solution* Assume that there exists a non-trivial irreducible character $\chi$ such that $\chi(g) = \chi(e)$, for some non-identity element $g \in G$. This yields that $g \in Ker\chi$, and so $Ker\chi \neq \{e\}$. If $T$ is a representation of $G$ with character $\chi$, then $Ker\chi = KerT$. Since $\chi$ is non-trivial and irreducible, it follows that $KerT \neq G$, and so $Ker\chi \neq G$. This yields that $Ker\chi$ is a non-trivial proper subgroup of $G$, which implies that $G$ is not simple.

Conversely, suppose that $G$ is not simple. Then, there is a normal subgroup $N$ of $G$ with $N \neq G$ and $N \neq \{e\}$. Now, by Theorem 10.59, there is an irreducible character $\chi$ of $G$ such that $Ker\chi$ is not $\{e\}$ or $G$. Since $Ker\chi \neq G$, it follows that $\chi$ is non-trivial; and taking $e \neq g \in Ker\chi$, we obtain $\chi(g) = \chi(e)$.   ∎

**Problem 10.115** Describe the commutator subgroup of a group in terms of the character table of $G$.

*Solution* We claim that the rows in the character table of $G$ with 1 in the first column are precisely the characters lifted from $G'$ to $G$. If $\chi$, a character of $G$, is lifted from $\widetilde{\chi}$, a character of $G/G'$, then $\chi = \widetilde{\chi} \circ \pi$ and so $\chi$ is linear, i.e., $\chi(e) = 1$. Conversely, if $\chi(e) = 1$, then we have to find a $\widetilde{\chi} : G/G' \to \mathbb{F}^*$ such that $\chi = \widetilde{\chi} \circ \pi$. The obvious choice would of course be $\widetilde{\chi} : G/G' \to \mathbb{F}^*$ with $\widetilde{\chi}(xG') = \chi(x)$, for all $x \in G$.

Since our definition of $\widetilde{\chi}$ involves a choice of coset representative, we have to show that this is well defined, i.e., the image is independent of the coset representative we chose: $\chi(x) = \chi(xg)$, for all $g \in G'$ or equivalently $\chi(g) = 1$, for all $g \in G'$. The last condition is just $G' \leq Ker\widetilde{\chi}$ and this is obvious because $G/Ker\widetilde{\chi} \leq \mathbb{F}^*$, and so is abelian, thus $G' \leq Ker\widetilde{\chi}$. Also, we observe that the characters lifted from the irreducible character of $G/G'$ are all distinct (clear from how a lifted character is defined), and so the character table of $G$ contains exactly $|G/G'|$ rows of dimension one. We further claim that

$$G' = \bigcap_{\chi(e)=1} Ker\chi.$$

In the earlier part, we have already shown that $G' \subseteq Ker\chi$, for all linear character $\chi$. So "$\subseteq$" is done. For the converse, recall that:

(1) By the earlier part, all irreducible characters of $G/G'$ are of the form $\chi'$, where $\chi' \circ \pi = \chi$ is a linear character of $G$;
(2) In general, if $x \in G$ satisfies $\chi(x) = 1$ for every irreducible character of $G$, then $x = e$.

Now, suppose that $g \in \bigcap_{\chi(e)=1} Ker\chi$. Then, $\chi(g) = 1$, for all linear character $\chi$ of $G$. By (1), we obtain $\chi'(gG') = \chi' \circ \pi(g) = \chi(g) = 1$, for all irreducible character $\chi'$ of $G/G'$. Now, by (2), we must have $gG' = G'$, and hence $g \in G'$. This completes the proof.   ∎

**Problem 10.116** Let $G$ be a finite group, and let $T: G \to GL_2(\mathbb{C})$ be a representation of $G$. Suppose that there are elements $x, y \in G$ such that the matrices $T(x)$ and $T(y)$ do not commute. Prove that $T$ is irreducible.

*Solution* Assume that $T: G \to GL_2(\mathbb{C})$ is reducible. Then, by the matrix form of Maschke's theorem, we have

$$T(x) = \begin{bmatrix} \lambda_x & 0 \\ 0 & \mu_x \end{bmatrix},$$

for some suitable $\lambda_x, \mu_x \in \mathbb{C}$ determined by $x$. Now, if this is true, then $T(x)T(y) = T(y)T(x)$, for all $x, y \in G$, because diagonal matrices commute but this implies $G$ is abelian. Consequently, we must have $T$ is irreducible.   ∎

**Problem 10.117** The dihedral group $D_n$ of order $2n$ has generators $a$, $b$ and relations $a^n = e$, $b^2 = e$ and $b^{-1}ab = a^{-1}$. Describe all of the irreducible complex characters of $D_n$.

*Solution* Suppose that $\omega = e^{2\pi i/n}$, Foe each integer $1 \le j < n/2$, take

$$A_j = \begin{bmatrix} \omega^j & 0 \\ 0 & \omega^{-j} \end{bmatrix} \text{ and } B_j = \begin{bmatrix} 0 & 1 \\ 1 & 0 \end{bmatrix}.$$

It is easy to see that $A_j^n = I_2$, $B_j^2 = I_2$ and $B_j^{-1}A_jB_j = A_j^{-1}$. Now, if we define $T_j :$ $G \to GL_{@}(\mathbb{C})$ by

$$T_j(a^r b^s) = A_j^r B_j^s \ (r, s \in \mathbb{Z}),$$

for all $1 \le j < n/2$, then $T_j$ is a representation of $G$. Now, by using Problem 10.116, we conclude that $T_j$ is irreducible, for all $1 \le j < n/2$. Moreover, if $i$ and $j$ are two distinct integers such that $1 \le i, j < n/2$, then $\omega^i \ne \omega^j$ and $\omega^i \ne \omega^{-j}$. This yields that $T_i(a)$ and $T_j(a)$ have different eigenvalues. Hence, does not exist matrix a $A$ such that $T_i(a) = A^{-1}(T_j(a))A$, and so $T_i$ and $T_j$ are not equivalent. Now, if $\psi_j$ is the character associated to $T_j$, then $\psi_j$ is irreducible character of $G_n$, for all $1 \le j < n/2$. Now, we consider two cases: $n$ odd and $n$ even.

*Case 1:* Let $n$ be odd. The conjugacy classes of $D_n$ are $\{e\}$, $\{a^r, a^{-r}\}$, for $1 \le r \le (n-1)/2$, and $\{a^s b \mid 0 \le s \le n-1\}$. The $(n-1)/2$ irreducible characters $\psi_1, \cdots, \psi_{(n-1)/2}$ each have degree 2. Since $G$ has $(n+3)/2$ irreducible characters, there are two more irreducible characters to be determined. Since $\langle a \rangle \trianglelefteq G$ and $G/\langle a \rangle \cong C_2$, it follows that there are two linear characters $\chi_1$ and $\chi_2$ of $G$ by lifting the irreducible characters of $G/\langle a \rangle$ to $G$. The characters $\chi_1$ and $\chi_2$ are given by $\chi_1 = 1$, trivial character of $G$, and

$$\chi_2(x) = \begin{cases} 1 & \text{if } x = a^k \text{ for some } k \\ -1 & \text{if } x = a^k b \text{ for some } k. \end{cases}$$

Now, we have determined all the irreducible characters of $D_n$ when $n$ is odd. Therefore, the character table of $D_n$, when $n$ is odd, is given in Table 10.7.

*Case 2:* Suppose that $n$ be even, and let $n = 2m$. The conjugacy classes of $D_n$ are $\{e\}$, $\{a^m\}$, $\{a^r, a^{-r}\}$, for $1 \le r \le m-1$, $\{a^s b \mid s$ is even$\}$, and $\{a^s b \mid s$ is odd$\}$. So, $G$ has $m+3$ irreducible characters, of which $m-1$ are given by $\psi_1, \ldots, \psi_{m-1}$. In order to determine the remaining four irreducible characters, note that $\langle a^2 \rangle \trianglelefteq G$ and

$$G/\langle a^2 \rangle = \{\langle a^2 \rangle, a\langle a^2 \rangle, b\langle a^2 \rangle, ab\langle a^2 \rangle\} \cong C_2 \times C_2.$$

Consequently, $G$ has four linear characters $\chi_1$, $\chi_2$, $\chi_3$ and $\chi_4$. Since these linear characters are the lifts of the irreducible characters of $G/\langle a^2 \rangle$, they are easy to obtain, and their values appear in the complete character table of $D_n$, when $n = 2m$ (Table 10.8). ∎

**Table 10.7** Character table of $D_n$ when $n$ is odd

| $x_i$ | $e$ | $a^r \ (1 \le r \le (n-1)/2)$ | $b$ |
|---|---|---|---|
| $\lvert C_{D_n}(x_i) \rvert$ | $2n$ | $n$ | $2$ |
| $\chi_1$ | $1$ | $1$ | $1$ |
| $\chi_2$ | $1$ | $1$ | $-1$ |
| $\psi_j \ (1 \le j \le (n-1)/2)$ | $2$ | $\omega^{jr} + \omega^{-jr}$ | $0$ |

**Table 10.8** Character table of $D_n$ when $n$ is even

| $x_i$ | $e$ | $a^m$ | $a^r \ (1 \le r \le m-1)$ | $b$ | $ab$ |
|---|---|---|---|---|---|
| $\lvert C_{D_n}(x_i) \rvert$ | $2n$ | $2n$ | $n$ | $4$ | $4$ |
| $\chi_1$ | $1$ | $1$ | $1$ | $1$ | $1$ |
| $\chi_2$ | $1$ | $1$ | $1$ | $-1$ | $-1$ |
| $\chi_3$ | $1$ | $(-1)^m$ | $(-1)^r$ | $1$ | $-1$ |
| $\chi_4$ | $1$ | $(-1)^m$ | $(-1)^r$ | $-1$ | $1$ |
| $\psi_j \ (1 \le j \le m-1)$ | $2$ | $2(-1)^j$ | $\omega^{jr} + \omega^{-jr}$ | $0$ | $0$ |

**Problem 10.118** Let $G$ be a finite group, $x \in G$, and let $H$ be a normal subgroup of $G$ such that $H \cap C_G(x) = \{e\}$. Show that

(1) $\lvert C_G(x) \rvert \le \lvert C_{G/H}(xH) \rvert$;
(2) If $\chi$ is an irreducible character of $G$ whose kernel does not contain $H$, then $\chi(x) = 0$.

*Solution* (1) Let $Irr(G) = \{\chi_1, \ldots, \chi_h\}$, the set of all irreducible characters of $G$. Then, we have

$$\lvert C_G(x) \rvert = \sum_{m=1}^{h} \chi_m(x) \chi_m(x^{-1}) = \sum_{m=1}^{h} \chi_m(x) \overline{\chi_m(x)}$$
$$= \sum_{m=1}^{h} \lvert \chi_m(x) \rvert^2 = \sum_{\chi \in Irr(G)} \lvert \chi(x) \rvert^2.$$

Since $Irr(G/H) = \{\chi \in Irr(G) \mid H \subseteq Ker\chi\}$, we can write

$$\lvert C_{G/H}(xH) \rvert = \sum_{\substack{\chi \in Irr(G) \\ H \subseteq Ker\chi}} \lvert \chi(x) \rvert^2 \le \sum_{\chi \in Irr(G)} \lvert \chi(x) \rvert^2 = \lvert C_G(x) \rvert. \qquad (10.16)$$

(2) We have

$$|C_G(x)| - |C_{G/H}(xH)| = \sum_{\substack{\chi \in Irr(G) \\ H \nsubseteq Ker\chi}} |\chi(x)|^2.$$

We define a function $\varphi$ from $C_G(x)$ to $C_{G/H}(xH)$ by $\varphi(g) = gH$, for all $g \in C_G(x)$. If $\varphi(g_1) = \varphi(g_2)$, for some $g_1, g_2 \in C_G(x)$, then $g_1 H = g_2 H$, which implies that $g_1^{-1}g_2 \in H$. Since $g_1^{-1}g_2 \in H \cap C_G(x) = \{e\}$, it follows that $g_1 = g_2$. This means that $\varphi$ is one to one, and so we must have

$$|C_G(x)| \le |C_{G/H}(xH)|. \tag{10.17}$$

From (10.16) and (10.17), we conclude that $|C_G(x)| = |C_{G/H}(xH)|$. This gives

$$\sum_{\substack{\chi \in Irr(G) \\ H \nsubseteq Ker\chi}} |\chi(x)|^2 = 0,$$

and so $\chi(x) = 0$, for all $\chi \in Irr(G)$ such that $H \nsubseteq Ker\chi$. ∎

## 10.11 Supplementary Exercises

1. Let $G$ be the group of matrices

$$\begin{bmatrix} 1 & x & y \\ 0 & 1 & z \\ 0 & 0 & 1 \end{bmatrix},$$

where $x, y, z$ are elements of the finite field $\mathbb{F}_5$. Classify irreducible representations of $G$ over $\mathbb{C}$.

2. Let $G$ be the finite abelian group $C_{n_1} \times \cdots \times C_{n_k}$. Prove that $G$ has a faithful representation of degree $k$. Can $G$ have a faithful representation of degree less than $k$?

3. Let $G$ be a group and $\mathbb{F}$ be a field of characteristic $p$. Suppose that $p \mid |G|$ and show that $\mathbb{F}G$ is not semisimple.

   *Hint:* $\left( \sum_{g \in G} g \right)^2 = 0$.

4. Let $G$ be a finite group which has a unique minimal normal subgroup and let $\mathbb{F}$ be a field whose characteristic does not divide $|G|$. Prove that $G$ has a faithful irreducible representation over $\mathbb{F}$.

5. Prove that an irreducible representation of a $p$-group over a field with characteristic $p$ has degree 1.

6. Find an example of irreducible characters $\chi$ and $\psi$ of a finite group $G$ such that $\chi\psi$ is not irreducible.

7. Prove that a cyclic group of order $n$ has a faithful irreducible representation of degree $\varphi(n)$ over the field $\mathbb{Q}$ (where $\varphi$ is Euler's function).

8. Prove that the sum of any row of the character table of $G$ is a nonnegative integer.

9. Prove that a finite $p$-group $G$ has a faithful irreducible representation over the field $\mathbb{C}$ if and only if the center of $G$ is cyclic.

10. Prove that for every finite simple group $G$, there exists a faithful irreducible $\mathbb{C}G$-module.

11. Suppose that $G = D_3 = \langle a, b \mid a^3 = b^2 = e, b^{-1}ab = a^{-1} \rangle$ and let $\omega = e^{2\pi i/3}$.

    (a) Prove that the two-dimensional subspace $W$ of $\mathbb{C}G$ defined by

    $$W = \langle 1 + \omega^2 a + \omega a^2, \ b + \omega^2 ab + \omega a^2 b \rangle$$

    is an irreducible $\mathbb{C}G$-module of the regular $\mathbb{C}G$-module;

    (b) Show that $a + a^{-1} \in Z(\mathbb{C}G)$;

    (c) Find $\lambda \in \mathbb{C}$ such that $x(a + a^{-1}) = \lambda x$, for all $x \in W$.

12. Prove that if $G$ is a simple group and $\chi$ is an irreducible character, then $\chi(e) \neq 2$.

13. Let $H$ be a normal subgroup of $G$ such that $G = \{hz \mid h \in H, \ z \in Z(G)\}$. Let $T$ be an irreducible representation of $G$. Prove that the restriction of $T$ to $H$ is also irreducible.

14. Let $\chi$ be a faithful complex character of the finite group $G$ with degree $n$. Denote by $r$ the number of distinct values assumed by $\chi$. Prove that each irreducible complex character occurs as a direct summand of at least one power $\chi^s$, $s = 0, 1, \ldots, r - 1$ (here $\chi^0$ is the trivial character). Deduce that the sum of degrees of the irreducible complex representations can not exceed $(n^r - 1)/((n - 1))$.

    *Hint:* Let $\psi$ be an irreducible complex character and show that not every $\langle \chi^s, \psi \rangle_G$ can be 0.

15. Let $\theta$ be a faithful complex character of $G$ with exactly $r$ distinct values $a_1 = \theta(e), a_2, \ldots, a_r$. Let $\chi$ be any irreducible complex character of $G$.

    (a) If $\langle \theta^n, \chi \rangle_G = 0$, for some integer $n \geq 0$, and $A_i = \{g \in G \mid \theta(g) = a_i\}$, show that

    $$0 = \sum_{j=1}^{r} a_j^n \sum_{g \in A_j} \overline{\chi(g)};$$

    (b) Show that for some $0 \leq n \leq r$, $\chi$ is a constituent of $\theta^n$.

16. Let $n \geq 4$ and put $A = \{1, \ldots, n\}$. Let $V$ be a complex vector space of dimension $n(n - 1)/2$ and with basis $\{v_{ab} \mid a, b \in A, \ a < b\}$. We also write $v_{ab} = v_{ba}$. The symmetric group $G = S_n$ acts on $V$ by putting $gv_{ab} = v_{ga,gb}$, for all $g \in G$. This makes $V$ into a $\mathbb{C}G$-module whose character will be written $\chi$. The linear transformation $V \to V$, where $v \mapsto gv$, is denoted by $t_g$. For $a, b, c, d \in A$ with $a \neq b$ and $c \neq d$ put

$$M(a, b, c, d) = \{g \in S_n \mid gv_{ab} = v_{ab}, \; gv_{cd} = v_{cd}\},$$

and $m(a, b, c, d) = |M(a, b, c, d)|$.

(a) Calculate the trace of $t_h$ if $n = 8$ and $h = (1\ 5\ 8)(2\ 4)$;
(b) Prove directly from the definitions

$$4 \cdot n! \cdot \langle \chi, \chi \rangle_G = \sum_{\substack{a,b,c,d \in A \\ a \neq b, c \neq d}} m(a, b, c, d);$$

(c) Assume that $a, b, c, d \in A$ are distinct. Prove that

$$m(a, b, c, d) = 4(n - 4)!,$$
$$m(a, b, a, b) = 2(n - 2)!,$$
$$m(a, b, a, c) = (n - 3)!;$$

(d) Use foregoing to prove $\langle \chi, \chi \rangle = 3$;
(e) Prove that there are distinct irreducible characters $\chi_1, \chi_2, \chi_3$ of $S_n$ such that $\chi = \chi_1 + \chi_2 + \chi_3$.

17. Show that no simple group can have an irreducible character of degree 2.
18. Let $G$ be a non-abelian group and let

$$k = \min\{\chi(e) \mid \chi \text{ is an irreducible character of } G \text{ and } \chi(e) > 1\}.$$

Show that

(a) If $|G'| \le k$, then $G' \subseteq Z(G)$;
(b) If $[G: G'] \le k$, then $G'$ is abelian;
(c) If $H \le G$ and $[G: H] \le k$, then $G' \subseteq H$.

19. Let $G$ be a simple group and suppose that $\chi$ is an irreducible character of $G$ with $\chi(e) = p$, a prime. Show that a Sylow $p$-subgroup of $G$ has order $p$.
20. Let $G$ be a finite group, $r$ be the number of conjugacy classes in $G$ and $s$ be the number of conjugacy classes in $G$ preserved by the involution $g \to g^{-1}$. Prove that the number of irreducible representations of $G$ over $\mathbb{R}$ is equal to $(r + s)/2$.

# References

1. S.S. Abhyankar, C. Christensen, Semidirect products: $x \mapsto ax + b$ as a first example. Math. Mag. **75**(4), 284–289 (2002)
2. M.A. Armstrong, *Groups and Symmetry, Undergraduate Texts in Mathematics* (Springer, New York, 1988)
3. R.B.J.T. Allenby, *Rings, Fields and Groups  An Introduction to Abstract Algebra*, 2nd edn. (Edward Arnold, London, 1991)
4. T.M. Apostol, *Introduction to Analytic Number Theory, Undergraduate Texts in Mathematics* (Springer, New York; Heidelberg, 1976)
5. H. Bacry, *Group Theory and Constellations* (Publibook, Paris, 2004)
6. T. Barnard, H. Neill, Discovering group theory, a transition to advanced mathematics, in *Textbooks in Mathematics* (CRC Press, Boca Raton, FL, 2017)
7. M.V. Bell, *Polya's Enumeration Theorem and its Applications*, Master's thesis, Primary supervisor: A.-M. Ernvall-Hytönen, Secondary supervisor: E. Elfving, University of Helsinki (2015)
8. O. Bogopolski, Introduction to group theory, in *EMS Textbooks in Mathematics*. Translated, revised and expanded from the 2002 Russian original (European Mathematical Society (EMS), Zürich, 2008)
9. C. Boyer, U. Merzbach, *A History of Mathematics*, 2nd edn. (Wiley, Hoboken, 1989)
10. P.M. Cohn, *Further Algebra and Applications* (Springer, London Ltd., London, 2003)
11. M.J. Collins, *Representations and Characters of Finite Groups*. Cambridge Studies in Advanced Mathematics, 22 (Cambridge University Press, Cambridge, 1990)
12. H.S.M. Coxeter, *Regular Polytopes* (Dover Publication. Inc., New York, 1973)
13. H.S.M. Coxeter, *Introduction to Geometry* (Wiley, Hoboken, 1969)
14. H.S.M. Coxeter, W.O.J. Moser, *Generators and Relations for Discrete Groups*, 4th edn. Ergebnisse der Mathematik und ihrer Grenzgebiete [Results in Mathematics and Related Areas], 14 (Springer, Berlin; New York, 1980)
15. M.R. Darafsheh, *Algebra I*, 3rd edn. (University of Tehran Press, Iran, 2010) (in Persian)
16. B. Davvaz, *A First Course in Group Theory* (Springer, Berlin, 2021)
17. B. Davvaz, *Polygroup Theory and Related Systems* (World Scientific Publishing Co., Pte. Ltd., Hackensack, NJ, 2013)
18. B. Davvaz, M.A. Iranmanesh, *Fundamentals of Group Theory* (Yazd University Press, Iran, 2005) (in Persian)
19. D.C. DeHovitz, The platonic solids—an exploration of the five regular polyhedra and the symmetries of three-dimensional space, Whitman College, Walla Walla, Washington (2016)

20. I.M.S. Dey, J. Wiegold, Generators for alternating and symmetric groups. J. Aust. Math. Soc. **12**, 63–68 (1971)
21. J.D. Dixon, *Problems in Group Theory* (Blaisdell Publishing Co. Ginn and Co., Waltham, Mass.-Toronto, Ont.-London, 1967)
22. J.D. Dixon, B. Mortimer, *Permutation Groups, Graduate Texts in Mathematics*, vol. 163 (Springer, New York, 1996)
23. N. Do, *Topics in Geometry* (McGill University, Montreal, 2010)
24. A. Doerr, K. Levasseur, *Applied Discrete Structures* (Department of Mathematical Sciences, University of Massachusetts Lowell, 2013)
25. L. Dornhoff, *Group Representation Theory. Part A: Ordinary Representation Theory*. Pure and Applied Mathematics, 7 (Marcel Dekker, Inc., New York, 1971)
26. C. Drutu, M. Kapovich, *Geometric Group Theory*, with an appendix by Bogdan Nica. American Mathematical Society Colloquium Publications, 63 (American Mathematical Society, Providence, RI, 2018)
27. H.B. Enderton, *Elements of Set Theory* (Academic Press [Harcourt Brace Jovanovich, Publishers], New York; London, 1977)
28. J.B. Fraleigh, *A First Course in Abstract Algebra*, 7th edn. (Pearson Education Limited, USA, 2014)
29. J.A. Gallian, *Contemporary Abstract Algebra*, 9th edn. (Cengage Learning, Boston, 2016)
30. J.A. Gallian, Classroom notes: group theory and the design of a letter facing machine. Am. Math. Monthly **84**(4), 285–287 (1977)
31. D. Guichard, *An Introduction to Combinatorics and Graph Theory* (Whitman College, 2020)
32. H.W. Guggenheimer, *Plane Geometry and its Groups* (Holden-Daay, 1967)
33. A. Hammel, Verifying the associative property for finite groups. Math. Teach. **61**(2), 136–139 (1968)
34. B. Hernandez, *Burnside's Lemma*, Undergraduate honors thesis, University of Redlands (2013)
35. R. Hartshorne, *Geometry: Euclid and Beyond* (Springer Science + Business Media, New York, 2000)
36. I.N. Herstein, *Topics in Algebra*, 2nd edn. (Xerox College Publishing, Lexington, Mass.-Toronto, Ont., 1975)
37. I.N. Herstein, *Abstract Algebra*, 3rd edn. with a preface by B. Cortzen, D.J. Winter (Prentice Hall Inc., Upper Saddle River, NJ, 1996)
38. V.E. Hill, *Groups, Representations and Characters* (Hafner Press, New York, 1975) (Binder)
39. K. Hoffman, R. Kunze, *Linear Algebra*, 2nd edn. (Prentice-Hall Inc., Englewood Cliffs, NJ, 1971)
40. J.F. Humphreys, *A Course in Group Theory* (Oxford Science Publications, The Clarendon Press, Oxford University Press, New York, 1996)
41. T.W. Hungerford, *Algebra*, Reprint of the 1974 original. Graduate Texts in Mathematics, 73 (Springer, New York; Berlin, 1980)
42. I.M. Isaacs, *Character Theory of Finite Groups* (AMS Chelsea Publishing, Providence, RI, 2006)
43. K. Iwasawa, Über die Einfachheit der speziellen projektiven Gruppen. Proc. Imp. Acad. Tokyo **17**, 57–59 (1941) (German)
44. G. James, M. Liebeck, *Representations and Characters of Groups*. Cambridge Mathematical Textbooks (Cambridge University Press, Cambridge, 1993)
45. D.L. Johnson, *Symmetries*. Springer Undergraduate Mathematics Series (Springer, London Ltd., London, 2001)
46. H. Kurzweil, B. Stellmacher, *The Theory of Finite Groups—An Introduction*. Translated from the 1998 German original (Universitext. Springer, New York, 2004)
47. T. Leinster, *Basic Category Theory*. Cambridge Studies in Advanced Mathematics, 143 (Cambridge University Press, Cambridge, 2014)
48. S.-Y.T. Lin, Y.-F. Lin, *Set Theory with Applications*, 2nd edn. (Marine Publishing Company Inc., 1981)

49. E.S. Lyapin, A.Y. Aizenshtat, M.M. Lesokhin, *Exercises in Group Theory*. Translated from the Russian by D.E. Zitarelli (Plenum Press, New York; Wolters-Noordhoff Publishing, Groningen, 1972)
50. R.C. Lyndon, *Groups and Geometry* (Cambridge University Press, Cambridge, England, 1985)
51. K. Mcgerty, *The Classical Groups* (University of Chicago, 2006)
52. J.H. McKay, Another proof of Cauchy's group theorem. Am. Math. Monthly **66**, 119 (1959)
53. P.M. Neumann, G.A. Stoy, E.C. Thompson, *Groups and Geometry* (Oxford Science Publications, The Clarendon Press, Oxford University Press, New York, 1994)
54. G. Polya, R.C. Reade, *Combinatorial Enumeration of Groups, Graphs, and Chemical Compounds* (Springer, New York, 1987)
55. D.J.S. Robinson, *A Course in the Theory of Groups*, 2nd edn. Graduate Texts in Mathematics, 80 (Springer, New York, 1996)
56. S. Roman, *Fundamentals of Group Theory—An Advanced Approach* (Birkhäuser/Springer, New York, 2012)
57. R.N. Umble, Z. Han, *Transformational Plane Geometry*, with a foreword by M.C. Bucur, L. Eisenhut. Textbooks in Mathematics (CRC Press, Boca Raton, FL, 2015)
58. H.E. Rose, *A Course on Finite Groups* (Springer, London Limited, London, 2009)
59. S. Ross, *A First Course in Probability*, 2nd edn. (Macmillan Co., New York; Collier Macmillan Ltd., London, 1984)
60. J.J. Rotman, *An Introduction to the Theory of Groups*, 4th edn. Graduate Texts in Mathematics, 148 (Springer, New York, 1995)
61. D. Schattschneider, The plane symmetry groups: their recognition and notation. Am. Math. Monthly **85**(6), 439–450 (1978)
62. D. Schattschneider, The 17 plane symmetry groups. Math. Gaz. **58**, 123–131 (1974)
63. W.R. Scott, *Group Theory*, 2nd edn. (Dover Publications Inc., New York, 1987)
64. S. Siksek, *Introduction to Abstract Algebra* (Mathematics Institute, University of Warwick, 2013)
65. S. Singh, *Modern Algebra* (Vikas Publishing House Pvt. Ltd., India, 1990)
66. M.K. Siu, *Which Latin Squares are Cayley Tables?* Am. Math. Monthly **98**(7), 625–627 (1991)
67. D.A. Suprunenko, *Matrix Groups*. Translated from the Russian. Translation edited by K.A. Hirsch. Translations of Mathematical Monographs, vol. 45 (American Mathematical Society, Providence, RI, 1976)
68. M. Suzuki, *Group Theory I*. Translated from the Japanese by the author. Grundlehren der Mathematischen Wissenschaften [Fundamental Principles of Mathematical Sciences], 247 (Springer, Berlin; New York, 1982)
69. T. Tsuzuku, *Finite Groups and Finite Geometries*. Translated from the Japanese by A. Sevenster, T. Okuyama [Tetsuro Okuyama]. Cambridge Tracts in Mathematics, 70 (Cambridge University Press, Cambridge; New York, 1982)
70. R.N. Umble, *Transformational Plane Geometry* (Millersville University of Pennsylvania, USA, 2012)
71. O. Viro, *Isometries*. Lecture Notes (Stony Brook University, 2014)

# Index

Printed in the United States
by Baker & Taylor Publisher Services